De Gruyter Studies in Mathematics 40

Nikolai A. Perestyuk
Viktor A. Plotnikov
Anatolii M. Samoilenko
Natalia V. Skripnik

Differential Equations with Impulse Effects

Multivalued Right-hand Sides with Discontinuities

De Gruyter

Mathematics Subject Classification 2010: 34A37, 34A60, 34C29, 34A30, 34A12.

ISBN 978-3-11-021816-9
e-ISBN 978-3-11-021817-6
ISSN 0179-0986

Library of Congress Cataloging-in-Publication Data

Differential equations with impulse effects : multivalued right-hand sides with discontinuities / by Nikolai A. Perestyuk ... [et al.].
p. cm. – (De Gruyter studies in mathematics ; 40)
Includes bibliographical references and index.
ISBN 978-3-11-021816-9 (alk. paper)
1. Impulsive differential equations. I. Perestyuk, N. A. (Nikolai Alekseevich)
QA377.D557 2011
515'.353–dc22
2011007994

Bibliographic information published by the Deutsche Nationalbibliothek

The Deutsche Nationalbibliothek lists this publication in the Deutsche Nationalbibliografie; detailed bibliographic data are available in the Internet at http://dnb.d-nb.de.

Typesetting: Da-TeX Gerd Blumenstein, Leipzig, www.da-tex.de
Printing and binding: Hubert & Co. GmbH & Co. KG, Göttingen
∞ Printed on acid-free paper

Printed in Germany

www.degruyter.com

To the Memory of
Viktor Aleksandrovich Plotnikov

Introduction

Significant interest in the investigation of systems with discontinuous trajectories is explained by the development of equipment in which significant role is played by impulsive control systems and impulsive computing systems. Impulsive systems are also encountered in numerous problems of natural sciences described by mathematical models with conditions reflecting the impulsive action of external forces with pulses whose duration can be neglected. It was discovered that the presence of a pulse action may significantly complicate the behavior of trajectories of these systems even in the case of quite simple differential equations.

Individual impulsive systems were studied by numerous researchers. Various examples of problems of this sort can be found in the works by N. N. Bogolyubov and N. M. Krylov [72], N. N. Bautin [16], B. S. Kalitin [60–62], A. E. Kobrinskii and A. A. Kobrinskii [68], N. A. Perestyuk and A. M. Samoilenko [142], and D. D. Bainov and A. B. Dishliev [11, 39].

In the works by N. N. Bogolyubov and N. M. Krylov [72], S. T. Zavalishchin and A. N. Sesekin [151, 152], and A. Halanay and D. Wexler [56], systems with pulse action were described by differential equations with generalized functions on the right-hand side. In these works, the differential equations describe pulses occurring at fixed moments of time, and the case where the times of pulse action depend on the phase vector is not investigated.

Another approach to the investigation of impulsive differential equations is based on the application of the classical methods of the theory of ordinary differential equations. As the first works in this direction, we can mention the works by A. D. Myshkis and A. M. Samoilenko [86, 87, 94], in which the general concepts of the theory of systems with pulse action are formulated from a new point of view and their basic specific features are investigated. Later, numerous works of many mathematicians were devoted to the analysis of the problems of stability of solutions of differential equations with pulse action, development of the theory of periodic and almost periodic solutions of impulsive systems, determination of invariant sets, construction of asymptotic expansions by the Krylov–Bogolyubov–Mitropol'skii method of small parameter, application of the method of comparison, solution of problems of the theory of optimal control, and investigation of impulsive systems with random perturbations [27,56,72,74,86–88,94,137,142,143,151,152]. The monographs [27,74,88,142,143] contain an extensive list of references in this field.

It is worth noting that the analysis of the dynamics of any real processes with the help of differential equations with univalent right-hand sides corresponds to the ideal model that does not take into account the action of random noises, errors of measurement in specifying the coefficients, and errors of specifying the functions on the right-hand sides of differential equations. If the probabilistic characteristics of the model are known, then the influence of random factors is taken into account by using stochastic differential equations. The theory of these equations is now rapidly developed and is extensively used in practice [1, 55, 64]. As a natural generalization of differential equations, we can mention differential inclusions capable of description of the dynamics of nondeterministic processes without using the probabilistic characteristics of the model. In numerous cases, this enables one to avoid the necessity of application of various a priori assumptions about these characteristics. The results of investigation of the model performed by the method of differential inclusions enable one to establish direct upper bounds for all results obtained by using probabilistic models, which is sometimes sufficient for applications.

The first investigations of differential equations with set-valued right-hand sides were carried out by S. Zaremba [158, 159] and A. Marchaud [80–83]. In these works, the authors made an attempt to extend the available results in the theory of differential equations to a more general case. Thus, S. Zaremba introduced the notion of differential equations in paratingents, and A. Marchaud proposed the notion of differential equations in contingents.

For the next 25 years, no works were published in this direction (we can mention only the works by A. D. Myshkis [92, 93]). This was explained by the absence of applications.

At the beginning of the 1960s, new fundamental results on the existence and properties of solutions of differential equations with set-valued right-hand sides (differential inclusions) were obtained in the cycles of works by T. Wazewski [157] and A. F. Filippov [49]. As one of the most important results obtained in the cited works, we can mention the established relationship between differential inclusions and problems of optimal control, which led to the extensive development of the theory of differential inclusions. The interest in the problems of control after the Second World War was connected with urgent needs of new technologies developed in the aviation, spacecraft engineering, and power-generating industry. This period was characterized by the appearance of new general methods for the solution of optimization problems of control, including the Pontryagin maximum principle, the Bellman method of dynamic programming, etc.

The principal results of the theory of differential equations with set-valued right-hand sides are presented in the works by A. F. Filippov [23, 48, 49, 51], T. Wazewski [157], V. I. Blagodatskikh [21–23], T. Donchev [41], M. Z. Zgurovskii, V. S. Mel'nik [153], A. I. Panasyuk and V. I. Panasyuk [101, 103], V. A. Plotnikov, A. V. Plotnikov, and A. N. Vityuk [115], A. A. Tolstonogov [145], O. P. Khapaev and M. M. Filatov [47], J.-P. Aubin and H. Frankovska [9], K. Deimling [36], and M. Kisielewicz [67]. The authors studied the problems of existence of solutions of differential inclusions and boundary-value problems, the problems of existence of monotone, bounded, and periodic solutions, stability of solutions, properties of solutions and integral funnels (compactness, connectedness, dependence on initial conditions and conditions on the right-hand side of the inclusion, and the relationship between the sets of solutions of the inclusions $\dot{x} \in F(t,x)$ and $\dot{x} \in \text{co}\, F(t,x)$), the problems of determination of the boundary of the set of attainability, conditions for the convexity of the set of solutions, the problems of averaging of differential inclusions, etc.

The investigation of properties of the integral funnels of differential inclusions is of high significance for the qualitative theory. In this connection, numerous researchers studied the properties of the set of attainability [6, 91, 101, 145] and various approximate methods for its construction, including the method of ellipsoids for linear systems [30, 73, 98], asymptotic methods [41, 115], and numerical methods [91, 96, 149].

In [145], it was shown that the integral funnel is a subset of the solution of the corresponding equation with Hukuhara derivative.

The first results in the theory of differential equations with Hukuhara derivative were obtained by F. S. de Blasi and F. Iervolino [25] and covered the problems related to the existence of solutions, their uniqueness, and continuous dependence on initial conditions and parameters. The possibility of application of the method of averaging to this class of problems was considered by M. Kisielewicz [66] and A. V. Plotnikov [109].

At present, the methods of the theory of differential equations with set-valued right-hand sides and differential equations with Hukuhara derivative are extensively used in the investigation of the dynamics of systems under the conditions of uncertainty, ambiguity, and incompleteness of information (so-called fuzzy systems) [75, 76].

The investigations of differential equations with discontinuous right-hand sides in the case of "sliding modes" carried out by A. F. Filippov [51], M. A. Aizerman [2], L. T. Ashchepkov [5], and V. I. Utkin [147] were also based on the theory of differential inclusions. Note that numerous important engineering problems related, e.g., to the motion of flying vehicles, propagation of seismic oscillations, development of shock and explosive processes, and control over manipulators can also be formulated in terms of discontinuous systems. Discontinuous systems are widely used in economics, chemical technology, theory of automated control, theory of systems with variable structure, and other fields of science.

The theory of impulsive differential equations and theory of differential inclusions were naturally developed in the works devoted to the investigation of differential inclusions with pulse action [7, 17–20, 43, 110–126, 156] dealing with the problems of existence of solutions of Cauchy and boundary-value problems, stability of solutions, existence of periodic solutions, and extendability and continuous dependence of solutions on the initial conditions and the right-hand sides of impulsive differential inclusions. Moreover, the hybrid control systems were also studied by the methods of impulsive differential inclusions.

Chapters 1 and 5 were written by A. M. Samoilenko and N. A. Perestyuk, Chapter 2 was written by V. A. Plotnikov, Chapters 3, 4, and 7 were written by N. V. Skripnik (née Plotnikova), and Chapter 6 was written by V. A. Plotnikov and N. V. Skripnik.

Notation

$\varnothing$	empty set
$\{x\}$	singleton set $x \in \mathbb{R}^n$
$\|x\|$	Euclidean norm of a vector $x \in \mathbb{R}^n$
$\|M\|$	spectral norm of a matrix
$B_r(a)$	closed ball of radius r centered at a point $a \in \mathbb{R}^n$
$S_r(a)$	sphere of radius r centered at a point $a \in \mathbb{R}^n$
$\mathrm{mes}(A)$	Lebesgue measure of a set A
$\mathrm{co}\, A$	convex hull of a set A
∂A	boundary of a set A
$\mathrm{int}\, A$	interior of a set A
$\overline{A}$	closure of a set A
$\rho(x, A)$	distance from a point x to a set A
$\mathrm{comp}(\mathbb{R}^n)$	space of nonempty compact subsets of $\mathbb{R}^n$ with Hausdorff metric
$\mathrm{conv}(\mathbb{R}^n)$	subspace of $\mathrm{comp}(\mathbb{R}^n)$ that consists of convex sets
$h(A, B)$	Hausdorff distance between sets A and B
$\vert A \vert$	modulus of a set A
$c(A, \psi)$	support function of a set A
$C[a, b]$	space of continuous functions with uniform metric on a segment $[a, b]$
$M[a, b]$	space of bounded functions with uniform metric on a segment $[a, b]$

Contents

Chapter 1
Impulsive Differential Equations

1.1 General Characterization of Systems of Impulsive Differential Equations

Description of a Mathematical Model. Let M be the phase space of a certain evolution process, i.e., the set of all possible states of this process. By $x(t)$ we denote a point that represents the state of this process at time t. We assume that the process is finite-dimensional, i.e., the description of its state at a fixed time requires a finite number, say n, of parameters. Under this assumption, the point $x(t)$ for a fixed t can be interpreted as an n-dimensional vector of the Euclidean space $\mathbb{R}^n$, and M can be regarded as a set from $\mathbb{R}^n$. The topological product $M \times \mathbb{R}$ of the phase space M and the real axis $\mathbb{R}$ is called the extended phase space of the evolution process under consideration. Assume that the law of evolution of the process is described by

(a) a system of differential equations

$$\frac{dx}{dt} = f(t, x), \quad x \in M,\ t \in \mathbb{R}, \tag{1.1}$$

(b) a certain set Γ_t given in the extended phase space, and

(c) an operator A_t given on the set Γ_t and mapping it onto the set $\Gamma'_t = A_t\Gamma_t$ of the extended phase space.

The process itself runs as follows: a representative point $P_t = (t, x(t))$ leaves a point (t_0, x_0) and moves along the curve $\{t, x(t)\}$ determined by the solution $x(t) = x(t, t_0, x_0)$ of the system of equations (1.1). The motion along this curve lasts up to a time $t = t_1 > t_0$ at which the point $(t, x(t))$ meets the set Γ_t (hits a point of the set Γ_t). At time $t = t_1$, the point P_t is "instantaneously" transferred by the operator A_t from the location $P_{t_1} = (t_1, x(t_1))$ to the location

$$P_{t_1}^+ = A_{t_1} P_{t_1} = (t_1, x^+(t_1)) \in \Gamma'_{t_1}$$

and then moves along the curve $\{t, x(t)\}$ described by the solution $x(t) = x(t, t_1, x^+(t_1))$ of the system of equations (1.1). The motion along the indicated curve lasts up to a time $t_2 > t_1$ at which the point P_t meets the set Γ_t again. At this time, the point P_t jumps "instantaneously" from the location $P_{t_2} = (t_2, x(t_2))$ to the location $P_{t_2}^+ = A_{t_2} P_{t_2} = (t_2, x^+(t_2))$ under the action of the operator A_t and moves further along the curve $\{t, x(t)\}$ described by the solution $x(t) = x(t, t_2, x^+(t_2))$ of the system of equations (1.1) up to a new contact with the set Γ_t, and so on.

In what follows, the collection of relations (a)–(c) characterizing the evolution of a process is called a system of differential equations with pulse action. The trajectory $\{t, x(t)\}$ of a point P_t in the extended phase space is called an integral curve, and the function $x = x(t)$ that defines this curve is called a solution of this system.

A system of differential equations with pulse action, i.e., the collection of relations (a)–(c), can be rewritten in a more compact form:

$$\frac{dx}{dt} = f(t, x), \quad (t, x) \notin \Gamma_t, \tag{1.2}$$

$$\Delta x|_{(t,x)\in\Gamma_t} = A_t x - x.$$

Thus, a solution $x = \varphi(t)$ of the system of equations (1.2) is a function that satisfies Eq. (1.1) outside the set Γ_t and has discontinuities of the first kind at the points of Γ_t with jumps

$$\Delta x = \varphi(t + 0) - \varphi(t - 0) = A_t\varphi(t - 0) - \varphi(t - 0). \tag{1.3}$$

A priori, solutions of Eqs. (1.2) may be of one of the following types:

(i) solutions not subjected to instantaneous changes; in this case, the integral curve of the system of equations (1.1) does not intersect the set Γ_t or intersects it at fixed points of the operator A_t;

(ii) solutions subjected to finitely many instantaneous changes; in this case, the integral curve intersects the set Γ_t at finitely many points that are not fixed points of the operator A_t;

(iii) solutions subjected to countably many instantaneous changes; in this case, the integral curve intersects the set Γ_t at countably many points that are not fixed points of the operator A_t.

Among the solutions whose integral curves pass through countably many points of Γ_t, we separate solutions that are absorbed by the set Γ_t (they remain in Γ_t beginning with a certain time $t_1 > t_0$) or have an accumulation point. The motion along a trajectory absorbed by the set Γ_t consists, beginning with a certain time $t_1 > t_0$, of successive transitions of the representative point P_t from the location (t_1, x_1) to the location $(t_1, A_{t_1}x_1)$, then from the latter to $(t_1, A_{t_1}^2 x_1)$, then to $(t_1, A_{t_1}^3 x_1)$, and so on. The motion along a trajectory having an accumulation point in Γ_t is a motion that meets and leaves the set Γ_t countably many times as time approaches a certain moment $t_1 > t_0$. Therefore, this motion cannot be extended to the time moment $t = t_1$.

The consideration of systems with pulse action meets the same problems as those for ordinary differential equations. However, some specific problems also arise. The character of these problems depends to a significant extent on properties of the operator A_t. For example, if A_t is not assumed to be one-to-one, then we encounter problems related to the study of motions for which the representative point can "instantaneously" split into several points at the times of contact with the set Γ_t. If the operator A_t is not assumed to be bijective, then we can consider problems related to motions for which independently moving points merge "instantaneously" into a single one at the time of contact with Γ_t.

Similar specific problems arise if we assume that the set $A_t \Upsilon_t$ is empty for some $\Upsilon_t \subset \Gamma_t$. This assumption allows one to consider "mortal" systems: a representative point P_t that hits Υ_t is transferred by the operator A_t to the empty set, i.e., it "dies" according to Vogel [150], and Υ_t serves as the set of "death" of trajectories. For systems of this type, it is natural to pose the problems of the mean lifetime of a moving point, the probability of its "death" in time $t_0 \leq t \leq T$, etc. Unfortunately, the wide variety of systems of differential equations that describe the evolution of a process between two successive times when a representative point hits the set Γ_t and the variety of sets Γ_t and mappings $A_t\colon \Gamma_t \to \Gamma_t'$ do not allow one to give a deep classification of systems of differential equations with pulse action according to their specific properties. Depending on the character of pulse action, three essentially different classes (types) of systems of equations under study can be distinguished:

(i) systems subjected to pulse action at fixed times;

(ii) systems subjected to pulse action at the times when a representative point P_t hits given surfaces $t = \tau_i(x)$ of the extended phase space;

(iii) discontinuous dynamical systems.

Prior to giving a brief description of these classes of systems, we present several examples that illustrate the variety of motions and trajectories in a system with pulse action and their essential dependence on the operator A_t and the set Γ_t.

Example 1. Assume that the phase space of a process is a straight line, the set Γ_t is given by the relation

$$\Gamma_t = \{(t, x) \in \mathbb{R}^2\colon x = \arctan(\tan t)\},$$

the operator A_t is defined by the equality

$$A_t(t, x) = (t, x^2 \operatorname{sign} x),$$

and the system of differential Eqs. (1.1) has the form

$$\frac{dx}{dt} = 0.$$

In other words, we consider the following system of differential equations with pulse action:

$$\frac{dx}{dt} = 0, \quad (t, x) \notin \Gamma_t, \tag{1.4}$$

$$\Delta x|_{(t,x)\in\Gamma_t} = x^2 \operatorname{sign} x - x.$$

We now study the integral curves and possible motions described by this system. In this system, every motion that starts at $t = 0$ from a point x_0, $|x_0| \geq \frac{\pi}{2}$, corresponds to the state of rest because the integral curve of this motion (the straight line $x = x_0$) does not hit the set Γ_t for any $t \geq 0$. The trajectory of each motion of this sort is the point x_0 (Figure 1). The motion that starts at $t = 0$ from a point x_0, $1 < |x_0| < \frac{\pi}{2}$, is subjected to finitely many pulse actions. The integral curve of this motion hits the set Γ_t finitely many times. For each motion of this sort, one can indicate the time $t_1 = t_1(x_0)$ beginning with which the integral curve stays in the set $|x| \geq \frac{\pi}{2}$, and, hence, this motion is not subjected to pulse action for $t > t_1(x_0)$. The trajectory of each motion of this type is a finite number of points. For example, the trajectory of the motion that starts at $t = 0$ from the point $x = \sqrt{2}$ consists of two points $x = \sqrt{2}$ and $x = 2$, whereas the trajectory of the motion that starts at $t = 0$ from the point $x = \sqrt[8]{2}$ consists of four points: $\{\sqrt[8]{2}, \sqrt[4]{2}, \sqrt{2}, 2\}$. The motion that starts at $t = 0$ from a point $x_0 \in (0, 1)$ is subjected to countably many pulse actions. The integral curve of this motion intersects the set Γ_t countably many times. In this case, one has $x(t, x_0) \to 0$ as $t \to \infty$. The trajectory of this motion consists of countably many points from the interval $(0, 1)$. For example, the trajectory of the motion that starts at $t = 0$ from the point $x = \frac{1}{2}$ is the set of points $x = \frac{1}{2^n}$, $n = 0, 1, 2, \ldots$. The integral curves that pass through the points $x = 0$ and $x = \pm 1$ also intersect the set Γ_t countably many times, but the motions corresponding to them are not subjected to pulse action and correspond to the state of rest. This is explained by the fact that the integral curves of these motions intersect the set Γ_t at fixed points of the operator A_t.

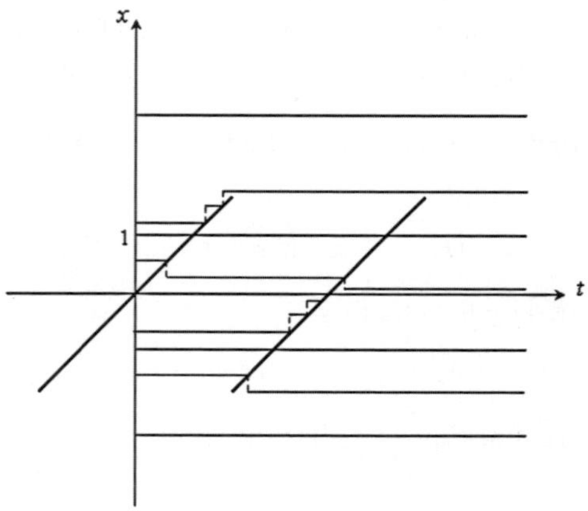

Figure 1. Integral curves (1.4) under different initial conditions.

The motions that start at $t = 0$ from points of the interval $(-1, 0)$ are subjected to countably many pulse actions on the segment $(\frac{3\pi}{4}, \pi)$. The sequence of times at which the motion is subjected to pulse action has the limit point $t = \pi$. Hence, the solution that corresponds to this motion cannot be extended to the interval $t \geq \pi$.

The example of these motions illustrates the phenomenon of beating of solutions of impulsive systems against the set Γ_t: on a small time interval, the integral curve hits the set Γ_t infinitely (countably) many times.

In addition to the variety of types of motions and integral curves, this example also shows that, in systems with pulse action, two integral curves can merge into a single one at a certain time. For example, the integral curves of motions that leave the points $x = \sqrt{2}$ and $x = 2$ at $t = 0$ merge into a single curve $x = 2$ at time $t = \sqrt{2}$.

Example 2. In the theory of optimal control, the following model problems are extensively studied: Find a control $u(t) \in U$ that minimizes the functional

$$I(u) = \int_0^T x^2 dt \tag{1.5}$$

on trajectories of the system

$$x^{(k)} = u, \quad x(0) = x^0, \quad x'(0) = x_1^0, \quad \dots, \quad x^{(k-1)}(0) = x_{k-1}^0. \tag{1.6}$$

For $k = 1$ and $U = [-1, 1]$, this example was first studied by L. I. Rozenoer in [130] for the illustration of the possibility of appearance of particular controls in the sense of the Pontryagin maximum principle. The optimal solution of the system has the form

$$x^*(t) = \begin{cases} x^0 - t \operatorname{sign} x^0, & 0 \le t \le |x^0|, \\ 0, & |x^0| \le t \le T, \end{cases} \tag{1.7}$$

$$u^*(t) = \begin{cases} -\operatorname{sign} x^0, & 0 \le t \le |x^0|, \\ 0, & |x^0| \le t \le T. \end{cases}$$

If $U = \{-1, 1\}$, then a solution does not exist in the class of absolutely continuous functions for $T > |x^0|$, and the so-called sliding mode begins at $t > |x^0|$.

The equation of motion along the optimal trajectory can be written in the form of an impulsive differential inclusion:

$$\begin{gathered} \dot{x} = u, \quad \dot{u} = 0, \quad x \neq 0, \\ \Delta x|_{x=0} = 0, \quad \Delta u|_{x=0} = \operatorname{sign} x^0, \\ x(0) = x^0, \quad u(0) = -\operatorname{sign} x^0. \end{gathered} \tag{1.8}$$

The solution of the impulsive differential Eq. (1.8) obviously coincides with (1.7); it is subjected to a single pulse action and then remains on the surface $x = 0$.

Problem (1.5), (1.6) corresponds to the motion of an object without regard for its inertia.

For $k = 2$ and $U = [-1, 1]$, this example was studied in detail in [53,54,79]. In this case, the surface of control switching has the form $x = -\gamma \dot{x}^2 \operatorname{sign} \dot{x}$, and the control has countably many switching points accumulated near the point τ_0. For $t > \tau_0$, the control satisfies the relation $u(t) \equiv 0$ (special mode). Note that if the problem is posed so that $x(0) = x^0$ and $x(T) = x^1$, then the second point of accumulation of switching points can appear for some $\tau_1 \in (\tau_0, T)$ (Figure 2) [1].

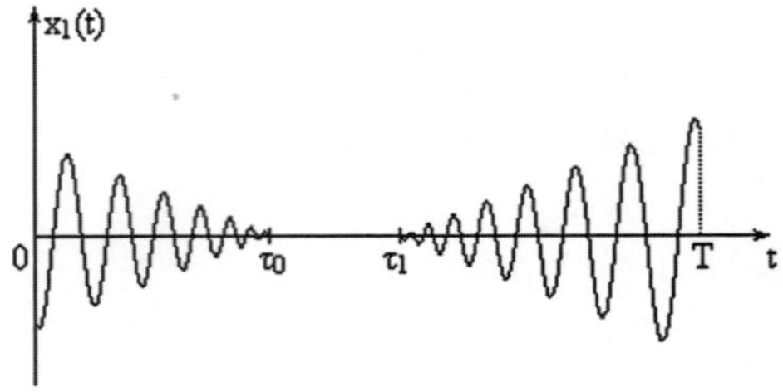

Figure 2. Integral curve (1.6) in the presence of the sliding mode.

This behavior of systems of optimal control is typical of a certain class of problems that take the inertia of an object into account. The equation of motion along the optimal trajectory can be written in the form of an impulsive differential equation:

$$\ddot{x} = u, \quad x(0) = x^0, \quad \dot{x}(0) = \dot{x}^0,$$

$$\dot{u} = 0, \quad u(0) = -\operatorname{sign} x^0, \quad x \neq -\gamma \dot{x}^2 \operatorname{sign} \dot{x}, \quad (x, \dot{x}) \neq 0,$$

$$\Delta x = 0, \quad \Delta \dot{x} = 0, \quad \Delta u = \begin{cases} -2u, & x = -\gamma \dot{x}^2 \operatorname{sign} \dot{x}, \\ -u, & x = \dot{x} = 0. \end{cases} \tag{1.9}$$

It is obvious that, for $T > \tau_0$, the solution of the impulsive differential Eq. (1.9) has the point of accumulation of switching points, and then the trajectory is located on the switching surface. If the problem is posed so that the final point $x(T) = x^1$ is also given, then there may exist the second point of accumulation of switching points $t = \tau_1 < T$.

Systems Subjected to Pulse Action at Fixed Times. If a real process described by the system of equations (1.1) is subjected to pulse action at fixed times, then the mathematical model of this process is given by the following system of differential equations with pulse action:

$$\frac{dx}{dt} = f(t, x), \quad t \neq \tau_i, \tag{1.10}$$

$$\Delta x|_{t=\tau_i} = I_i(x).$$

In this system, the set Γ_t is a sequence of hyperplanes $t = \tau_i$ of the extended phase space, where $\{\tau_i\}$ is a given (finite or infinite) sequence of times. In this case, it is sufficient to define the operator A_t only for $t = \tau_i$. In other words, it is sufficient to consider only its restriction to the hyperplanes $t = \tau_i$, $A_{t_i}: M \to M$. The most convenient is to consider the sequence of operators $A_i: M \to M$ defined as follows:

$$A_i: x \to A_i x = x + I_i(x). \tag{1.11}$$

Definition 1. The solution of Eqs. (1.10) is defined as a piecewise-continuous function $\varphi(t)$ with discontinuities of the first kind at the points $t = \tau_i$ for which the following conditions are satisfied:

(1) $\varphi'(t) = f(t, \varphi(t))$ for all $t \neq \tau_i$;

(2) for $t = \tau_i$, the following jump condition is satisfied:

$$\Delta\varphi|_{t=\tau_i} = \varphi(\tau_i + 0) - \varphi(\tau_i - 0) = I_i(\varphi(\tau_i - 0)). \tag{1.12}$$

In what follows, the value of the function $\varphi(t)$ at a point t' is understood as $\lim_{t\uparrow t'} \varphi(t)$, i.e., if τ_i is a point of discontinuity of $\varphi(t)$, then we assume that $\varphi(t)$ is left-continuous and

$$\varphi(\tau_i) = \varphi(\tau_i - 0) = \lim_{t\uparrow\tau_i} \varphi(t). \tag{1.13}$$

Following [94], we present several general theorems on properties of solutions of the systems of Eqs. (1.10). We assume that the function $f(t, x)$ is defined in the entire space $(t, x) \in \mathbb{R}^{n+1}$ (the case where it is defined in a certain domain of this space can be considered by analogy). We also assume that the solutions of the system of equations (1.1) possess the following properties:

(i) *extendability:* every solution $x(t)$ is a continuous function defined on an interval (a, b), $-\infty \leq a < b \leq \infty$, which is individual for every solution; in this case, if $a > -\infty$ ($b < \infty$), then $\|x(a+0)\| = \infty$ ($\|x(b-0)\| = \infty$, respectively);

(ii) *local character:* if a function $x(t)$, $a < t < b$, satisfies condition (i) and, for any $t_0 \in (a, b)$, there exists $\varepsilon > 0$ such that the function $x(t)$ coincides with a certain solution on each of the intervals $(t_0 - \varepsilon, t_0)$ and $(t_0, t_0 + \varepsilon)$, then $x(t)$ is also a solution;

(iii) *solvability of the Cauchy problem:* for any t_0 and x_0, there exists at least one solution $x(t)$, $a < t < b$, for which $a < t_0 < b$ and $x(t_0) = x_0$.

These conditions are satisfied, in particular, for system (1.1) whose right-hand side is continuous or satisfies the Carathéodory conditions. Generally speaking, the operators A_i are not assumed to be one-to-one, i.e., for any $x \in \mathbb{R}^n$, $i \in K$, $A_i x$ is a certain (possibly empty) subset of $\mathbb{R}^n$.

The definition of impulsive system and the assumptions concerning solutions of system (1.1) yield the following statement:

Theorem 1 ([142]). *If the solutions of the system of equations* (1.1) *satisfy conditions* (i)–(iii), *then, for any* $t_0 \in \mathbb{R}$ *and* $x_0 \in \mathbb{R}^n$, *there exists at least one solution* $x(t)$, $a < t < b$, *of the impulsive system* (1.10) *for which* $a < t_0 \leq b$ *and either* $x(t_0) = x_0$ $(a \geq -\infty,\ b \leq \infty)$ *(for* $t_0 < b$*) or* $x(t_0 - 0) = x_0$ *(for* $t_0 = b$*). In this case, the following assertions are true:*

(a) *if* $a > -\infty$, *then either* $\|x(a+0)\| = \infty$ *or* $a = \tau_i$, $x(a+0)$ *exists (as a finite limit), and* $x(a+0) \notin A_i\mathbb{R}^n$;

(b) *if* $b < \infty$, *then either* $\|x(b-0)\| = \infty$ *or* $b = \tau_j$, $x(b-0)$ *exists, and* $A_j x(b-0) = \varnothing$.

A solution $x(t)$ of this type cannot be extended.

For any $M \subseteq \mathbb{R}^n$, we denote by $g(t, t_0)M$ the set of values of $x(t)$ for all solutions of system (1.1) for which $x(t_0) \in M$. Then an analogous set for solutions of system (1.10) takes the form $G(t, t_0)M$, where the mapping G is defined by the following relation for $t > t_0$:

$$G(t, t_0)M = g(t, \tau_i)A_i g(\tau_i, \tau_{i-1})A_{i-1} \cdots A_j g(\tau_j, t_0)M \tag{1.14}$$
$$(\tau_i < t < \tau_{i+1},\ i = j-1, j, \ldots,\ j = \min\{i : \tau_i \geq t_0\}).$$

Furthermore, if there are only finitely many times τ_i for $t > t_0$ and $\tau_m = \max\{\tau_i\}$, then relation (1.14) with $i = m$ holds for $\tau_m < t < \infty$ (in what follows, we do not make a special mention of this fact). For the construction of a solution of system (1.10) in the case where t decreases, i.e., for $t < t_0$, an analogous formula is valid, in which A_i should be replaced by the naturally introduced mappings A_i^{-1}. The introduction of the operator $G(t, t_0)$ of shift along the trajectories of a system with pushes allows one to reformulate, in an obvious manner, the conditions of boundedness, stability, etc., of solutions of this system in terms of properties of this operator.

As a remark on Theorem 1, we note that if one additionally assumes that $A_i x \neq \varnothing$, then $\|x(b-0)\| = \infty$ for $b < \infty$. If one assumes instead that $A_i\mathbb{R}^n = \mathbb{R}^n$ $(i \in K)$, then $\|x(a+0)\| = \infty$ for $a > -\infty$. The case where $A_i x^* = \varnothing$ corresponds, according to Vogel, to the "death" of a trajectory that hits the point x^* at time τ_i. Thus, the set $\{x : A_j x = \varnothing\}$ serves as the "set of death" of trajectories at time τ_j.

For example, a solution $x = \varphi(t)$, $\varphi(0) = 0$, of the impulsive equation

$$\frac{dx}{dt} = 1, \quad t \neq \tau_i, \quad \Delta x|_{t=\tau_i} = \ln(1-x), \tag{1.15}$$

where $\tau_i = i$, $i = 1, 2, \ldots$, cannot be extended to the interval $[0, 2]$, and the time $t = 2$ is the time of death of this solution. Indeed, for $0 \leq t < 2$, this solution is determined by the equality $x = \varphi(t) = t$ (for $t = \tau_1 = 1$, one has $\varphi(\tau_1) = 1$, and, therefore, this solution does not have a discontinuity at $t = \tau_1$ because $\ln(2-\varphi(1)) = 0$). For $t = \tau_2 = 2$, one has $\varphi(2) = 2$, and the function $\ln(2-x)$ is not defined at the point $x = \varphi(2)$. Thus, this solution dies at time $t = \tau_2$.

Theorem 2 ([142]). *For the uniqueness of a solution of the Cauchy problem for the impulsive system* (1.10) *with arbitrary initial data in the case where t increases, it is necessary and sufficient that system* (1.1) *possess this property for any $t_0 \neq \tau_i$ and that, for any $t_0 = \tau_i$ and $x_0 \in A_i\mathbb{R}^n$, each of the sets $A_i x$ contain at most one element. For the uniqueness of a solution of the Cauchy problem for system* (1.10) *in the case where t decreases, it is necessary and sufficient that system* (1.1) *possess this property for any t_0 and that each of the sets $A_i^{-1}x$ contain at most one point.*

Thus, even if a solution of the Cauchy problem for system (1.1) is unique, solutions of an impulsive system can split or merge in the course of their extension under the action of the operators A_i. For the unbounded extendability of all solutions of system (1.10) forward (back) in time, it is necessary and sufficient that the solutions of system (1.1) possess this property and $A_i x \neq \varnothing$ (respectively, $A_i\mathbb{R}^n = \mathbb{R}^n$) for all $i \in K$.

One should not think that if a solution of the Cauchy problem for the system of equations (1.1) cannot be extended, say, to the interval $[t_0, t_0 + h]$, $h > 0$, then a solution of the corresponding Cauchy problem for the system of equations (1.10) cannot also be extended to this interval. For example, a solution $x = \varphi(t)$, $\varphi(0) = 0$, of the equation

$$\frac{dx}{dt} = 1 + x^2$$

cannot be extended to the interval $[0, \pi/2]$ (this solution goes to infinity in finite time: $\varphi(t) = \tan t \to \infty$ as $t \uparrow \pi/2$). However, considering a solution $x = \varphi(t)$, $\varphi(0) = 0$, of the impulsive equation

$$\frac{dx}{dt} = 1 + x^2, \quad t \neq \tau_i, \quad \Delta x|_{t=\tau_i} = -1, \quad \tau_i = \frac{\pi i}{4},$$

we conclude that this solution is extendable for all $t \geq 0$. It is easy to prove that this solution is periodic with period $\pi/4$ for $t \geq 0$. For $t \in (0, \pi/4]$, this function is determined by the equality $\varphi(t) = \tan t$, i.e.,

$$\varphi(t) = \tan\left(t - \frac{\pi i}{4}\right) \quad \text{for } t \in (\tau_i, \tau_{i+1}].$$

We assume that the solutions of system (1.1) also possess the following property:

(iv) *local compactness:* for any t_0 and x_0, there exists $\varepsilon > 0$ such that if $|\bar{t}_0 - t_0| < \varepsilon$ and $\|\bar{x}_0 - x_0\| \leq \varepsilon$, then any solution $x(t)$ for which $x(\bar{t}_0) = \bar{x}_0$ exists on the interval $[t_0 - \varepsilon, t_0 + \varepsilon]$, and the set of these solutions for fixed t_0, x_0, and ε is compact (in itself) in the metric of $C[t_0 - \varepsilon, t_0 + \varepsilon]$.

We also assume that the mappings A_i are upper semicontinuous.

Theorem 3 ([142]). *Suppose that, under the assumptions introduced above, for given t_0 and $x(t_0)$ and a nonempty compact set $K \subset \mathbb{R}^n$ all solutions of system* (1.1) *for* *which* $x(t_0) \in K$ *exist on a certain interval* $t_0 \le t \le T$, $T < \infty$. *Then, for some* $\varepsilon > 0$, *any solution* $\overline{x}(t)$ *satisfying the condition* $\rho(\overline{x}(\overline{t}_0), K) \le \varepsilon$, $|\overline{t}_0 - t_0| \le \varepsilon$, *exists on the entire interval* $t_0 - \varepsilon \le t \le T$, *and the set of these solutions for fixed* t_0, K, T, *and* ε *is compact in the metric of uniform deviations for discontinuous functions. If* $t_0 = \tau_i$, *then the assertion presented above is valid under the additional condition* $\overline{t}_0 \le t_0$.

Note that if $T = \tau_i$, then we can take the segment $[t_0-\varepsilon, T+\varepsilon]$ instead of $[t_0-\varepsilon, T]$. To extend Theorem 3 to the segment $T_1 \le t \le t_0$, $T_1 < t_0$, one should assume that the mappings A_i^{-1} are upper semicontinuous and take the segment $[T_1 - \varepsilon, t_0 + \varepsilon]$, $t_0 \ne \tau_i$, or $[T_1 - \varepsilon, t_0]$, $t_0 = \tau_i$, instead of $[t_0 - \varepsilon, T]$.

Corollary 1. *Under the additional assumption that a solution of the Cauchy problem for system* (1.1) *is unique and the mappings* A_i *are bijective, the solution* $x(t, t_0, x_0)$ *of the impulsive system* (1.10) *depends continuously on* $t_0 \ne \tau_i$ *and* x_0 *on every closed interval of the axis* t *on which it is defined; for* $t_0 = \tau_i$, *this dependence is left continuous.*

Corollary 2. *Under the conditions of Theorem* 3, *the set* $G(t, t_0)K$, $t_0 \le t \le T$, *is compact for every* t *and depends continuously on* $t \ne \tau_i$; *furthermore, for* $t = \tau_i$, *this dependence is left continuous and*

$$G(\tau_i + 0, t_0)K = A_i G(\tau_i, t_0)K.$$

The dependence of $G(t, t_0)K$ *on* K *is upper semicontinuous uniformly in* t. *If system* (1.1) *possesses the Knezer property of connectedness of a section of an integral funnel, all sets* $A_i x$ *are connected, and* K *is connected, then the set* $G(t, t_0)K$ *is connected for every* $t \in [t_0, T]$.

We now present sufficient conditions that must be satisfied by the system of Eqs. (1.10) in order that its solutions depend continuously on the initial data and right-hand sides. For what follows, we need the lemmas presented below.

Lemma 1 ([142]). *Suppose that a nonnegative piecewise-continuous function* $u(t)$ *satisfies the following inequality for* $t \ge t_0$:

$$u(t) \le C + \int_{t_0}^{t} v(s)u(s)ds + \sum_{t_0 \le \tau_i < t} \beta_i u(\tau_i),$$

where $C \ge 0$, $\beta_i \ge 0$, $v(t)$ *is a positive continuous function, and* τ_i *are the points of discontinuity of the first kind of the function* $u(t)$. *Then the following estimate holds for the function* $u(t)$:

$$u(t) \le C \prod_{t_0 \le \tau_i < t} (1 + \beta_i) \exp\left(\int_{t_0}^{t} v(s)ds\right).$$

Lemma 2 ([142]). *Suppose that a nonnegative piecewise-continuous function $u(t)$ satisfies the following inequality for $t \geq t_0$:*

$$u(t) \leq C + \int_{t_0}^{t} \gamma u(s)ds + \sum_{t_0 \leq \tau_i < t} \beta u(\tau_i),$$

where $C \geq 0$, $\beta \geq 0$, $\gamma > 0$, and τ_i are the points of discontinuity of the first kind of the function $u(t)$. Then the following estimate holds for the function $u(t)$:

$$u(t) \leq C(1+\beta)^{i(t_0,t)} e^{\gamma(t-t_0)},$$

where $i(t_0, t)$ is the number of points τ_i on the interval $[t_0, t)$.

Lemma 3 ([142]). *Suppose that a nonnegative piecewise-continuous function $u(t)$ satisfies the following inequality for $t \geq t_0$:*

$$u(t) \leq \alpha + \int_{t_0}^{t} [\beta + \gamma u(s)]ds + \sum_{t_0 \leq \tau_i < t} [\beta + \gamma u(\tau_i)],$$

where $\alpha \geq 0$, $\beta \geq 0$, $\gamma > 0$, and τ_i are the points of discontinuity of the first kind of the function $u(t)$. Then the following estimate is true:

$$u(t) \leq \left(\alpha + \frac{\beta}{\gamma}\right)(1+\gamma)^{i(t_0,t)} e^{\gamma(t-t_0)} - \frac{\beta}{\gamma}.$$

Using these lemmas, we estimate the variation of solutions of the system of Eqs. (1.10) corresponding to the variation of the initial data and right-hand sides of this system.

Assume that the functions $f(t,x)$ and $I_i(x)$ are continuous in their variables for $x \in M$ and $t \in I = [t_0, t_0 + T]$ and satisfy the Lipschitz condition with respect to x uniformly in $t \in I$ and $i \in K$, i.e.,

$$\begin{aligned} \|f(t,x') - f(t,x'')\| &\leq L\|x' - x''\|, \\ \|I_i(x') - I_i(x'')\| &\leq L\|x' - x''\|. \end{aligned} \tag{1.16}$$

Parallel with Eq. (1.10), we consider the system of equations

$$\begin{aligned} \frac{dy}{dt} &= f(t,y) + R(t,y), \quad t \neq \tau_i, \\ \Delta y|_{t=\tau_i} &= I_i(y) + R_i(y), \end{aligned} \tag{1.17}$$

where the functions $R(t,y)$ and $R_i(y)$ are such that solutions of system (1.17) exist.

Assume that the inequalities

$$\|R(t,y)\| \leq \eta \quad \text{and} \quad \|R_i(y)\| \leq \eta \tag{1.18}$$

hold for all $x \in M$ and $t \in I$. Consider solutions $x(t, x_0)$ of the system of equations (1.10) and solutions $y(t, x_0)$ of the system of equations (1.17). Assume that these solutions are defined for $t_0 \leq t \leq t_0 + T$ and

$$\|x_0 - y_0\| < \delta. \tag{1.19}$$

Theorem 4 ([142]). *If the functions that determine the systems of Eqs.* (1.10) *and* (1.17) *satisfy inequalities* (1.16) *and* (1.18), *then the solutions* $x(t, x_0)$ *and* $y(t, y_0)$ *of systems* (1.10) *and* (1.17) *whose initial values satisfy inequality* (1.19) *admit the following estimate for all* $t_0 \leq t \leq t_0 + T$*:*

$$\|x(t, x_0) - y(t, y_0)\| < \left(\delta + \frac{\eta}{L}\right)(1 + L)^{i(t_0,t)} e^{L(t-t_0)} - \frac{\eta}{L}. \tag{1.20}$$

Proof. The solution $x(t, x_0)$, $x(t_0, x_0) = x_0$, of system (1.10) for $t \in (\tau_i, \tau_{i+1}]$ coincides with one of solutions of the system of ordinary differential equations

$$\frac{dx}{dt} = f(t, x).$$

Since every solution $x = \varphi(t)$ of the latter system can be represented in the form

$$\varphi(t) = \varphi(\tau) + \int_\tau^t f(s, \varphi(s)) ds,$$

for $t \in (\tau_i, \tau_{i+1}]$ we have

$$x(t, x_0) = x(\tau_i, x_0) + I_i(x(\tau_i, x_0)) + \int_{\tau_i}^t f(s, x(s, x_0)) ds.$$

Hence, for $t \in [t_0, t_0 + T]$, the solution $x(t, x_0)$ admits the representation

$$x(t, x_0) = x_0 + \sum_{t_0 \leq \tau_i < t} I_i(x(\tau_i, x_0)) + \int_{t_0}^t f(s, x(s, x_0)) ds. \tag{1.21}$$

An analogous representation is valid for the solution $y(t, y_0)$ of the system of equations (1.17), i.e.,

$$y(t, y_0) = y_0 + \sum_{t_0 \leq \tau_i < t} [I_i(y(\tau_i, y_0)) + R_i(y(\tau_i, y_0))] + \int_{t_0}^t [f(s, y(s, y_0)) + R(s, y(s, y_0))] ds. \tag{1.22}$$

Therefore, for the norm of the difference of solutions $x(t, x_0) - y(t, y_0)$, we have

$$\begin{aligned}
&\|x(t, x_0) - y(t, y_0)\| \\
&\leq \|x_0 - y_0\| + \sum_{t_0 \leq \tau_i < t} [\|I_i(x(\tau_i, x_0)) - I_i(y(\tau_i, y_0))\| + \|R_i(y(\tau_i, y_0))\|] \\
&\quad + \int_{t_0}^{t} [\|f(s, x(s, x_0)) - f(s, y(s, y_0))\| + \|R(s, y(s, y_0))\|] ds.
\end{aligned}$$

Taking inequalities (1.16) and (1.17) into account, we obtain

$$\begin{aligned}
\|x(t, x_0) - y(t, y_0)\| \leq \|x_0 - y_0\| &+ \sum_{t_0 \leq \tau_i < t} [\eta + L\|x(\tau_i, x_0) - y(\tau_i, y_0)\|] \\
&+ \int_{t_0}^{t} [\eta + L\|x(s, x_0) - y(s, y_0)\|] ds.
\end{aligned}$$

In other words, the norm of the difference $\|x(t, x_0) - y(t, y_0)\|$ satisfies the conditions of Lemma 3 if we set

$$u(t) = \|x(t, x_0) - y(t, y_0)\|, \quad a = \|x_0 - y_0\|, \quad \beta = \eta, \quad \gamma = L.$$

By virtue of Lemma 3, we have

$$\|x(t, x_0) - y(t, y_0)\| \leq \left(\|x_0 - y_0\| + \frac{\eta}{L}\right)(1 + L)^{i(t_0,t)} e^{L(t - t_0)} - \frac{\eta}{L}. \tag{1.23}$$

The required inequality (1.20) follows directly from (1.23) because the initial values x_0 and y_0 satisfy condition (1.19). Theorem 4 is proved. □

We note some special cases of Theorem 4. Let $\eta = 0$. Then $x(t, x_0)$ and $y(t, y_0)$ are solutions of the same system of equations (1.10), but with different initial conditions. For these solutions, estimate (1.20) takes the form

$$\|x(t, x_0) - y(t, y_0)\| \leq \delta(1 + L)^{i(t_0,t)} e^{L(t - t_0)},$$

i.e., for all $t \in [t_0, t_0 + T]$ we have

$$\|x(t, x_0) - y(t, y_0)\| \leq \delta(1 + L)^p e^{LT}, \tag{1.24}$$

where p is the number of points τ_i on the interval $[t_0, t_0 + T]$.

It follows from inequality (1.24) that, for any $\varepsilon > 0$, one can find a number $\delta = \delta(\varepsilon) = \varepsilon(1 + L)^{-p} e^{-LT}$ such that if $\|x_0 - y_0\| < \delta$, then $\|x(t, x_0) - y(t, y_0)\| < \varepsilon$ for all $t \in [t_0, t_0 + T]$. This means that if inequalities (1.16) are satisfied, then the solutions of system (1.10) depend continuously on the initial conditions. Moreover, it follows from estimate (1.23) that the solutions of the system of equations (1.10) satisfy the Lipschitz condition with respect to x_0 uniformly in $t \in [t_0, t_0 + T]$:

$$\|x(t, x_0) - y(t, y_0)\| \leq (1 + L)^p e^{LT} \|x_0 - y_0\|. \tag{1.25}$$

If $\delta = 0$ and $\eta \neq 0$, then we have the case of permanently acting perturbations. Therefore, estimate (1.20) takes the form

$$\|x(t, x_0) - y(t, y_0)\| \leq \frac{\eta}{L}((1 + L)^{i(t_0,t)} e^{L(t-t_0)} - 1),$$

i.e.,

$$\|x(t, x_0) - y(t, y_0)\| \leq \frac{\eta}{L}((1 + L)^p e^{LT} - 1) \tag{1.26}$$

for all $t \in [t_0, t_0 + T]$.

It follows from (1.26) that, for any number $\varepsilon > 0$, one can find $\eta = \eta(\varepsilon) = \varepsilon L((1 + L)^p e^{LT} - 1)^{-1}$ such that if inequalities (1.18) are satisfied, then

$$\|x(t, x_0) - y(t, y_0)\| < \varepsilon \quad \text{for all } t \in [t_0, t_0 + T].$$

This statement expresses the property of continuity of solutions of the impulsive system (1.10) in some functional space of right-hand sides. In particular, if the right-hand sides of system (1.10) depend continuously on a certain parameter λ, then the obtained estimates yield the continuity of the solutions with respect to this parameter.

Systems with Nonfixed Times of Pulse Action. The systems subjected to pulse action at the times when the integral curve crosses given surfaces of the extended phase space are more complicated than the systems subjected to pulse action at fixed times. If the integral curve is subjected to pulse action when it hits the surface defined by the equation $\Phi(t, x) = 0$ in the extended phase space, then the indicated impulsive system of differential equations can be written in the form

$$\frac{dx}{dt} = f(t, x), \quad \Phi(t, x) \neq 0, \tag{1.27}$$
$$\Delta x|_{\Phi(t,x)=0} = I(t, x)|_{\Phi(t,x)=0}.$$

In this case, the sets Γ_t and Γ'_t used in the definition of impulsive system are defined as

$$\Gamma_t = \{(t, x)\colon \Phi(t, x) = 0\},$$
$$\Gamma'_t = \{(t, x)\colon \Phi(t, A_t^{-1}(t, x)) = 0\},$$

and the operator $A_t\colon \Gamma_t \to \Gamma'_t$ acts according to the rule

$$A_t\colon (t, x) \to (t, x + I(t, x)).$$

With regard for the form of the function $\Phi(t, x)$, we can distinguish the following cases:

(I) $\Phi(t, x)$ does not explicitly depend on x, i.e., the equation of the impulsive surface takes the form $\Phi(t) = 0$. In this case, the times of pulse action are fixed and correspond to the roots of the equation $\Phi(t) = 0$. The following qualitatively different cases are possible in this case:

(1) finitely many pulses on a finite interval, e.g., $\Phi_1(t) = \sin t$;

(2) countably many pulses with one accumulation point, e.g.,

$$\Phi_2(t) = \begin{cases} \sin \frac{1}{T_0 - t}, & t \neq T_0, \\ 0, & t = T_0; \end{cases}$$

(3) countably many pulses with several accumulation points, e.g.,

$$\Phi_3(t) = \begin{cases} \prod_{i=1}^{k} \sin \frac{1}{T_i - t}, & t \neq T_i,\ i = \overline{1, k}, \\ 0, & t = T_i,\ i = \overline{1, k}; \end{cases}$$

(4) countably many pulses with an accumulation point that is the limit point of accumulation points, e.g.,

$$\Phi_4(t) = \begin{cases} \sin \frac{1}{\Phi_2(t)}, & \Phi_2(t) \neq 0, \\ 0, & \Phi_2(t) = 0. \end{cases}$$

(II) $\Phi(t, x)$ does not explicitly depend on t, i.e., the equation of the impulsive surface has the form $\Phi(x) = 0$.

(III) $\Phi(t, x)$ depends explicitly on t and x. The simplest case here is the case where the equation $\Phi(t, x) = 0$ has finitely many simple roots $\tau_i(x)$, though all situations analogous to those in case (I) are also possible.

In what follows, we assume that the equation $\Phi(t, x) = 0$ is solvable with respect to t; furthermore, we assume that it has countably many solutions if the system is studied on the entire real axis or for $t \geq t_0$, and finitely many solutions if the system is considered on a finite time interval. We denote these solutions by $t = \tau_i(x)$ and enumerate them by the set of integers (or its subset) so that $\tau_i(x) \to \infty$ as $i \to \infty$, and $\tau_i(x) \to -\infty$ as $i \to -\infty$.

The restriction of the operator A'_t to the hypersurface $t = \tau_i(x)$ defines an operator acting according to the rule

$$x \to A_{\tau_i(x)} x \equiv x^+ = x + I(\tau_i(x), x) \equiv x + I_i(x). \tag{1.28}$$

Therefore, system (1.27) can also be written in the form

$$\begin{aligned} \frac{dx}{dt} &= f(t, x), \quad t \neq \tau_i(x), \\ \Delta x|_{t=\tau_i(x)} &= I_i(x). \end{aligned} \tag{1.29}$$

In contrast to systems of the form (1.10), despite the fact that the solutions of system (1.29) are piecewise-continuous functions, the points of discontinuity depend on a solution, i.e., every solution has its own points of discontinuity in this case. This substantially complicates the investigation of these systems of equations.

One of the difficulties consists of beating of solutions against the surface $t = \tau_i(x)$. Exactly due to beating, a solution quite often cannot be extended to a sufficient interval. This phenomenon has already been mentioned in Example 1. Moreover, beating is quite often the main reason for a solution to leave the domain of definition of an impulsive system or the domain where it is studied.

As an illustration, one may consider the following impulsive system:

$$\frac{dx}{dt} = -x, \quad t \neq \tau_i(x), \quad \Delta x|_{t=\tau_i(x)} = \alpha x,$$

where $\alpha \in \mathbb{R}$ and $\tau_i(x) = \arctan x + i\pi$, $i = 0, 1, 2, \ldots$. If $\alpha > 0$, then the integral curve of any solution $x = \varphi(t)$, $\varphi(0) = x_0 \leq 0$, of this impulsive system of equations crosses every curve $t = \tau_i(x)$ only once (Figure 3); the integral curve of the solution $x = \varphi(t)$, $\varphi(0) = x_0 > 0$, hits the curve $t = \arctan x$ countably many times. In other words, the solution $x = \varphi(t)$ beats against impulsive surfaces. In the case under consideration, the integral curve of each solution goes to infinity in time less than $\frac{\pi}{2}$.

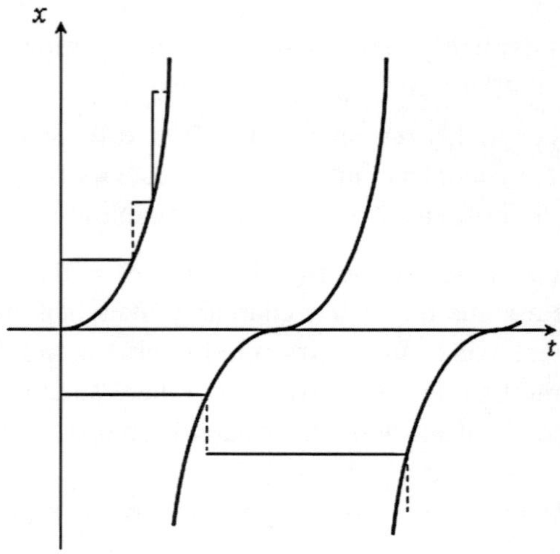

Figure 3. Case $\alpha > 0$.

However, if $-1 < \alpha < 0$ in this example, then the integral curve of the solutions $x = \varphi(t)$ for which $\varphi(0) = x_0 \geq 0$ hits each line $t = \tau_i(x)$ once (Figure 4).

The solutions $x = \varphi(t)$ for which $\varphi(0) = x_0 < 0$ undergo beating: the integral curve of each of these solutions hits the curve $t = \pi + \arctan x$ countably many times and approaches the straight line $x = 0$ as $t \uparrow \pi$. In this case, each solution $x = \varphi(t)$,

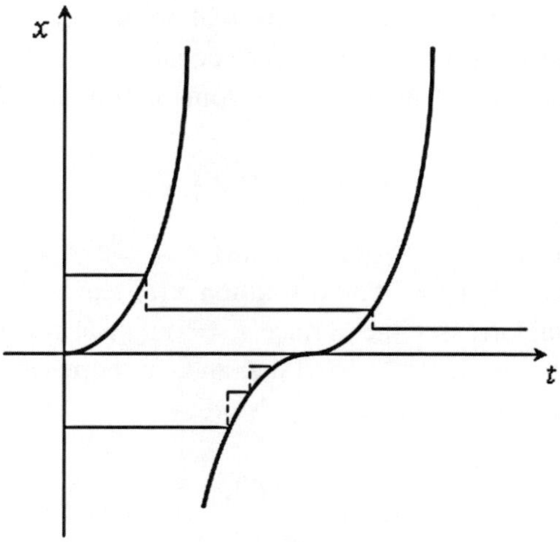

Figure 4. Case $-1 < \alpha < 0$.

$\varphi(0) = x_0 < 0$, cannot be extended to the interval $[0, a]$, where $a \geq \pi$, and the integral curve of each solution of this type tends to the point $(\pi, 0)$ as $t \uparrow \pi$.

In what follows, we consider impulsive systems of differential equations without beating. In other words, we consider equations whose solutions cross each hypersurface $t = \tau_i(x)$ only once. A sufficient condition for the absence of beating is given by the following lemma:

Lemma 4 ([142]). *Let the functions $f(t, x)$, $I_i(x)$, and $\tau_i(x)$ that determine the system of equations* (1.29) *be continuous for $(t, x) \in I \times D$, let $\tau_i(x)$ be continuously differentiable with respect to $x \in D$, and let*

$$\max_{(t,x)\in I\times D} \|f(t, x)\| \leq C, \quad \max_{x\in D} \left\| \frac{\partial \tau_i(x)}{\partial x} \right\| \leq N, \tag{1.30}$$

where D is a certain compact set of the phase space and C and N are positive numbers. Also assume that the inequality

$$\max_{0\leq\sigma\leq 1} \left(\frac{\partial \tau_i(x + \sigma I_i(x))}{\partial x}, I_i(x) \right) \leq 0 \tag{1.31}$$

holds for all $x \in D$.

Then one can find a positive number N_0 such that, for all $N \leq N_0$, the integral curve of any solution $x(t)$ of the system of equations (1.29) *belonging to the domain D for $t_0 \leq t \leq t_0 + T$ ($[t_0, t_0 + T] \subseteq I$) crosses each hypersurface $t = \tau_i(x)$ from the interval $[t_0, t_0 + T]$ only once.*

Proof. To prove the lemma, it suffices to show that, for sufficiently small values of the constant N, any solution $x(t)$ of the system of equations (1.29) that leaves the point $x_0 + I_i(x_0)$ at $t = \tau_i(x_0) + 0$ and lies in the domain D at least for $\tau_i(x_0) < t < \bar{t}_i$, where

$$\bar{t}_i = \max_{x \in D} \tau_i(x),$$

does not hit the surface $t = \tau_i(x)$ for $t > \tau_i(x_0)$.

Assume the contrary. Suppose that a solution $x(t)$ leaves the point $x_0 + I_i(x_0)$ at $t = \tau_i(x_0) + 0$ and crosses the surface $t = \tau_i(x)$ at a certain point (t_i^*, x^*), $t_i^* = \tau_i(x^*)$. It is obvious that $t_i^* > \tau_i(x_0)$, and, furthermore, the solution $x(t)$ is continuous on the interval $\tau_i(x_0) < t < t_i^*$. Moreover,

$$x^* = x_0 + I_i(x_0) + \int_{\tau_i(x_0)}^{t_i^*} f(s, \varphi(s))ds. \tag{1.32}$$

Consider the difference $t_i^* - \tau_i(x_0)$. We have

$$\begin{aligned} t_i^* - \tau_i(x_0) &= \tau_i(x^*) - \tau_i(x_0) \\ &= \tau_i(x^*) - \tau_i(x_0 + I_i(x_0)) + \tau_i(x_0 + I_i(x_0)) - \tau_i(x_0) \\ &= \int_0^1 \left(\frac{\partial \tau_i}{\partial x}(x_0 + I_i(x_0) + \sigma h), h \right) d\sigma \\ &\quad + \int_0^1 \left(\frac{\partial \tau_i}{\partial x}(x_0 + \sigma I_i(x_0)), I_i(x_0) \right) d\sigma, \end{aligned} \tag{1.33}$$

where

$$h = \int_{\tau_i(x_0)}^{t_i^*} f(s, \varphi(s))ds.$$

Inequalities (1.30) imply that the first term on the right-hand side of equality (1.33) admits the following estimate by virtue of the Cauchy–Schwarz inequality:

$$\int_0^1 \left(\frac{\partial \tau_i}{\partial x}(x_0 + I_i(x_0) + \sigma h), h \right) d\sigma \le NC(t_i^* - \tau_i(x_0)). \tag{1.34}$$

Therefore,

$$(1 - NC)(t_i^* - \tau_i(x_0)) \le \int_0^1 \left(\frac{\partial \tau_i}{\partial x}(x_0 + \sigma I_i(x_0)), I_i(x_0) \right) d\sigma. \tag{1.35}$$

To complete the proof the lemma, it suffices to choose N_0 from the condition $CN_0 \le 1$ because inequality (1.35) becomes contradictory in this case by virtue of condition (1.31). Lemma 4 is proved. □

The lemma below gives conditions under which there is no beating of solutions of the system of equations (1.29) against the surfaces $t = \tau_i(x)$ in the case where the functions $\tau_i(x)$ are not continuously differentiable.

Lemma 5 ([142]). *Suppose that the functions $f(t,x)$ and $I_i(x)$ in the system of equations* (1.29) *satisfy the conditions of the previous lemma and the functions $\tau_i(x)$ satisfy the Lipschitz condition*

$$|\tau_i(x') - \tau_i(x'')| \leq N\|x' - x''\|, \quad x', x'' \in D. \tag{1.36}$$

Also assume that the following inequalities hold for all $x \in D$ and i:

$$\tau_i(x) \geq \tau_i(x + I_i(x)). \tag{1.37}$$

Then one can find $N_0 > 0$ such that, for all $N \leq N_0$, the integral curve of any solution $x(t)$ of the system of equations (1.29) *that belongs to the domain D for $t_0 \leq t \leq t_0 + T$ crosses each hypersurface $t = \tau_i(x)$ from the interval $[t_0, t_0 + T]$ only once.*

Proof. The proof of this lemma is carried out by analogy with the proof of the previous lemma with relation (1.34) replaced by the following:

$$\begin{aligned}|\tau_i(x^*) - \tau_i(x_0 + I_i(x_0))| &\leq N \int_{\tau_i(x_0)}^{\tau_i(x^*)} \|f(s, x(s))\| ds \\ &\leq NC(\tau_i(x^*) - \tau_i(x_0)).\end{aligned}$$

Therefore,

$$\begin{aligned}t_i^* - \tau_i(x_0) &= \tau_i(x^*) - \tau_i(x_0) \\ &= \tau_i(x^*) - \tau_i(x_0 + I_i(x_0)) + \tau_i(x_0 + I_i(x_0)) - \tau_i(x_0) \\ &\leq NC(\tau_i(x^*) - \tau_i(x_0)) + \tau_i(x_0 + I_i(x_0)) - \tau_i(x_0),\end{aligned}$$

i.e.,

$$(1 - NC)(t_i^* - \tau_i(x_0)) \leq \tau_i(x_0 + I_i(x_0)) - \tau_i(x_0). \tag{1.38}$$

If N is so small that $1 - NC > 0$, then the last inequality becomes contradictory by virtue of (1.37), which completes the proof of the lemma. □

We have proved earlier that the solutions of the system of equations (1.10) subjected to pulse action at fixed times depend continuously on the initial conditions if the functions $f(t,x)$ and $I_i(x)$ satisfy the Lipschitz condition. Furthermore, the continuous dependence of the solutions on the initial conditions is uniform in $t \in [t_0, t_0 + T]$.

The solutions of the impulsive systems (1.29) do not possess this property, which is well illustrated by the following example:

$$\frac{dx}{dt} = 0, \quad t \neq \tau_i(x), \quad \Delta x|_{t=\tau_i(x)} = b, \quad \tau_i(x) = 2i - x, \quad b > 0.$$

Consider two solutions of this system of equations that leave the points $x = 1$ and $x = 1 + \alpha$ at $t = 0$, where α is a number arbitrarily small in modulus. Denote these solutions by $\varphi(t)$ and $\psi(t)$, respectively, i.e., $\varphi(0) = 1$ and $\psi(0) = 1 + \alpha$. No matter how small α is on the interval $[0, 1]$, there is no continuous dependence on x_0 because this interval contains the interval $[1-\alpha, 1]$ if $\alpha > 0$ (or $[1, 1+|\alpha|]$ if $\alpha < 0$), on which $|\varphi(t) - \psi(t)| = b + |\alpha|$, although the difference $|\varphi(t) - \psi(t)|$ is arbitrarily small outside this interval whenever α is sufficiently small.

However, if sufficiently small neighborhoods of the times when the integral curve crosses the surfaces $t = \tau_i(x)$ are removed from the interval $[t_0, t_0 + T]$, then the continuous dependence of solutions on the initial data is uniform with respect to the remaining values of the independent variable.

Assume that the functions $f(t, x)$, $I_i(x)$, and $\tau_i(x)$ in Eqs. (1.29) are continuous for $(t, x) \in I \times D$ and the following inequalities hold for all $t \in I$ and $x, x', x'' \in D$:

$$\begin{aligned} &\|f(t,x)\| \leq C, && \|f(t,x') - f(t,x'')\| \leq L\|x' - x''\|, \\ &\|I_i(x') - I_i(x'')\| \leq L\|x' - x''\|, && |\tau_i(x') - \tau_i(x'')| \leq N\|x' - x''\|. \end{aligned} \tag{1.39}$$

Let $x(t, x_0)$ and $x(t, y_0)$ be two solutions of Eqs. (1.29) that belong to the domain D for all $t \in [t_0, t_0 + T]$. Assume that each of these solutions crosses each hypersurface $t = \tau_i(x)$ only once, and denote the times when these solutions cross the surfaces $t = \tau_i(x)$ by $\tau_i^{x_0}$ and $\tau_i^{y_0}$, respectively. The following lemma is true:

Lemma 6 ([142]). *If the conditions presented above are satisfied and* $NC < 1$, *then*

$$\|x(t, x_0) - x(t, y_0)\| \leq \left(1 + \frac{L}{1 - NC}\right)^p e^{LT} \|x_0 - y_0\| \tag{1.40}$$

for all $t \in (\overline{\tau}_i, \underline{\tau}_{i+1}]$, *where* $\underline{\tau}_i = \min(\tau_i^{x_0}, \tau_i^{y_0})$ *and* $\overline{\tau}_i = \max(\tau_i^{x_0}, \tau_i^{y_0})$.

Proof. Without loss of generality, we assume that the hyperplanes $t = t_0$ and $t = t_0 + T$ do not cross the hypersurfaces $t = \tau_i(x)$. Let us represent the solutions $x(t, x_0)$ and $x(t, y_0)$ indicated in the conditions of the lemma in the integral form:

$$x(t, x_0) = x_0 + \int_{t_0}^{t} f(s, x(s, x_0))ds + \sum_{t_0 < \tau_i^{x_0} < t} I_i(x(\tau_i^{x_0}, x_0)),$$

$$x(t, y_0) = y_0 + \int_{t_0}^{t} f(s, x(s, y_0))ds + \sum_{t_0 < \tau_i^{y_0} < t} I_i(x(\tau_i^{y_0}, y_0)).$$

Estimating the difference of these solutions, we get

$$\begin{aligned}
&\|x(t,x_0)-x(t,y_0)\| \\
&\le \|x_0-y_0\| + L\int_{t_0}^{t}\|x(s,x_0)-x(s,y_0)\|ds \\
&\quad + \left\| \sum_{t_0<\tau_i^{x_0}<t} I_i(x(\tau_i^{x_0},x_0)) - \sum_{t_0<\tau_i^{y_0}<t} I_i(x(\tau_i^{y_0},y_0)) \right\|. \qquad (1.41)
\end{aligned}$$

If $t \in [t_0, t_0+T]\setminus\bigcup_i[\underline{\tau}_i,\overline{\tau}_i]$, where the union is taken over i for which $\tau_i^{x_0}, \tau_i^{y_0} \in [t_0, t_0+T]$, then

$$\begin{aligned}
&\left\| \sum_{t_0<\tau_i^{x_0}<t} I_i(x(\tau_i^{x_0},x_0)) - \sum_{t_0<\tau_i^{y_0}<t} I_i(x(\tau_i^{y_0},y_0)) \right\| \\
&\le L \sum_{t_0<\underline{\tau}_i<t} (\|x(\underline{\tau}_i,x_0)-x(\underline{\tau}_i,y_0)\| + \|x(\tau_i^{x_0},z_0)-x(\tau_i^{y_0},z_0)\|),
\end{aligned}$$

where $z_0 = x_0$ if $\underline{\tau}_i = \tau_i^{y_0}$, and $z_0 = y_0$ if $\underline{\tau}_i = \tau_i^{x_0}$. The second term on the right-hand side of the last inequality admits the estimate

$$\begin{aligned}
\|x(\tau_i^{x_0},z_0)-x(\tau_i^{y_0},z_0)\| &\le \int_{\underline{\tau}_i}^{\overline{\tau}_i}\|f(s,x(s,z_0))\|ds \\
&\le C|\tau_i^{x_0}-\tau_i^{y_0}| \le \frac{NC}{1-NC}\|x(\underline{\tau}_i,x_0)-x(\overline{\tau}_i,y_0)\|
\end{aligned}$$

because

$$\begin{aligned}
|\tau_i^{x_0}-\tau_i^{y_0}| &\le N\|x(\tau_i^{x_0},x_0)-x(\tau_i^{y_0},y_0)\| \\
&\le N\|x(\underline{\tau}_i,x_0)-x(\underline{\tau}_i,y_0)\| + N\|x(\tau_i^{x_0},z_0)-x(\tau_i^{y_0},z_0)\| \\
&\le N\|x(\underline{\tau}_i,x_0)-x(\underline{\tau}_i,y_0)\| + NC|\tau_i^{x_0}-\tau_i^{y_0}|.
\end{aligned}$$

Relation (1.41) yields

$$\begin{aligned}
\|x(t,x_0)-x(t,y_0)\| &\le \|x_0-y_0\| + L\int_{t_0}^{t}\|x(s,x_0)-x(s,y_0)\|ds \\
&\quad + \frac{L}{1-NC}\sum_{t_0<\underline{\tau}_i<t}\|x(\underline{\tau}_i,x_0)-x(\underline{\tau}_i,y_0)\|.
\end{aligned}$$

Hence, by virtue of Lemma 2, we get

$$\|x(t,x_0)-x(t,y_0)\| \le \left(1+\frac{L}{1-NC}\right)^p e^{LT}\|x_0-y_0\|$$

for all $t \in [t_0, t_0+T]\setminus\bigcup_i[\underline{\tau}_i,\overline{\tau}_i]$, where p is the number of the points τ_i (or, which is the same, $\overline{\tau}_i$) on the interval $[t_0, t_0+T]$. Lemma 6 is proved. □

Theorem 5 ([142]). *Suppose that the functions $f(t,x)$, $I_i(x)$, and $\tau_i(x)$ from the system of equations* (1.29) *satisfy inequalities* (1.39), *relation* (1.37) *is true, and $NC < 1$. If a solution $x(t,x_0)$ of system* (1.29) *is determined for $t \in [t_0, t_0+T]$, then the continuous dependence of this solution on the initial condition x_0 holds in the following sense: for any $\varepsilon > 0$, one can find $\delta = \delta(\varepsilon) > 0$ such that, for any other solution $x(t,y_0)$ of system* (1.29), *the inequality $\|x_0 - y_0\| < \delta$ implies that*

$$\|x(t,x_0) - x(t,y_0)\| < \varepsilon \tag{1.42}$$

for all $t \in [t_0, t_0+T]$ that satisfy the condition $|t - \tau_i^{x_0}| > \varepsilon$, where $\tau_i^{x_0}$ are the times when the integral curve of the solution $x(t,x_0)$ crosses the hypersurfaces $t = \tau_i(x)$.

Proof. The conditions of Theorem 5 rule out beating of solutions of system (1.29) against the surfaces $t = \tau_i(x)$ and guarantee that the conditions of Lemma 6 are satisfied. According to Lemma 6, estimate (1.40) holds for two solutions $x(t,x_0)$ and $x(t,y_0)$ of Eqs. (1.29). We take an arbitrary $\varepsilon > 0$ and choose $\delta_1 > 0$ so small that

$$\max_{\|x_0 - y_0\| \le \delta_1} |\tau_i^{x_0} - \tau_i^{y_0}| < \varepsilon$$

for all i such that $\tau_i^{x_0} \in [t_0, t_0+T]$. We choose δ as follows:

$$\delta = \min\left(\delta_1, \varepsilon\left(1 + \frac{L}{1-NC}\right)^{-1} e^{-LT}\right),$$

where p is the number of points $\tau_i^{x_0}$ that belong to the interval $[t_0, t_0+T]$. If y_0 is such that $\|x_0 - y_0\| < \delta$, then, according to (1.40), the solutions $x(t,x_0)$ and $x(t,y_0)$ satisfy inequality (1.42) for all $t \in [t_0, t_0+T]$ such that $|t - \tau_i(x_0)| > \varepsilon$. Theorem 5 is proved. □

We indicate one more general property of solutions of the system of equations (1.29) that is not inherent in the solutions of equations subjected to pulse action at fixed times.

In Eqs. (1.10), two different solutions can merge into a single one beginning with a certain time of pulse action $t = \tau_i$ because the mapping $g_i : x \to x + I_i(x)$ is not bijective. If a point y is the preimage of at least two points x_1 and x_2 under the indicated mapping, then the integral curves that hit the points x_1 and x_2 at time $t = \tau_i$ merge into a single integral curve for $t > \tau_i$. For the system of equations (1.29), the described phenomenon is also possible in the case where the mapping g_i is bijective.

Let $\sigma_i(x) = \tau_i(g_i^{-1}(x))$, where g_i^{-1} is the mapping inverse to g_i. Assume that $\tau_i(x) < \sigma_i(x)$ for all x from the domain of definition of system (1.29). Each solution of Eqs. (1.29) that starts at a point (t_0, x_0) from the domain $\tau_i(x) < t < \sigma_i(x)$ and crosses the surface $t = \sigma_i(x)$ for $t > t_0$, say, at a point $(\sigma_i(\overline{x}), \overline{x})$, merges for $t > \sigma_i(\overline{x})$ with the solution that hits the point $g_i^{-1}(\overline{x})$ at $t = \sigma_i(\overline{x})$.

Analyzing the location of the surfaces $t = \tau_i(x)$ and $t = \sigma_i(x)$, one can establish the presence or absence of beating in the system of equations (1.29). It is known that the beating of solutions against the surface $t = \tau_i(x)$ occurs if the inequality $\sigma_i(x) < \tau_i(x)$ has solutions in the domain of definition of the impulsive system (1.29).

Discontinuous Dynamical Systems. A discontinuous dynamical system is defined by relations (a)–(c) in which the differential Eq. (1.1) does not explicitly depend on t, the sets Γ_t and Γ'_t are subsets of the phase space, and $A_t = A$ for all $t \in \mathbb{R}$. Thus, a discontinuous dynamical system can be represented in the form

$$\frac{dx}{dt} = f(x), \quad (t, x) \notin \Gamma, \tag{1.43}$$
$$\Delta x|_{x \in \Gamma} = Ax - x = I(x).$$

To obtain interesting effects in the systems under consideration, it is necessary that a moving point hit the set Γ fairly often. Some examples of these systems were considered in detail in [142].

1.2 Linear Systems

General Properties of Solutions of Linear Systems. We now establish some properties of solutions of homogeneous linear systems of differential equations with pulse action

$$\frac{dx}{dt} = A(t)x, \quad t \neq \tau_i, \quad \Delta x|_{t=\tau_i} = B_i x, \tag{1.44}$$

and inhomogeneous linear systems.

Here, $A(t)$ is an $n \times n$ matrix continuous on the interval I, B_i are constant matrices, and $\tau_i \in I$ are fixed times enumerated in the increasing order ($\tau_i < \tau_{i+1}$) by the set of integers or its subset.

One of the main results of the theory of linear systems of the form (1.44) is the following one:

Theorem 6 ([142]). *Suppose that the interval* $[t_0, t_0 + h] \subset I$ *contains finitely many points* τ_i. *Then, for any* $x_0 \in \mathbb{R}^n$, *the solution* $x(t, x_0)$, $x(t_0, x_0) = x_0$, *of the system of equations* (1.44) *exists for all* $t \in [t_0, t_0 + h]$. *Furthermore, if the matrices* $E + B_i$ *are not degenerate for all* i *such that* $\tau_i \in [t_0, t_0 + h]$, *then* $x(t, x_0) \neq x(t, y_0)$ *for all* $t \in [t_0, t_0 + h]$ *whenever* $x_0 \neq y_0$.

Proof. Let $t_0 < \tau_j < \tau_{j+1} < \cdots < \tau_{j+k} \leq t_0 + h$. By virtue of the Picard–Cauchy theorem, a solution $x = \varphi_j(t)$ of the system

$$\frac{dx}{dt} = A(t)x \tag{1.45}$$

with initial condition $\varphi_j(t_0) = x_0$ exists and is unique on the interval $[t_0, \tau_j]$ for any $x_0 \in \mathbb{R}^n$. For $t \in [t_0, \tau_j]$, we set $x(t, x_0) = \varphi_j(t)$. By virtue of system (1.44), for $t = \tau_j$ we have

$$x(\tau_j + 0, x_0) = (E + B_j)x(\tau_j, x_0) = x_j^+.$$

According to the Picard–Cauchy theorem, there exists a unique solution $x = \varphi_{j+1}(t)$, $\varphi_{j+1}(\tau_j) = x_j^+$, of the system of equations (1.45) on the interval $[\tau_j, \tau_{j+1}]$. Therefore, one can extend the solution $x(t, x_0)$ of the original impulsive system (1.44) to the time $t = \tau_{j+1}$ by setting

$$x(t, x_0) = \varphi_{j+1}(t) \quad \text{for } \tau_j < t \le \tau_{j+1}.$$

For $t = \tau_{j+1} + 0$, we have

$$x(\tau_{j+1} + 0, x_0) = (E + B_{j+1})x(\tau_{j+1}, x_0) = x_{j+1}^+.$$

Denoting the solution of system (1.45) with initial condition $\varphi_{j+2}(\tau_{j+1}) = x_{j+1}^+$ by $\varphi_{j+2}(t)$, we can extend the solution $x(t, x_0)$ of system (1.44) to the time τ_{j+2} by setting $x(t, x_0) = \varphi_{j+2}(t)$ for $\tau_{j+1} < t \le \tau_{j+2}$, etc.

Since, according to the conditions of the theorem, the interval $[t_0, t_0 + h]$ contains finitely many points τ_j, we can thus construct a solution $x(t, x_0)$ of system (1.44) on the entire interval.

We have indicated a method for the construction of a solution $x(t, x_0)$ under the assumption that $t_0 < \tau_j$. If $t_0 = \tau_j$, then the construction of $x(t, x_0)$ is the same as for $t_0 < \tau_j$ with the difference that, by setting

$$x(t_0 + 0) = x_j^+ = (E + B_j)x_0,$$

we construct the function $\varphi_{j+1}(t)$ as a solution of system (1.45) satisfying the initial condition $\varphi_{j+1}(\tau_j) = x_j^+$.

To prove the second assertion of the theorem, we note that if $x(\tau_i + 0, x_0) \ne x(\tau_i + 0, y_0)$, then, by virtue of the Picard–Cauchy theorem, one has $x(t, x_0) \ne x(t, y_0)$ for all $\tau_i < t \le \tau_{i+1}$, $i = \overline{j, j+k}$. Since

$$x(\tau_i + 0, x_0) - x(\tau_i + 0, y_0) = (E + B_i)(x(\tau_i, x_0) - x(\tau_i, y_0)),$$

the relation $x(\tau_i, x_0) \ne x(\tau_i, y_0)$ yields $x(\tau_i + 0, x_0) \ne x(\tau_i + 0, y_0)$ under the condition of nondegeneracy of the matrix $E + B_i$. Hence, $x(t, x_0) \ne x(t, y_0)$ for all $t \in [t_0, t_0 + h]$ whenever $x_0 \ne y_0$ and $\det(E + B_i) \ne 0$ for $i = \overline{j, j+k}$. Theorem 6 is proved. □

If $[t_0 - h, t_0] \subset I$ and the matrices $E + B_i$ are nondegenerate for all i such that $\tau_i \in [t_0 - h, t_0]$, then the solution $x(t, x_0)$ can be uniquely extended to the interval $[t_0 - h, t_0]$. If the matrix $E + B_i$ is degenerate at one of the indicated points $\tau_i = \tau_{i_1}$, then the solution $x(t, x_0)$ can be uniquely extended to the left only up to the time τ_{i_1}. The extension of this solution to the interval $[t_0 - h, t_0]$ is either impossible, or possible, but in an ambiguous way.

Indeed, let τ_{i_1} be the time closest from the left to t_0 such that the matrix $E + B_i$ with its number is degenerate. Let r be the rank of the matrix $E + B_{i_1}$. The solution $x(t, x_0)$ can be uniquely extended to the interval $(\tau_{i_1}, t_0]$. Denote $x^+ = x(\tau_{i_1} + 0, x_0)$. The linear operator in $\mathbb{R}^n$ determined by the matrix $E + B_{i_1}$ projects the space $\mathbb{R}^n$ to a linear subspace $\mathbb{R}^r$ of this space. If x^+ is contained in the set of images of the indicated operator, then the solution $x(t, x_0)$ can be extended to the left of the point τ_{i_1}. Moreover, this extension is not unique because the equation $(E + B_{i_1})x = x^+$ has an infinite set of solutions. If x^+ does not belong to the set of images of the indicated linear operator, then the equation $(E + B_{i_1})x = x^+$ does not have solutions. Hence, the solution $x(t, x_0)$ cannot be extended to the left of the point τ_{i_1}.

Thus, the degeneracy of a certain matrix $E + B_i$ with index i such that $\tau_i \in [t_0 - h, t_0]$ leads to the situation where a part of solutions $x(t, x_0)$, $x(t_0, x_0) = x_0$, cannot be extended to the left of the point τ_i, and each of the remaining solutions splits into a set of solutions at time τ_i.

Let τ_i be the time closest from the right to t_0 such that the matrix $E + B_i$ is degenerate. Each of the solutions $x(t, x_0)$ of the original system with initial condition $x(t_0, x_0) = x_0$ can be extended to the entire interval $[t_0, t_0 + h]$, but many of these solutions merge into a single one at time $t = \tau_i$. The set of points $x(\tau_i, x_0)$, $x_0 \in \mathbb{R}^n$, forms the space $\mathbb{R}^n$, and the set of points $(E + B_i)x(\tau_i, x_0)$ forms a subspace of this space with dimension $n - r$, where r is the rank of the matrix $E + B_i$. Furthermore, new solutions defined for $t > \tau_i$ are "created" at time τ_i, namely, the solutions $x(t, y)$, $x(\tau_i + 0, y) = y$, for which the initial point is such that the algebraic system of equations $(E + B_i)x = y$ is unsolvable.

In what follows, we restrict ourselves to the investigation of systems (1.44) for which the following conditions are satisfied:

(1) any compact interval $[a, b] \subset I$ contains finitely many points τ_i;

(2) for all i such that $\tau_i \in I$, the matrices $E + B_i$ are not degenerate.

Under these assumptions, the following statement is true:

Theorem 7 ([142]). *The set of all solutions Υ of the linear homogeneous system of differential equations with pulse action* (1.44) *on the interval* $[a, b]$ *forms an n-dimensional vector space.*

Definition 2. A basis of the linear space of solutions Υ is called a fundamental system of solutions of system (1.44).

Theorem 7 yields the following important corollaries:

(1) the system of equations (1.44) has a fundamental system of n solutions $\varphi_1(t)$, $\varphi_2(t), \ldots, \varphi_n(t)$;

(2) any solution of the system of equations (1.44) is a linear combination of solutions of the fundamental system;

(3) any $n+1$ solutions of Eqs. (1.44) are linearly dependent.

Let $X(t)$ denote a matrix whose columns are solutions of system (1.44) that form a fundamental system of solutions. The matrix $X(t)$ is called a fundamental matrix of system (1.44). It is obvious that, for any constant vector c, the function

$$x(t) = X(t)c \tag{1.46}$$

is a solution of system (1.44). If c passes through the entire space $\mathbb{R}^n$, then the family of functions (1.46) forms a space. It follows from the definition of the matrix $X(t)$ that it satisfies the following matrix equation with pulse action:

$$\frac{dX}{dt} = A(t)X, \quad t \neq \tau_i, \quad \Delta X|_{t=\tau_i} = B_i X. \tag{1.47}$$

It is also obvious that any nondegenerate solution of the matrix system (1.47) is a fundamental matrix of the system of equations (1.44). All nondegenerate solutions of system (1.47) are given by the formula $X(t) = X_0(t)C$, where $X_0(t)$ is a nondegenerate solution of system (1.47) and C is an arbitrary nondegenerate matrix. The nondegenerate solution $X(t)$ of system (1.47) that satisfies the condition $X(t_0) = E$ is called the matrizant of system (1.44) and is denoted by $X(t, t_0)$.

Let $U(t, s)$ be a solution of the Cauchy matrix problem

$$\frac{dU}{dt} = A(t)U, \quad U(t,s) = E, \tag{1.48}$$

i.e., the matrizant of system (1.45). Then any solution $X(t)$ of the matrix system (1.47) admits the representation

$$X(t) = U(t, \tau_{j+k})(E + B_{j+k})U(\tau_{j+k}, \tau_{j+k-1}) \cdots (E + B_j)U(\tau_j, t_0)X(t_0),$$
$$\tau_{j-1} < t_0 \leq \tau_j < \tau_{j+k} < t \leq \tau_{j+k+1}. \tag{1.49}$$

In particular, for the matrizant $X(t, t_0)$, we have

$$X(t, t_0) = U(t, \tau_{j+k})(E + B_{j+k})U(\tau_{j+k}, \tau_{j+k-1}) \cdots (E + B_j)U(\tau_j, t_0),$$
$$\tau_{j-1} < t_0 \leq \tau_j < \tau_{j+k} < t \leq \tau_{j+k+1},$$

or

$$X(t,t_0) = U(t,\tau_{j+k})(E + B_{j+k}) \times \prod_{\nu=k}^{1} U(\tau_{j+\nu},\tau_{j+\nu-1})(E + B_{j+\nu-1})U(\tau_j,t_0). \tag{1.50}$$

By virtue of the Liouville–Ostrogradskii formula, relation (1.49) yields

$$\begin{aligned}\det X(t) &= \det U(t,\tau_{j+k})\det(E + B_{j+k}) \prod_{\nu=k}^{1} \det U(\tau_{j+\nu},\tau_{j+\nu-1}) \\ &\quad \times \det(E + B_{j+\nu-1}) \det U(\tau_j,t_0)\det X(t_0) \\ &= e^{\int_{\tau_{j+k}}^{t} \operatorname{Sp} A(s)ds} \det(E + B_{j+k}) \\ &\quad \times \prod_{\nu=k}^{1} e^{\int_{\tau_{j+\nu-1}}^{\tau_{j+\nu}} \operatorname{Sp} A(s)ds} \det(E + B_{j+\nu-1}) \\ &\quad \times e^{\int_{t_0}^{\tau_j} \operatorname{Sp} A(s)ds} \det X(t_0),\end{aligned}$$

i.e.,

$$\det X(t) = \det X(t_0) e^{\int_{t_0}^{t} \operatorname{Sp} A(s)ds} \prod_{\nu=1}^{k+1} \det(E + B_{j+\nu-1}), \tag{1.51}$$
$$\tau_{j-1} < t_0 \le \tau_j < \tau_{j+k} < t \le \tau_{j+k+1}.$$

The condition of nondegeneracy of the matrices $E + B_i$ and relation (1.51) imply that the matrix $X(t)$ is nondegenerate if the matrix $X(t_0)$ is nondegenerate.

If the matrix $X(t)$ is nondegenerate, then the inverse matrix $X^{-1}(t)$ is determined by the relation

$$\begin{aligned}X^{-1}(t) &= X^{-1}(t_0)U^{-1}(\tau_j,t_0)(E + B_j)^{-1} \cdots \\ &\quad \times U^{-1}(\tau_{j+k},\tau_{j+k-1})(E + B_{j+k})^{-1}U^{-1}(t,\tau_{j+k}) \\ &= X^{-1}(t_0)U^{-1}(\tau_j,t_0) \prod_{\nu=1}^{k} (E + B_{j+\nu-1})^{-1}U^{-1}(\tau_{j+\nu},\tau_{j+\nu-1}) \\ &\quad \times (E + B_{j+k})^{-1}U^{-1}(t,\tau_{j+k}),\end{aligned}$$
$$\tau_{j-1} < t_0 \le \tau_j < \tau_{j+k} < t \le \tau_{j+k+1},$$

and

$$X(t)X^{-1}(s) = U(t, \tau_{j+k}) \prod_{\nu=k}^{m+1} (E + B_{j+\nu})U(\tau_{j+\nu}, \tau_{j+\nu-1})(E + B_{j+m})U(\tau_{j+m}, s),$$

$$\tau_{j+m-1} < s \le \tau_{j+m} < \tau_{j+k} < t \le \tau_{j+k+1}.$$

In particular, for the matrizant $X(t, t_0)$, we have

$$X^{-1}(t, t_0) = U^{-1}(\tau_j, t_0) \prod_{\nu=1}^{k} (E + B_{j+\nu-1})^{-1} U^{-1}(\tau_{j+\nu}, \tau_{j+\nu-1})$$
$$\times (E + B_{j+k})^{-1} U^{-1}(t, \tau_{j+k}),$$
$$X(t, t_0)X^{-1}(s, t_0) = U(t, \tau_{j+k}) \prod_{\nu=k}^{m+1} (E + B_{j+\nu})U(\tau_{j+\nu}, \tau_{j+\nu-1})$$
$$\times (E + B_{j+m})U(\tau_{j+m}, s) = X(t, s), \tag{1.52}$$
$$\tau_{j-1} < t_0 \le \tau_j \le \tau_{j+m-1} < s \le \tau_{j+m} < \tau_{j+k} < t \le \tau_{j+k+1}.$$

If $\tau_i < s \le t \le \tau_{i+1}$, then $X(t, t_0)X^{-1}(s, t_0) = U(t, s)$. Also note that any solution of system (1.44) $x(t, x_0)$, $x(t_0, x_0) = x_0$, can be written with the help of the matrizant $X(t, t_0)$ in the form

$$x(t, x_0) = X(t, t_0)x_0. \tag{1.53}$$

The system of equations

$$\frac{dx}{dt} = A(t)x + f(t), \quad t \ne \tau_i, \quad \Delta x|_{t=\tau_i} = B_i x + a_i, \tag{1.54}$$

where the matrices $A(t)$ and B_i and times τ_i are the same as in system (1.44), $f(t)$ is a function continuous (piecewise continuous) on the interval I, and a_i are constant vectors, is called a linear inhomogeneous system of differential equations with pulse action. The relationship between solutions of the inhomogeneous system (1.54) and the corresponding homogeneous system (1.44) is described by the following theorem:

Theorem 8 ([142]). *If $x = \varphi(t)$ is a solution of system* (1.44) *and $x = \psi(t)$ is a solution of system* (1.54), *then the function $x = \varphi(t) + \psi(t)$ is a solution of system* (1.54). *Conversely, if $x = \varphi_1(t)$ and $x = \varphi_2(t)$ are solutions of the inhomogeneous system* (1.54), *then the function $x = \varphi_1(t) - \varphi_2(t)$ is a solution of the system of equations* (1.44).

In what follows, we use a linear change of dependent variables in systems (1.44) and (1.54).

Theorem 9 ([142]). *Let $S(t)$ be a nondegenerate matrix continuously differentiable for $t \in [a,b] \setminus \{\tau_i\}$. Then the linear change*

$$x = S(t)y \tag{1.55}$$

reduces system (1.54) *to the form*

$$\begin{aligned} \frac{dy}{dt} &= S^{-1}(t)\left[A(t)S(t) - \frac{dS}{dt}\right]y + S^{-1}(t)f(t), \quad t \neq \tau_i, \\ \Delta y|_{t=\tau_i} &= S^{-1}(\tau_i+0)(-\Delta S + B_i S)y|_{t=\tau_i} + S^{-1}(\tau_i+0)a_i. \end{aligned} \tag{1.56}$$

In particular, if $S(t)$ is a fundamental matrix $X(t)$ of the system of equations (1.44), then the change of variables (1.55) is called a "variation of constants" because it is realized by the replacement of the constant vector c in (1.46) by a variable vector $y(t)$. Then system (1.54) reduces to the system

$$\frac{dy}{dt} = X^{-1}(t)f(t), \quad t \neq \tau_i, \quad \Delta y|_{t=\tau_i} = X^{-1}(\tau_i+0)a_i, \tag{1.57}$$

which can easily be integrated.

With regard for the relation $X(\tau_i+0) = (E+B_i)X(\tau_i)$, the condition of jump in Eqs. (1.57) can be written in the form

$$\Delta y = X^{-1}(\tau_i)(E+B_i)^{-1}a_i. \tag{1.58}$$

For $t \geq t_0$, Eqs. (1.57) yield

$$y(t) = c + \int_{t_0}^{t} X^{-1}(s)f(s)ds + \sum_{t_0 \leq \tau_i < t} X^{-1}(\tau_i)(E+B_i)^{-1}a_i, \tag{1.59}$$

where $c = y(t_0)$ is a constant vector.

Let $X(t)$ be a fundamental matrix of system (1.44) in which the matrices $E + B_i$ are nondegenerate. Then any solution of the system of equations (1.54) for $t \geq t_0$ is given by the relation

$$x(t) = X(t)\left[c + \int_{t_0}^{t} X^{-1}(s)f(s)ds + \sum_{t_0 \leq \tau_i < t} X^{-1}(\tau_i)(E+B_i)^{-1}a_i\right]. \tag{1.60}$$

In particular, if $X(t) = X(t,t_0)$ is the matrizant of system (1.44), then any solution $x(t,x_0)$, $x(t_0,x_0) = x_0$, of system (1.54) for $t \geq t_0$ can be represented in the form

$$x(t,x_0) = X(t,t_0)x_0 + \int_{t_0}^{t} X(t,s)f(s)ds + \sum_{t_0 \leq \tau_i < t} X(t,\tau_i)a_i. \tag{1.61}$$

The last two relations show that if the solutions of the corresponding homogeneous system are known, then the solutions of system (1.54) are determined in quadratures.

Stability of Solutions of Linear Impulsive Systems. We now study the problem of stability of solutions of the homogeneous system with pulse action

$$\frac{dx}{dt} = A(t)x, \quad t \neq \tau_i, \quad \Delta x|_{t=\tau_i} = B_i x, \tag{1.62}$$

where the matrix $A(t)$ is continuous (piecewise continuous) and bounded for $t \geq t_0$, the matrices B_i, $i = 1, 2, \ldots$, are uniformly bounded with respect to $i \in N$, and the times τ_i are enumerated by the set of natural numbers $t_0 \leq \tau_1 < \tau_2 < \cdots < \tau_i < \tau_{i+1} < \cdots$ and are such that $\tau_i \to \infty$ as $i \to \infty$.

Let $X(t, t_0)$ be the matrizant of system (1.62). Since the difference of any two solutions $x(t, x_0) - x(t, y_0)$ of system (1.62) can be represented in the form

$$x(t, x_0) - x(t, y_0) = X(t, t_0)(x_0 - y_0), \tag{1.63}$$

we conclude that the stability or instability of solutions of system (1.62) depends on the behavior of the matrizant $X(t, t_0)$ as $t \to \infty$.

If the matrizant $X(t, t_0)$ is bounded for $t \geq t_0$, i.e., there exists a constant $M \geq 0$ such that the estimate

$$\|X(t, t_0)\| \leq M < \infty$$

holds for all $t \geq t_0$, then, for all $t \geq t_0$ and any solution $x(t, x_0)$ of system (1.62), the following inequality is true:

$$\|x(t, x_0) - x(t, y_0)\| \leq \|X(t, t_0)\| \cdot \|x_0 - y_0\| \leq M\|x_0 - y_0\|.$$

By virtue of this inequality, we have $\|x(t, x_0) - x(t, y_0)\| < \varepsilon$ for $t \geq t_0$ whenever

$$\|x_0 - y_0\| < \delta = \frac{\varepsilon}{M}.$$

This means that the solution $x(t, x_0)$ is stable.

Assume that

$$\lim_{t \to \infty} \|X(t, t_0)\| = 0.$$

In this case, the matrix $X(t, t_0)$ is bounded for $t \geq t_0$, and, hence, the solution $x(t, x_0)$ is stable. Furthermore, it follows from relation (1.63) that

$$\lim_{t \to \infty} \|x(t, x_0) - x(t, y_0)\| = 0$$

for any solution $x(t, y_0)$, i.e., the solution $x(t, x_0)$ is asymptotically stable.

Assume that the matrix $X(t, t_0)$ is unbounded for $t \geq t_0$. In other words, there exists an infinite increasing sequence of numbers $t_0 \leq t_1 < t_2 < \cdots$ such that

$$\lim_{k \to \infty} \|X(t_k, t_0)\| = \infty.$$

In this case, the elements of the matrix $X(t, t_0)$ contain at least one element $q_{\alpha\beta}(t)$ for which

$$\lim_{k\to\infty} |q_{\alpha\beta}(t_k)| = \infty.$$

Consider the solution $x(t, x_0^*)$ of system (1.62) that passes at $t = t_0$ through the point x_0^* with the coordinates

$$x_{10}^* = x_{10}, \quad x_{20}^* = x_{20}, \quad \ldots, \quad x_{\beta 0}^* \neq x_{\beta 0},$$
$$x_{\beta+1\,0}^* = x_{\beta+1\,0}, \quad \ldots, \quad x_{n0}^* = x_{n0}.$$

For this solution, we have

$$x_\alpha(t, x_0^*) - x_\alpha(t, x_0) = q_{\alpha\beta}(t)(x_{\beta 0}^* - x_{\beta 0}).$$

Therefore,

$$\lim_{k\to\infty} |x_\alpha(t_k, x_0^*) - x_\alpha(t_k, x_0)| = \infty.$$

No matter how small the difference $x_{\beta 0}^* - x_{\beta 0}$ is in modulus, the function $x_\alpha(t, x_0^*) - x_\alpha(t, x_0)$ is unbounded for $t \to \infty$, and, hence, the difference $x(t, x_0^*) - x(t, x_0)$ is also unbounded. This means that the solution $x(t, x_0)$ of system (1.62) is unstable.

We have proved that the boundedness of the matrizant $X(t, t_0)$ for all $t \geq t_0$ is a sufficient condition for the stability, the equality $\lim_{t\to\infty} X(t, t_0) = 0$ is a sufficient condition for the asymptotic stability, and the unboundedness of the matrix $X(t, t_0)$ for $t \geq t_0$ is a sufficient condition for the instability of any solution of the system of equations (1.62).

One can also prove that the conditions presented above are not only sufficient but also necessary for the stability, asymptotic stability, and instability of any solution of the linear system (1.62), respectively.

Thus, the following theorem is true:

Theorem 10 ([142]). *For the stability of a solution $x(t, x_0)$ of the linear impulsive system* (1.62), *it is necessary and sufficient that the matrizant $X(t, t_0)$ (and, hence, any fundamental matrix) of this system be bounded for $t \geq t_0$; for its asymptotic stability, it is necessary and sufficient that the matrizant satisfy the condition*

$$\lim_{t\to\infty} X(t, t_0) = 0;$$

and for its instability, it is necessary and sufficient that the matrizant be unbounded for $t \geq t_0$.

Since the matrix $X(t, t_0)$ does not depend on the initial value of a solution $x(t, x_0)$ of system (1.62), all solutions of the linear impulsive system (1.62) are simultaneously stable or unstable. Therefore, the linear system (1.62) is called stable, asymptotically stable, or unstable depending on whether its solutions are stable, asymptotically stable, or unstable.

For example, the first-order equation

$$\frac{dx}{dt} = a(t)x, \quad t \neq \tau_i, \quad \Delta x|_{t=\tau_i} = b_i x$$

is stable, asymptotically stable, or unstable depending on whether the expression

$$\int_{t_0}^{t} a(s)ds + \sum_{t_0 \leq \tau_i < t} \ln|1 + b_i|$$

is bounded, tends to $-\infty$ as $t \to \infty$, or is unbounded for $t \geq t_0$.

Assume that the matrices $A(t)$ and B_i in the system of equations (1.62) are representable in the form $A(t) = A + P(t)$ and $B_i = B + I_i$, where A and B are constant matrices. Then system (1.62) can be rewritten in the form

$$\frac{dx}{dt} = Ax + P(t)x, \quad t \neq \tau_i, \quad \Delta x|_{t=\tau_i} = Bx + I_i x. \tag{1.64}$$

Together with system (1.64), we consider the system

$$\frac{dx}{dt} = Ax, \quad t \neq \tau_i, \quad \Delta x|_{t=\tau_i} = Bx. \tag{1.65}$$

The following theorem is true:

Theorem 11 ([142]). *If the solutions of the system of equations* (1.65) *are stable, then the solutions of the system of equations* (1.64) *are also stable, provided that*

$$\int_{t_0}^{\infty} \|P(t)\|dt < \infty \quad \textit{and} \quad \prod_{\tau_i \geq t_0} (1 + \|I_i\|) < \infty. \tag{1.66}$$

Proof. The matrizant $X(t, t_0)$ of the system with constant coefficients (1.65) takes the form

$$X(t, t_0) = e^{A(t-\tau_i)} \prod_{t_0 \leq \tau_j < t} (E + B)e^{A(\tau_j - \tau_{j-1})}, \quad \tau_0 = t_0. \tag{1.67}$$

By virtue of the nondegeneracy of the matrix $E + B$, the matrix $X(t, t_0)$ is nondegenerate and

$$X(t, t_0)X^{-1}(s, t_0) = e^{A(t-\tau_i)} \times \prod_{s \leq \tau_j < t} (E + B)e^{A(\tau_j - \tau_{j-1})}(E + B)e^{A(\tau_{k+1} - s)}, \tag{1.68}$$

$$\tau_i < t \leq \tau_{i+1}, \quad \tau_k < s < \tau_{k+1}, \quad k < i.$$

It follows from the stability of solutions of system (1.65) and from representation (1.68) that there exists a positive number K such that

$$\|X(t,t_0)\| \le K, \quad \|X(t,t_0)X^{-1}(s,t_0)\| \le K, \quad t_0 \le s \le t. \tag{1.69}$$

Note that any solution $x(t,x_0)$, $x(t_0,x_0) = x_0$, of system (1.64) admits the representation

$$x(t,x_0) = X(t,t_0)x_0 + \int_{t_0}^{t} X(t,s)P(s)x(s,x_0)ds + \sum_{t_0 \le \tau_i < t} X(t,\tau_i)I_i x(\tau_i,x_0). \tag{1.70}$$

Therefore, for any two solutions $x(t,x_0)$ and $x(t,y_0)$ of system (1.64), with regard for (1.69) we get

$$\begin{aligned}\|x(t,x_0) - x(t,y_0)\| \le K\|x_0 - y_0\| &+ \int_{t_0}^{t} K\|P(s)\| \cdot \|x(s,x_0) - x(s,y_0)\|ds \\ &+ \sum_{t_0 \le \tau_i < t} K\|I_i\| \cdot \|x(\tau_i,x_0) - x(\tau_i,y_0)\|.\end{aligned}$$

Hence, by virtue of Lemma 1, we obtain the following estimate for all $t \ge t_0$:

$$\|x(t,x_0) - x(t,y_0)\| \le K \prod_{t_0 \le \tau_i < t} (1 + K\|I_i\|)e^{\int_{t_0}^{t} K\|P(s)\|ds}\|x_0 - y_0\|. \tag{1.71}$$

The convergence of the product $\prod_{\tau_i \ge t_0}(1 + \|I_i\|)$ yields the convergence of the product $\prod_{t_0 \le \tau_i < t}(1 + K\|I_i\|)$ by virtue of the fact that $1 + K\|I_i\| \le (1 + \|I_i\|)^K$. Therefore, relation (1.71) yields

$$\|x(t,x_0) - x(t,y_0)\| \le K_1\|x_0 - y_0\|, \quad t \ge t_0, \tag{1.72}$$

where

$$K_1 = \prod_{\tau_i \ge t_0} (1 + K\|I_i\|)e^{\int_{t_0}^{\infty} K\|P(s)\|ds}.$$

The stability of solutions of system (1.64) follows from inequality (1.72). Theorem 11 is proved. □

Criterion for Stability in the First Approximation. We now study the stability of solutions of the nonlinear impulsive system of differential equations

$$\frac{dx}{dt} = f(t,x), \quad t \ne \tau_i, \quad \Delta x|_{t=\tau_i} = I_i^0(x). \tag{1.73}$$

The problem of stability of a solution $x = \varphi(t)$ of Eqs. (1.73) reduces to the study of the stability of the trivial solution of some other system. To construct this system we change the variables in Eqs. (1.73) by setting $x = y + \varphi(t)$. As a result, the system of equations (1.73) is transformed into the system

$$\frac{dy}{dt} = F(t, y), \quad t \neq \tau_i, \quad \Delta y|_{t=\tau_i} = I_i^1(y), \tag{1.74}$$

where

$$F(t, y) = f(t, y + \varphi(t)) - f(t, \varphi(t)), \quad F(t, 0) \equiv 0,$$

and

$$I_i^1(y) = I_i^0(y + \varphi(\tau_i)) - I_i^0(\varphi(\tau_i)), \quad I_i^1(0) = 0,$$

and the solution $x = \varphi(t)$ is transformed into the solution $y = 0$.

Thus, without loss of generality, we assume that system (1.73) has the trivial solution $x = 0$ and study its stability.

We represent the functions $f(t, x)$ and $I_i^0(x)$ as follows:

$$f(t, x) = A(t)x + g(t, x), \quad I_i^0(x) = B_i x + I_i(x),$$

where $A(t)$ and B_i are matrices, and $g(t, x)$ and $I_i(x)$ satisfy the conditions $g(t, 0) = 0$ and $I_i(0) = 0$, respectively. We now rewrite the system of equations (1.73) in the form

$$\frac{dx}{dt} = A(t)x + g(t, x), \quad t \neq \tau_i, \quad \Delta x|_{t=\tau_i} = B_i x + I_i(x). \tag{1.75}$$

Together with Eqs. (1.75), we consider the linear system of equations

$$\frac{dx}{dt} = A(t)x, \quad t \neq \tau_i, \quad \Delta x|_{t=\tau_i} = B_i x. \tag{1.76}$$

We call it the system of the first approximation with respect to system (1.75). Assume that the times of pulse action are enumerated by the set of natural numbers in natural order and

$$\tau_{i+1} - \tau_i \geq \theta \tag{1.77}$$

for some $\theta > 0$.

Theorem 12 ([142]). *Suppose that, for all t and s, $t_0 \leq s \leq t$, the matrizant $X(t, s)$ of the system of equations* (1.76) *admits the estimate*

$$\|X(t, s)\| \leq K e^{-\nu(t-s)}, \quad K \geq 1, \ \nu > 0, \tag{1.78}$$

and the functions $g(t, x)$ and $I_i(x)$ satisfy the inequalities

$$\|g(t, x)\| \leq a\|x\|, \quad \|I_i(x)\| \leq a\|x\| \tag{1.79}$$

for all $t \geq t_0$, $i = 1, 2, \ldots$, and $\|x\| \leq h$, $h > 0$. Then, for sufficiently small a, the trivial solution of Eq. (1.75) *is asymptotically stable.*

Proof. Each solution of Eq. (1.75) can be represented in the form

$$x(t, x_0) = X(t, t_0)x_0 + \int_{t_0}^{t} X(t, s)g(s, x(s, x_0))ds + \sum_{t_0 \le \tau_i < t} X(t, \tau_i)I_i(x(\tau_i, s_0)). \tag{1.80}$$

Hence, with regard for inequalities (1.78) and (1.79), we get

$$\|x(t, x_0)\| \le Ke^{-\nu(t-t_0)}\|x_0\| + \int_{t_0}^{t} Ke^{-\nu(t-s)}a\|x(s, x_0)\|ds + \sum_{t_0 \le \tau_i < t} Ke^{-\nu(t-\tau_i)}a\|x(\tau_i, x_0)\|,$$

or

$$e^{\nu(t-t_0)}\|x(t, x_0)\| \le K\|x_0\| + \int_{t_0}^{t} Kae^{\nu(s-t_0)}\|x(s, x_0)\|ds + \sum_{t_0 \le \tau_i < t} Kae^{\nu(\tau_i - t_0)}\|x(\tau_i, x_0)\|. \tag{1.81}$$

By virtue of Lemma 2, we obtain

$$e^{\nu(t-t_0)}\|x(t, x_0)\| \le K\|x_0\|(1 + Ka)^{i(t_0,t)}e^{Ka(t-t_0)}.$$

Since the times τ_i satisfy inequality (1.77), relation (1.81) yields

$$\|x(t, x_0)\| \le Ke^{-(\nu - Ka - \frac{1}{\theta}\ln(1+Ka))(t-t_0)}\|x_0\|. \tag{1.82}$$

Thus, if a is so small that

$$\nu - Ka - \frac{1}{\theta}\ln(1 + Ka) > 0,$$

then any solution $x(t, x_0)$, $\|x_0\| < \frac{h}{K}$, of Eq. (1.75) is defined for all $t \ge t_0$, and

$$\lim_{t\to\infty} \|x(t, x_0)\| = 0.$$

In other words, the trivial solution of Eq. (1.75) is asymptotically stable. □

Linear Periodic Systems with Pulse Action. A linear homogeneous system of differential equations with pulse action

$$\frac{dx}{dt} = A(t)x, \quad t \ne \tau_i, \quad \Delta x|_{t=\tau_i} = B_i x \tag{1.83}$$

is called periodic with period T (or T-periodic) if the matrix $A(t)$ is T-periodic and one can find a natural number p such that

$$B_{i+p} = B_i, \quad \tau_{i+p} = \tau_i + T \tag{1.84}$$

for all $i \in \mathbb{Z}$.

We assume that the matrix $A(t)$ is continuous (piecewise continuous with discontinuities of the first kind at $t = \tau_i$), the matrices $E + B_i$ are nondegenerate, and the times τ_i are enumerated by the set of integers $\mathbb{Z}$ so that $0 \le \tau_1 < \cdots < \tau_p < T$.

Let $X(t)$ be the fundamental matrix of the periodic system (1.83) such that $X(0) = E$, i.e., the matrizant of this system. By virtue of the periodicity of system (1.83), it is easy to prove that $X(t+T)$ is also a fundamental matrix of system (1.83) and

$$X(t+T) = X(t)X(T), \tag{1.85}$$

where

$$X(T) = U(T, \tau_p) \prod_{\nu=p-1}^{1} (E + B_{\nu+1})U(\tau_{\nu+1}, \tau_\nu)(E + B_1)U(\tau_1, 0)$$

is the matrix of monodromy and $U(t,s)$, $U(s,s) = E$, is the matrizant of the differential system from (1.83).

The eigenvalues of the matrix $X(T)$ are called the multipliers of system (1.83). The following theorem is true:

Theorem 13 ([142]). *For any multiplier ρ, there exists a nontrivial solution $x = \varphi(t)$ of the periodic system of Eqs.* (1.83) *that satisfies the condition*

$$\varphi(t+T) = \rho\varphi(t). \tag{1.86}$$

Conversely, if relation (1.86) *is satisfied for some nontrivial solution $x = \varphi(t)$ and some number ρ, then ρ is a multiplier of this system of equations.*

Proof. As a solution $\varphi(t)$, we take the solution of system (1.83) for which the vector $\varphi(0)$ is the eigenvector of the matrix of monodromy corresponding to the eigenvalue ρ. We have

$$X(T)\varphi(0) = \rho\varphi(0), \quad \varphi(t) = X(t)\varphi(0).$$

Hence,

$$\varphi(t+T) = X(t+T)\varphi(0) = X(t)X(T)\varphi(0) = X(t)\rho\varphi(0) = \rho\varphi(t),$$

i.e., condition (1.86) is satisfied.

Assume that equality (1.86) is satisfied for some nontrivial solution $\varphi(t) = X(t)\varphi(0)$, i.e.,

$$X(t+T)\varphi(0) = \rho X(t)\varphi(0), \quad X(t)X(T)\varphi(0) = \rho X(t)\varphi(0),$$

whence

$$(X(T) - \rho E)\varphi(0) = 0.$$

This implies that ρ is a root of the characteristic equation $\det(X(T) - \rho E) = 0$, i.e., it is a multiplier of system (1.83). Theorem 13 is proved. □

Theorem 13 yields the following important corollary: The linear T-periodic system of equations (1.83) has a nontrivial kT-periodic solution if and only if the k-th power of at least one of its multipliers is equal to 1.

We now consider conditions for the existence of periodic solutions of linear inhomogeneous periodic systems of differential equations with pulse action, i.e., systems of the form

$$\frac{dx}{dt} = A(t)x + f(t), \quad t \neq \tau_i, \quad \Delta x|_{t=\tau_i} = B_i x + a_i, \tag{1.87}$$

where $A(t)$ and $f(t)$ are continuous (piecewise continuous) T-periodic matrix and vector functions, respectively, B_i are constant matrices, a_i are constant vectors, and the times τ_i are such that

$$B_{i+p} = B_i, \quad a_{i+p} = a_i, \quad \tau_{i+p} = \tau_i + T \tag{1.88}$$

for some natural p and all $i \in \mathbb{Z}$. It is also assumed that $\det(E + B_i) \neq 0$, and $0 \leq \tau_1 < \tau_2 < \cdots < \tau_p < T$.

Let $X(t,s)$, $X(s,s) = E$, be the matrizant of the homogeneous system corresponding to (1.87). Any solution $x(t, x_0)$, $x(0, x_0) = x_0$, of the system of equations (1.87) can be represented in the form

$$x(t, x_0) = X(t,0)x_0 + \int_0^t X(t,s)f(s)ds + \sum_{0 \leq \tau_i < t} X(t, \tau_i)a_i. \tag{1.89}$$

Among these solutions, the T-periodic solutions are those for which x_0 satisfies the equation

$$(E - X(T,0))x_0 = \int_0^T X(T,s)f(s)ds + \sum_{i=1}^{p} X(T, \tau_i)a_i. \tag{1.90}$$

Assume that $\det(E - X(T,0)) \neq 0$. This condition is equivalent to the assumption that the homogeneous system corresponding to Eqs. (1.87) does not have nontrivial T-periodic solutions, i.e., none of its multipliers is equal to 1. In this case, Eq. (1.90) has the unique solution

$$x_0 = (E - X(T,0))^{-1}\Bigg[\int_0^T X(T,s)f(s)ds + \sum_{i=1}^{p} X(T, \tau_i)a_i\Bigg]. \tag{1.91}$$

Hence, the system of equations (1.87) has the unique T-periodic solution

$$x^*(t) = X(t,0)(E - X(T,0))^{-1}\Big[\int_0^T X(T,s)f(s)ds + \sum_{i=1}^{p} X(T,\tau_i)a_i\Big]$$
$$+ \int_0^t X(t,s)f(s)ds + \sum_{0\le\tau_i<t} X(t,\tau_i)a_i, \quad t \ge 0.$$

Denoting

$$G(t,s) = \begin{cases} X(t,0)(E - X(T,0))^{-1}X^{-1}(s,0), & 0 \le s \le t \le T, \\ X(t+T,0)(E - X(T,0))^{-1}X^{-1}(s,0), & 0 \le t < s \le T, \end{cases} \tag{1.92}$$

we can represent the periodic solution in the form

$$x^*(t) = \int_0^T G(t,s)f(s)ds + \sum_{i=1}^{p} G(t,\tau_i)a_i. \tag{1.93}$$

The function $G(t,s)$ is called the Green function of the problem of periodic solutions of system (1.87). Note the following properties of the function $G(t,s)$:

(1) if $t \ne \tau_i$, then $G(s+0,s) - G(s,s) = E$;

(2) $G(0,s) = G(T,s)$;

(3) for $t \ne s$, the function $G(t,s)$ satisfies the impulsive system of equations

$$\frac{dG(t,s)}{dt} = A(t)G(t,s), \quad t \ne \tau_i, \ \Delta G|_{t=\tau_i} = B_i G(\tau_i,s),$$

so that

$$G(\tau_i + 0, s) = (E + B_i)G(\tau_i, s);$$

(4) $G(t,\tau_i + 0) = G(t,\tau_i)(E + B_i)^{-1}$.

These properties of the function $G(t,s)$ can easily be verified with regard for properties of the function $X(t,0)$. Also note that properties (1)–(4) uniquely define the Green function $G(t,s)$.

Let

$$\max_{t,s\in[0,T]} \|G(t,s)\| = K.$$

It follows from (1.93) that $x^*(t)$ admits the estimate

$$\|x^*(t)\| \le K\Big(\int_0^T \|f(t)\|dt + \sum_{i=1}^{p} \|a_i\|\Big). \tag{1.94}$$

Thus, we have proved the following statement:

Theorem 14 ([142]). *If the homogeneous system corresponding to* (1.87) *does not have nontrivial T-periodic solutions, then the system of equations* (1.87) *has the unique T-periodic solution $x^*(t)$ for any T-periodic function $f(t)$ and any periodic sequence a_i ($a_{i+p} = a_i$, $i \in \mathbb{Z}$). This solution satisfies estimate* (1.94).

Consider the case where the homogeneous system corresponding to system (1.87) has nontrivial T-periodic solutions.

The initial values of these solutions are determined from the equations

$$(E - X(T,0))x_0 = 0. \tag{1.95}$$

Assume that the homogeneous system has $k \leq n$ linearly independent solutions. Then the algebraic system (1.95) has exactly k linearly independent solutions, i.e., the rank of the matrix $E - X(T,0)$ is equal to $n - k$. The matrizant of the conjugate system

$$\frac{dx}{dt} = -A^T(t)x, \quad t \neq \tau_i, \quad \Delta x|_{t=\tau_i} = -(E + B_i^T)^{-1} B_i x \tag{1.96}$$

is the matrix $Y(t,s) = (X^T(t,s))^{-1}$. Therefore, the initial conditions of nontrivial T-periodic solutions of the conjugate system must satisfy the condition

$$(E - (X^T(T,0))^{-1})y_0 = 0 \quad \text{or} \quad (X(T,0) - E)^T y_0 = 0. \tag{1.97}$$

The rank of the matrix $(X(T,0)-E)^T$ is equal to the rank of the matrix $X(T,0)-E$, i.e., it is equal to $n-k$. Hence, the system of algebraic Eqs. (1.97) has k linearly independent solutions, which correspond to k linearly independent T-periodic solutions of the conjugate system (1.96).

Theorem 15 ([142]). *Suppose that the linear homogeneous T-periodic system corresponding to Eqs.* (1.87) *has k linearly independent T-periodic solutions $\varphi_1(t), \dots, \varphi_k(t)$, $k = \overline{1,n}$.*

The system of equations (1.87) *has T-periodic solutions if and only if the conditions*

$$\int_0^T (\psi_j(t), f(t))dt + \sum_{i=1}^{p} (\psi_j(\tau_i), a_i) = 0, \quad j = \overline{1,k}, \tag{1.98}$$

are satisfied. Here, $\psi_1(t), \dots, \psi_k(t)$ are linearly independent T-periodic solutions of the conjugate system of equations (1.96). *In this case, the T-periodic solutions of system* (1.87) *form a k-parameter family of solutions.*

Proof. Let $x(t)$ be a T-periodic solution of the inhomogeneous system of equations (1.87). It follows from the condition of T-periodicity that the initial data $x(0) = x_0$ satisfy the condition

$$(E - X(T,0))x_0 = \int_0^T X(T,0)X^{-1}(s,0)f(s)ds + \sum_{i=1}^{p} X(T,0)X^{-1}(\tau_i,0)a_i. \tag{1.99}$$

Let $\psi(t)$ be a nontrivial T-periodic solution of the conjugate system of equations (1.96). Then $(E - X(T,0))^T\psi(0) = 0$ and, hence,

$$\begin{aligned} 0 &= ((E - X(T,0))^T\psi(0), x_0) = (\psi(0), (E - X(T,0))x_0) \\ &= \left(X^T(T,0)\psi(0), \int_0^T X^{-1}(s,0)f(s)ds + \sum_{i=1}^{p} X^{-1}(\tau_i,0)a_i\right) \\ &= \int_0^T ((X^T(s,0))^{-1}\psi(0), f(s))ds + \sum_{i=1}^{p}((X^T(\tau_i,0))^{-1}\psi(0), a_i) \\ &= \int_0^T (\psi(s), f(s))ds + \sum_{i=1}^{p}(\psi(\tau_i), a_i). \end{aligned}$$

The necessity is proved.

Let us prove the sufficiency of conditions (1.98). Assume that equalities (1.98) are true.

If y_0 is the eigenvector of the matrix $X^T(T,0)$ that corresponds to the multiplier $\rho = 1$, then the solution $y(t)$, $y(0) = y_0$, of the conjugate system (1.96) is T-periodic and

$$y(t) = Y(t,0)y_0 = \sum_{j=1}^{k} c_j\psi_j(t).$$

Therefore,

$$\begin{aligned} 0 &= \int_0^T \left(\sum_{j=1}^{k} c_j\psi_j(t), f(t)\right)dt + \sum_{i=1}^{p}\sum_{j=1}^{k}(c_j\psi_j(\tau_i), a_i) \\ &= \int_0^T (Y(t,0)y_0, f(t))dt + \sum_{i=1}^{p}(Y(\tau_i,0)y_0, a_i) \end{aligned}$$

$$= \int_0^T (X^T(T,0)y_0, X^{-1}(t,0)f(t))dt$$

$$+ \sum_{i=1}^{p} (X^T(T,0)y_0, X^{-1}(\tau_i,0)a_i)$$

$$= \left(y_0, \int_0^T X(T,0)X^{-1}(t,0)f(t)dt + \sum_{i=1}^{p} X(T,0)X^{-1}(\tau_i,0)a_i \right).$$

Thus, the system of equations $(X(T,0)-E)^T y_0 = 0$ is equivalent to the system

$$(X(T,0)-E)y_0 = 0,$$

$$\left(y_0, \int_0^T X(T,0)X^{-1}(t,0)f(t)dt + \sum_{i=1}^{p} X(T,0)X^{-1}(\tau_i,0)a_i \right) = 0,$$

and, hence, the ranks of the matrices of these systems are identical. Therefore, denoting the row vector

$$b = \left(\int_0^T X(T,0)X^{-1}(t,0)f(t)dt + \sum_{i=1}^{p} X(T,0)X^{-1}(\tau_i,0)a_i \right)^T,$$

for the rank of the system of equations (1.99) we get

$$\operatorname{rang}((X(T,0)-E)b^T) = \operatorname{rang}\begin{pmatrix} (X(T,0)-E)^T \\ b \end{pmatrix}$$

$$= \operatorname{rang}(X(T,0)-E)^T = \operatorname{rang}(X(T,0)-E) = n-k.$$

By virtue of the Kronecker–Capelli theorem, the system of equations (1.99), which determines the initial conditions for T-periodic solutions of the inhomogeneous system (1.87), is consistent and has exactly k linearly independent solutions. Theorem 15 is proved. □

Chapter 2

Impulsive Differential Inclusions

2.1 Differential Inclusions with Fixed Times of Pulse Action

Consider a differential inclusion with pulse action at fixed times:

$$\dot{x} \in F(t, x), \quad t \neq \tau_i, \; x(t_0) = x_0 \in X_0, \tag{2.1}$$

$$\Delta x|_{t=\tau_i} \in I_i(x), \quad i = \overline{1, k}. \tag{2.2}$$

Here, $t \in I = [t_0, T]$ is time, $x \in \mathbb{R}^n$ is a phase vector, $F: I \times \mathbb{R}^n \to \operatorname{conv}(\mathbb{R}^n)$, $\tau_i \in I$ are the times of pulse action, and $I_i: \mathbb{R}^n \to \operatorname{conv}(\mathbb{R}^n)$.

Definition 1 ([23]). A vector function $x(t)$ defined on an interval or a segment $J \subset I$ is called a solution of the impulsive differential inclusion (2.1), (2.2) if it is absolutely continuous, satisfies inclusion (2.1) almost everywhere on intervals that do not contain τ_i, and satisfies the condition of jump (2.2) for $t = \tau_i$.

Definition 2 ([23]). The integral funnel of a point (t_0, x_0) (or of the set K) is the set of points lying on the graphs of all solutions that pass through this point (respectively, through the points of the set K). The cross-section $t = t_1$ of the funnel of a point (t_0, x_0) is the set of attainability at time t_1, i.e., the set of points $\{x(t_1)\}$ that can be reached at time t_1 by moving along various solutions leaving the point x_0 at time t_0.

If we omit the assumption that the set $F(t, x)$ is convex, then the funnel and the set of attainability can be nonclosed (see Appendix 2).

Now consider the problem of the connectedness of the integral funnel and the set of attainability of system (2.1), (2.2), depending on properties of the functions $F(t, x)$ and $I_i(x)$ and the set K.

Assume that $I_i(x)$ map any connected compact set onto a connected compact set. Then, in view of [23], the following statement is true:

Theorem 1 ([113]). *Suppose that the following conditions are satisfied in the domain $Q = I \times D$, where D is a compact set in $\mathbb{R}^n$:*

(1) *a set-valued mapping $F(t, x)$ is upper semicontinuous in x and measurable in t, and $|F(t, x)| \leq g(t)(1 + \|x\|)$, where $g(t)$ is summable on I;*

(2) *$I_i(x)$ map a connected compact set onto a connected compact set;*

(3) *the set K is compact and connected.*

Then the set $H(K)$ of all solutions of system (2.1), (2.2) *with various initial conditions $(t_0, x_0) \in K$ is compact in the metric of $C(I)$ and is connected on the time interval between pulses.*

The set of attainability is connected and compact.

By analogy with the notion of R-solution for differential inclusions without pulses [23, 101], we introduce the notion of R-solution for impulsive differential inclusions and present conditions for the existence and uniqueness of R-solutions.

Definition 3. A set-valued mapping $R\colon I \to \mathrm{comp}(\mathbb{R}^n)$ is called an R-solution of the differential inclusion (2.1), (2.2) if $R(t)$ satisfies the initial condition $R(t_0) = X_0$ and is absolutely continuous on the time interval between two successive pulses, for almost all t one has

$$\frac{1}{\Delta} h\Big(R(t+\Delta), \bigcup_{x \in R(t)} \Big\{ x + \int_t^{t+\Delta} F(s,x)ds \Big\} \Big) \to 0 \quad (\Delta \downarrow 0),$$

and the following relation holds at the times of pulse action:

$$R(\tau_i + 0) = \bigcup_{x \in R(\tau_i)} \{x + I_i(x)\}.$$

Here and in what follows, the integral of a set-valued mapping is understood in the sense of Aumann [10] (unless otherwise stated).

Theorem 2. *Suppose that the following conditions are satisfied in the domain $Q = I \times D,\ D \in \mathrm{comp}(\mathbb{R}^n)$:*

(1) *a set-valued mapping $F(t,x)$ is continuous in x and measurable in t, and $|F(t,x)| \le g(t)(1 + \|x\|)$, where $g(t)$ is summable on I;*

(2) *$I_i(x)$ map a connected compact set onto a connected compact set.*

Then, for any compact set $X_0 \subset \mathbb{R}^n$, there exists an R-solution of the differential inclusion (2.1), (2.2)*. The integral funnel is the graph of the R-solution $R(t)$.*

Assume, in addition, that the following condition is satisfied for any $r > 0$ and almost all t $(t \ne \tau_i)$: If $\|x - y\| \le r$, then

$$h(F(t,x), F(t,y)) \le \omega(t,r),$$

where $\omega(t,r)$ is the Kamke function. Then the R-solution is unique.

Now consider the problem of continuous dependence of solutions of impulsive differential inclusions on the initial data and right-hand sides.

Consider differential inclusion with multivalued pulses:

$$\dot{x} \in F_1(t,x), \quad t \neq \tau_i, \; x(t_0) \in X_0, \tag{2.3}$$

$$\Delta x|_{t=\tau_i} \in I_1^i(x) \tag{2.4}$$

and

$$\dot{y} \in F_2(t,y), \quad t \neq \tau_i, \; y(t_0) \in Y_0, \tag{2.5}$$

$$\Delta y|_{t=\tau_i} \in I_2^i(y), \quad i = \overline{1,k}, \tag{2.6}$$

where $F_j \colon I \times \mathbb{R}^n \to \mathrm{conv}(\mathbb{R}^n)$ and $I_j^i \colon \mathbb{R}^n \to \mathrm{conv}(\mathbb{R}^n)$, $j = 1,2$.

Theorem 3 ([114]). *Suppose that the following conditions are satisfied in the domain* $Q = I \times D$, $D \in \mathrm{comp}(\mathbb{R}^n)$*:*

(1) *set-valued mappings* $F_j(t,x)$, $j = 1,2$, *are measurable in* t *and satisfy the Lipschitz condition with respect to* x *with summable function* $l(t)$*;*

(2) *set-valued mappings* $I_j^i(x)$ *satisfy the Lipschitz condition with constant* μ*;*

(3) *the following estimates are true:*

$$h(F_1(t,x), F_2(t,x)) \le \eta, \quad h(I_1^i(x), I_2^i(x)) \le \eta, \quad h(X_0, Y_0) \le \delta_0. \tag{2.7}$$

Then, for any solution $y(t)$ *of inclusion* (2.3), (2.4)*, there exists a solution* $x(t)$ *of inclusion* (2.5), (2.6) *such that*

$$\|x(t) - y(t)\| \le A\delta_0 + B\eta,$$

and vice versa. The constants A *and* B *are independent of* η *and* δ_0.

Proof. Let $y(t)$ be a solution of inclusion (2.5), (2.6) and let $\tau_1 > t_0$ be the time of the first pulse. Then, by virtue of (2.7), we have

$$\rho(\dot{y}(t), F_1(t, y(t))) \le h(F_2(t, y(t)), \quad F_1(t, y(t)) < \eta.$$

According to the Filippov theorem [51], a solution $x(t)$ of inclusion (2.3), (2.4) exists on the interval $[t_0, \tau_1]$ and is such that

$$\|x(t) - y(t)\| \le \delta_0 e^{m(t)} + \eta \int_{t_0}^{t} e^{m(t)-m(s)} ds,$$

where

$$m(t) = \int_{t_0}^{t} l(s) ds.$$

Hence,

$$\delta_1^- = \|x(\tau_1) - y(\tau_1)\| \le \delta_0 e^{m(\tau_1)} + \eta \int_{t_0}^{\tau_1} e^{m(\tau_1)-m(s)} ds.$$

According to (2.6), we have $y(\tau_1 + 0) \in y(\tau_1) + I_2^1(y(\tau_1))$.
We choose $x(\tau_1 + 0)$ so that

$$\begin{aligned}
\delta_1^+ &= \|x(\tau_1 + 0) - y(\tau_1 + 0)\| = \min_{x \in x(\tau_1) + I_1^1(x)} \|x - y(\tau_1 + 0)\| \\
&\le h(x(\tau_1) + I_1^1(x(\tau_1)), y(\tau_1) + I_2^1(y(\tau_1))) \\
&\le \|x(\tau_1) - y(\tau_1)\| + h(I_1^1(x(\tau_1)), I_2^1(y(\tau_1))) \\
&= \delta_1^- + h(I_1^1(x(\tau_1)), I_2^1(y(\tau_1))) \\
&\le \delta_1^- + h(I_1^1(x(\tau_1)), I_2^1(x(\tau_1))) + h(I_2^1(x(\tau_1)), I_2^1(y(\tau_1))) \\
&\le \delta_1^- + \mu\|x(\tau_1) - y(\tau_1)\| + \eta = \delta_1^- + \mu\delta_1^- + \eta.
\end{aligned}$$

Hence, $\delta_1^+ \le \delta_1^-(1 + \mu) + \eta$.
By analogy, for any $i = \overline{0, k}$ and $t \in (\tau_i, \tau_{i+1}]$, we get

$$\begin{aligned}
\delta_{i+1}^+ &\le \delta_{i+1}^-(1 + \mu) + \eta, \\
\delta_{i+1}^- &\le \delta_i^+ e^{m(\tau_{i+1})-m(\tau_i)} + \eta \int_{\tau_i}^{\tau_{i+1}} e^{m(\tau_{i+1})-m(s)} ds,
\end{aligned}$$

$$\begin{aligned}
\|x(t) - y(t)\| &\le \delta_i^+ e^{m(t)-m(\tau_i)} + \eta \int_{\tau_i}^{t} e^{m(t)-m(s)} ds \\
&\le \delta_i^+ e^{m(\tau_{i+1})-m(\tau_i)} + \eta \int_{\tau_i}^{\tau_{i+1}} e^{m(\tau_{i+1})-m(s)} ds \\
&\le (\delta_i^- (1 + \mu) + \eta)\, e^{m(\tau_{i+1})-m(\tau_i)} + \eta\, e^{m(\tau_{i+1})} \int_{\tau_i}^{\tau_{i+1}} e^{-m(s)} ds \\
&\le \left[(1 + \mu)\left(\delta_{i-1}^+ e^{m(\tau_i)-m(\tau_{i-1})} + \eta e^{m(\tau_i)} \int_{\tau_{i-1}}^{\tau_i} e^{-m(s)} ds\right) + \eta\right] \\
&\quad \times e^{m(\tau_{i+1})-m(\tau_i)} + \eta\, e^{m(\tau_{i+1})} \int_{\tau_i}^{\tau_{i+1}} e^{-m(s)} ds \\
&\le (1 + \mu)\delta_{i-1}^+ e^{m(\tau_{i+1})-m(\tau_{i-1})} + \eta e^{m(\tau_{i+1})-m(\tau_i)} \\
&\quad + \eta e^{m(\tau_{i+1})}\left[(1 + \mu) \int_{\tau_{i-1}}^{\tau_i} e^{-m(s)} ds + \int_{\tau_i}^{\tau_{i+1}} e^{-m(s)} ds\right] \le \cdots
\end{aligned}$$

$$\le (1+\mu)^i \delta_0\, e^{m(\tau_i+1)} + \eta \sum_{j=0}^{i-1} (1+\mu)^j e^{m(\tau_i+1)-m(\tau_{i-j})}$$

$$+ \eta\, e^{m(\tau_i+1)} \sum_{j=0}^{i} (1+\mu)^j \int_{\tau_{i-j}}^{\tau_{i-j+1}} e^{-m(s)}\, ds$$

$$\le (1+\mu)^k \delta_0\, e^{m(T)}$$

$$+ \eta e^{m(T)} \sum_{j=0}^{k} (1+\mu)^j \left(1 + \int_{\tau_{k-j}}^{\tau_{k-j+1}} e^{-m(s)}\, ds\right).$$

Denoting $A = (1+\mu)^k\, e^{m(T)}$ and

$$B = e^{m(T)} \sum_{j=0}^{k} (1+\mu)^j \left(1 + \int_{\tau_{k-j}}^{\tau_{k-j+1}} e^{-m(s)}\, ds\right),$$

we finally obtain

$$\|x(t) - y(t)\| \le A\delta_0 + B\eta, \quad t \in I. \qquad \square$$

By analogy, we can prove the following generalization of Theorem 3:

Theorem 4 ([114]). *Suppose that the following conditions are satisfied in the domain* $Q = I \times D$, $D \in \operatorname{comp}(\mathbb{R}^n)$*:*

(1) *set-valued mappings* $F_j(t,x)$, $j = 1,2$, *are measurable in* t*, and the following inequality holds for* $\|x - y\| \le r$ *and almost all* t*:*

$$h(F_j(t,x), F_j(t,y)) \le \omega(t,r),$$

where $\omega(t,r)$ *is the Kamke function;*

(2) *set-valued mappings* $I_j^i(x)$ *satisfy the Lipschitz condition with constant* μ*;*

(3) *the following estimates are true:*

$$h(F_1(t,x), F_2(t,x)) \le \eta(t),\ h(I_1^i(x), I_2^i(x)) \le \bar{\eta}, \quad h(X_0, Y_0) \le \delta_0,$$

where $\eta(t)$ *is summable on* I*.*

Then, for any solution $y(t)$ *of inclusion* (2.5), (2.6)*, a solution* $x(t)$ *of inclusion* (2.3), (2.4) *exists and is such that*

$$\|x(t) - y(t)\| \le r(t), \quad t \in I,$$

and vice versa. Here, $r(t)$ *is the upper (maximum) solution of the problem*

$$\dot{r}(t) = \omega(t, r(t)) + \eta(t), \quad t \in (\tau_i, \tau_{i+1}],\ \tau_0 = t_0,\ \tau_{k+1} = T,$$

$$r(\tau_i + 0) = (1+\mu)\, r(\tau_i) + \bar{\eta}, \quad r(t_0) = \delta_0,\ i = \overline{0,k}.$$

Definition 4 ([114]). We say that a solution of the differential inclusion (2.1), (2.2) is extendable (weakly extendable) to I if any solution (at least one solution) $x(t)$ of the differential inclusion is extendable to I.

Example 1. Consider the control problem

$$\dot{x}(t) = u(t)(1 + x^2(t)), \quad x(0) = 0, \tag{2.8}$$

where $t \geq 0$ is time, $x \in \mathbb{R}$ is a phase vector, and the control $u(t) \in [0, 1]$ is a measurable function.

It is obvious that Eq. (2.8) can be represented in the form of a differential inclusion:

$$\dot{x}(t) \in [0, 1] \times (1 + x^2(t)), \quad x(0) = 0. \tag{2.9}$$

If $u(t)$ is an arbitrary given measurable function, then solution (2.8) takes the form

$$x(t) = \tan \int_0^t u(s)\, ds. \tag{2.10}$$

Note that solution (2.10) is defined on $[0, \infty)$ if and only if

$$\int_0^t u(s)\, ds < \frac{\pi}{2}$$

for any fixed $t < \infty$. If $u(t)$ is such that

$$\int_0^t u(s)\, ds = \frac{\pi}{2}$$

for some t, then the corresponding solution cannot be extended to the interval $[0, \infty)$. It is obvious that the set of solutions is weakly extendable. The set of solutions of (2.8) is not bounded on the interval $[0, \infty)$.

Using a pulse control, one can obtain the set of solutions that are strongly extendable and bounded.

Let a pulse action be given in the form

$$\Delta x|_{t=\frac{k\pi}{2}} = -x, \quad k = 1, 2, \dots. \tag{2.11}$$

The set of solutions of (2.9) and (2.11) has the form

$$X(t) = \left\{ \tan \int_{\frac{(k-1)\pi}{4}}^t u(s)\, ds,\ u(s) \in [0, 1], t \in \left(\frac{(k-1)\pi}{4}, \frac{k\pi}{4} \right], k = 1, 2, \dots \right\},$$

i.e., it is strongly extendable and bounded.

2.2 Differential Inclusions with Nonfixed Times of Pulse Action

Consider differential inclusions with pulse actions at nonfixed times:

$$\dot{x} \in F(t, x), \quad t \neq \tau_i(x), \; x(t_0) = x_0, \tag{2.12}$$

$$\Delta x|_{t=\tau_i(x)} \in I_i(x), \tag{2.13}$$

where $t \in I = [t_0, T]$, $x \in D$ (D is a domain in $\mathbb{R}^n$), $F: I \times D \to \operatorname{conv}(\mathbb{R}^n)$, $\tau_i: D \to \mathbb{R}$, $i = \overline{1, k}$, and $I_i: D \to \operatorname{comp}(\mathbb{R}^n)$.

Definition 5 ([114]). A piecewise absolutely continuous vector function $x(t)$ defined on a segment or interval $J \subset I$ is called a solution of inclusion (2.12), (2.13) if it satisfies (2.12) almost everywhere for $t \in J$, $t \neq \tau_i(x(t))$, $i = \overline{1, k}$, and has discontinuities of the first kind at the points $t = \tau_i(x(t))$ with jumps

$$\Delta x(t) \in I_i(x(t)).$$

Definition 6 ([114]). Solutions of the differential inclusion (2.12), (2.13) are called extendable (weakly extendable) to I if every solution (at least one solution) $x(t)$ of the differential inclusion is extendable to I.

Definition 7 ([36, 128]). A tangent cone $T_D(x)$ to a nonempty set $D \subset \mathbb{R}^n$ at a point $x \in D$ is defined as follows:

$$T_D(x) = \left\{ y \in \mathbb{R}^n \mid \lim_{s \to +0} s^{-1} \inf_{z \in D} |x + sy - z| = 0 \right\}.$$

If $D \in \operatorname{conv}(\mathbb{R}^n)$, then the set $T_D(x)$ is closed and convex, and $0 \in T_D(x)$.

Consider sufficient conditions for the absence of beating of solutions against the surface $t = \tau_i(x)$, i.e., conditions guaranteeing that any solution of a differential inclusion crosses the impulsive surface at most once.

Theorem 5 ([114]). *Suppose that the following conditions are satisfied in the domain $Q = I \times D$:*

(1) *a set-valued mapping $F(t, x)$ is bounded by a constant C ($|F(t, x)| \leq C$), is continuous in t, satisfies the Lipschitz condition with respect to x with constant λ, and is such that $F(t, x) \subset T_D(x)$;*

(2) *set-valued mappings $x + I_i(x): D \to \operatorname{conv}(D)$ satisfy the Lipschitz condition with constant λ, $i = \overline{1, k}$;*

(3) *surfaces* $\tau_i(x)$ *are smooth, disjoint, and such that, for every* $z_i \in I_i(x)$, *one has*

$$\max_{0\le\sigma\le 1}\left(\frac{\partial\tau_i(x+\sigma z_i)}{\partial x}, z_i\right)\le 0, \qquad \left\|\frac{\partial\tau_i(x)}{\partial x}\right\|\le N; \tag{2.14}$$

(4) $NC < 1$.

Then every solution $x(t)$, $x_0 \in D$, *of the differential inclusion* (2.12), (2.13) *is extendable to* I *and crosses the surfaces* $t = \tau_i(x)$ *at most once.*

Proof. Conditions (1) and (2) guarantee that the solutions remain in the domain D. Assume that there exists a solution $x(t)$ that hits the surface $t = \tau_i(x)$ at time $t_* = \tau_i(x(t_*))$ and crosses the same surface at another point $(t^*, x(t^*))$, where $t^* = \tau_i(x(t^*))$.

First, we consider the case where (t_*, t^*) is the interval of continuity of $x(t)$. Then

$$x(t^*) = x(t_*) + z_i + \int_{t_*}^{t^*} u(s)\,ds,$$

where $z_i \in I_i(x(t_*))$ and $u(s) \in F(s, x(s))$ is a measurable selector.

Consider the difference

$$\begin{aligned} t^* - t_* &= \tau_i(x(t^*)) - \tau_i(x(t_*)) \\ &= \tau_i(x(t^*)) - \tau_i(x(t_*) + z_i) + \tau_i(x(t_*) + z_i) - \tau_i(x(t_*)) \\ &= \int_{t_*}^{t^*}\left(\frac{\partial\tau_i(x(s))}{\partial x}, u(s)\right)ds + \int_0^1\left(\frac{\partial\tau_i(x(t_*)+\sigma z_i)}{\partial x}, z_i\right)d\sigma. \end{aligned} \tag{2.15}$$

By virtue of the Cauchy–Schwarz inequality and conditions (1) and (3) of Theorem 5, we obtain the following estimate:

$$\int_{t_*}^{t^*}\left(\frac{\partial\tau_i(x(s))}{\partial x}, u(s)\right)ds \le NC(t^* - t_*). \tag{2.16}$$

Using (2.15) and (2.16), we get

$$(1 - NC)(t^* - t_*) \le \int_0^1\left(\frac{\partial\tau_i(x(t_*)+\sigma z_i)}{\partial x}, z_i\right)d\sigma. \tag{2.17}$$

Since $NC < 1$ and $t^* > t_*$, inequality (2.17) contradicts condition (3) of Theorem 5. Therefore, in the case under consideration, each solution crosses each surface $t = \tau_i(x)$ at most once. It follows from conditions (1) and (2) of Theorem 5 that every local solution can be extended to the interval I.

Now consider the case where a solution $x(t)$ crosses other surfaces on the interval (t_*, t^*), i.e., there exist the moments of time $t_{i_j} = \tau_{i_j}(x(t_{i_j}))$, $i_1 = i$, $j = \overline{1, p}$,

$p \le k$. We have proved that, on the interval $(t_{i_{j-1}}, t_{i_j}]$, which is the interval of continuity of the function $x(t)$, one has $\tau_{i_{j-1}}(x(t)) < t$, and, obviously,

$$t_{i_j} = \tau_{i_j}(x(t_{i_j})) > \tau_{i_{j-1}}(x(t_{i_j})).$$

We now prove that $\tau_{i_j}(x(t)) > \tau_{i_{j-1}}(x(t))$ on the interval $(t_{i_{j-1}}, t_{i_j}]$. Assume the contrary. Then there exists a continuous function $y(t)\colon [0,1] \to D$, $y(0) = x(t)$, $y(1) = x(t_{i_j})$, such that

$$\tau_{i_j}(y(0)) \le \tau_{i_{j-1}}(y(0)), \quad t_{i_j} = \tau_{i_j}(y(1)) > \tau_{i_{j-1}}(y(1)).$$

It follows from these inequalities that there exists $\xi \in [0,1)$ such that $\tau_{i_j}(y(\xi)) = \tau_{i_{j-1}}(y(\xi))$, which contradicts the fact that the surfaces $t = \tau_j(x)$ are disjoint.

Since $\tau_{i_{j+1}}(x(t_{i_j})) > \tau_{i_j}(x(t_{i_j})) > \tau_{i_{j-1}}(x(t_{i_j}))$, we conclude, by analogy, that $\tau_{i_{j+1}}(x(t)) > \tau_{i_{j-1}}(x(t))$ on the interval $(t_{i_j}, t_{i_{j+1}}]$. Using the method of complete mathematical induction, one can show that $\tau_{i_p}(x(t)) > \tau_i(x(t))$ on the interval $(t_{i_{p-1}}, t_{i_p}]$. The interval $(t_{i_p}, t^*]$ is the interval of continuity of the solution $x(t)$. Taking the inequality $t_{i_p} > \tau_i(x_{i_p})$ into account and setting $t_* = x(t_{i_p})$ and $z_i = z_{i_p}$, we conclude that inequality (2.17) is not true.

Theorem 5 is proved. □

Remark 1. The condition $F(t,x) \subset T_D(x)$ can be replaced by the following one:

For every $x_0 \in D' \subset D$, there exists a constant $\rho > 0$ for which all solutions of (2.12), (2.13) *belong to the domain D on the interval $[t_0, T]$ together with their ρ-neighborhoods.*

In this case, the conditions $x + I_i(x) \subset D$ are redundant. Note that the aforementioned conditions are not equivalent.

Remark 1 is also valid for the subsequent theorems.

Remark 2. If the conditions of Theorem 5

$F(t,x) \subset T_D(x)$, $x + I_i(x)\colon D \to \mathrm{conv}(D)$, and, for every $z \in I_i(x)$, inequalities (2.14) *are true*

are replaced by the conditions

$F(t,x) \bigcap T_D(x) \ne \varnothing$, $I_i(x)\colon D \to \mathrm{conv}(\mathbb{R}^n)$, $x + I_i(x) \bigcap D \ne \varnothing$, and there exists $z \in I_i(x)$, $x + z \in D$, for which inequalities (2.14) *are true,*

then one can show that there exists a solution that crosses the impulsive surfaces at most once and that the set of solutions of (2.12), (2.13) is weakly extendable.

The theorem below establishes conditions for the absence of beating of solutions of system (2.12), (2.13) against the surfaces $t = \tau_i(x)$ in the case where the functions $\tau_i(x)$ are not continuously differentiable.

Theorem 6 ([114]). *Suppose that conditions* (1), (2), *and* (4) *of Theorem* 5 *are satisfied for the system of equations* (2.12), (2.13), *and condition* (3) *is replaced by the following one:*

(3′) *the surfaces* $t = \tau_i(x)$ *are disjoint, the functions* $\tau_i(x)$ *satisfy the Lipschitz condition with constant* N, *and, for all* $x \in D$ *and* $z_i \in I_i(x)$, *one has*

$$\tau_i(x) \geq \tau(x + z_i). \tag{2.18}$$

Then every solution $x(t)$, $x_0 \in D$, *of the differential inclusion* (2.12), (2.13) *can be extended to* I *and crosses the surfaces* $t = \tau_i(x)$ *at most once.*

Proof. This theorem is proved by analogy with the previous theorem. The difference is that (2.15) is replaced by

$$\begin{aligned} t^* - t_* &= \tau_i(x(t^*)) - \tau_i(x(t_*)) \\ &= \tau_i(x(t^*)) - \tau_i(x(t_*) + z_i) + \tau_i(x(t_*) + z_i) - \tau_i(x(t_*)) \\ &\leq N \left\| \int_{t_*}^{t^*} u(s)ds \right\| + \tau_i(x(t_*) + z_i) - \tau_i(x(t_*)) \\ &\leq N \int_{t_*}^{t^*} \|u(s)\|ds + \tau_i(x(t_*) + z_i) - \tau_i(x(t_*)) \\ &\leq NC(t^* - t_*) + \tau_i(x(t_*) + z_i) - \tau_i(x(t_*)), \end{aligned}$$

whence

$$(1 - NC)(t^* - t_*) \leq \tau_i(x(t_*) + z_i) - \tau_i(x(t_*)).$$

By virtue of conditions (3′) and (4) of Theorem 6, the last inequality becomes contradictory. □

Remark 3. It is obvious that if $\tau_i(x)$ possesses the Lipschitz property, then inequalities (2.14) can be written with the use of the subdifferential $\partial\tau_i(x)$ [128].

Consider the problem of continuous dependence of solutions on initial data. Let $x(t, x_0)$ and $x(t, y_0)$ be two solutions of (2.12), (2.13) with initial conditions $x(t_0, x_0) = x_0$ and $x(t_0, y_0) = y_0$. Let $\sigma_i(x_0)$ and $\sigma_i(y_0)$ be the corresponding times when the surface $t = \tau_i(x)$ is crossed and let

$$\tau_i^- = \min\{\sigma_i(x_0), \sigma_i(y_0)\}, \quad \tau_i^+ = \max\{\sigma_i(x_0), \sigma_i(y_0)\}.$$

Theorem 7 ([114]). *Suppose that the conditions of Theorem* 5 *are satisfied in the domain* Q.

Then, for every solution $x(t, x_0)$*, there exist a constant* $\delta > 0$ *and a solution* $x(t, y_0)$ *such that, if* $\|x_0 - y_0\| < \delta$*, then the following inequality holds for* $t \in I$*,* $t \notin [\tau_i^-, \tau_i^+]$*,* $i = \overline{1,k}$*:*

$$\|x(t, x_0) - x(t, y_0)\| \le \left(1 + \frac{\lambda}{1 - NC}\right)^k e^{\lambda(T - t_0)} \|x_0 - y_0\|. \tag{2.19}$$

Proof. Assume that the solution $x(t, y_0)$ crosses the surfaces $t = \tau_i(x)$ in the same order as the solution $x(t, x_0)$. In what follows, we prove that this is indeed the case for all sufficiently small $\delta > 0$.

We represent the solutions $x(t, x_0)$ and $x(t, y_0)$ as follows:

$$x(t, x_0) = x_0 + \int_{t_0}^{t} u(s)\, ds + \sum_{t_0 \le \sigma_i(x_0) < t} z^i,$$

$$x(t, y_0) = y_0 + \int_{t_0}^{t} v(s)\, ds + \sum_{t \le \sigma_i(y_0) < t} w^i,$$

where $u(s) \in F(s, x(s, x_0))$ and $v(s) \in F(s, x(s, y_0))$ are measurable functions, $z^i \in I_i(x(\sigma_i(x_0), x_0))$, and $w^i \in I_i(x(\sigma_i(y_0), y_0))$.

Let us estimate the distance between the indicated solutions:

$$\|x(t, x_0) - x(t, y_0)\| \le \|x_0 - y_0\| + \int_{t_0}^{t} \|u(s) - v(s)\|\, ds$$
$$+ \left\| \sum_{t_0 \le \sigma_i(x_0) < t} z^i - \sum_{t_0 \le \sigma_i(y_0) < t} w^i \right\|. \tag{2.20}$$

If $t \in [t_0, T] \setminus \bigcup_i [\tau_i^-, \tau_i^+]$, then, for every measurable function $u(s) \in F(s, x(s, x_0))$, there exists a measurable function $v(s) \in F(s, x(s, y_0))$ such that

$$\int_{t_0}^{t} \|u(s) - v(s)\|\, ds \le \int_{t_0}^{t} h(F(s, x(s, x_0)), F(s, x(s, y_0)))\, ds$$
$$\le \lambda \int_{t_0}^{t} \|x(s, x_0) - x(s, y_0)\|\, ds.$$

Similarly, for every vector $z^i \in I_i(x(\sigma_i(x_0), x_0))$, there exists a vector $w^i \in I_i(x(\sigma_i(y_0), y_0))$ such that

$$\left\| \sum_{t_0 \le \sigma_i(x_0) < t} z^i - \sum_{t_0 \le \sigma_i(y_0) < t} w^i \right\| \le \sum_i \|z^i - w^i\|$$
$$\le \sum_i h(I_i(x(\sigma_i(x_0), x_0)), I_i(x(\sigma_i(y_0), y_0)))$$

$$\leq \sum_i \lambda \|x(\sigma_i(x_0), x_0) - x(\sigma_i(y_0), y_0)\|$$
$$\leq \lambda \sum_i (\|x(\sigma_i(x_0), z_0) - x(\sigma_i(y_0), z_0)\|$$
$$+ \|x(\tau_i^-, x_0) - x(\tau_i^-, y_0)\|),$$

where $z_0 = x_0$ if $\tau_i^- = \sigma_i(y_0)$, and $z_0 = y_0$ if $\tau_i^- = \sigma_i(x_0)$.

If $r(s) \in F(s, x(s, z_0))$, then

$$\|x(\sigma_i(x_0), z_0) - x(\sigma_i(y_0), z_0)\| \leq \int_{\tau_i^-}^{\tau_i^+} \|r(s)\| \, ds \leq C(\tau_i^+ - \tau_i^-).$$

Since

$$\begin{aligned}
\tau_i^+ - \tau_i^- &= |\sigma_i(x_0) - \sigma_i(y_0)| \\
&= |\tau_i(x(\sigma_i(x_0), x_0)) - \tau_i(x(\sigma_i(y_0), y_0))| \\
&\leq N \|x(\sigma_i(x_0), x_0) - x(\sigma_i(y_0), y_0)\| \\
&\leq N(\|x(\sigma_i(x_0), z_0) - x(\sigma_i(y_0), z_0)\| + \|x(\tau_i^-, x_0) - x(\tau_i^-, y_0)\|) \\
&\leq NC(\tau_i^+ - \tau_i^-) + N\|x(\tau_i^-, x_0) - x(\tau_i^-, y_0)\|,
\end{aligned}$$

we have

$$\tau_i^+ - \tau_i^- \leq \frac{N}{1 - NC} \|x(\tau_i^-, x_0) - x(\tau_i^-, y_0)\|.$$

Hence, inequality (2.20) takes the form

$$\|x(t, x_0) - x(t, y_0)\| \leq \|x_0 - y_0\| + \lambda \int_{t_0}^{t} \|x(s, x_0) - x(s, y_0)\| \, ds$$
$$+ \frac{\lambda}{1 - NC} \sum_{t_0 \leq \tau_i^- < t} \|x(\tau_i^-, x_0) - x(\tau_i^-, y_0)\|,$$

and, according to Lemma 2 in Chapter 1, the following estimate is true:

$$\|x(t, x_0) - x(t, y_0)\| \leq \left(1 + \frac{\lambda}{1 - NC}\right)^k e^{\lambda(T - t_0)} \|x_0 - y_0\|, \tag{2.21}$$
$$t \in [t_0, T] \setminus \bigcup_i [\tau_i^-, \tau_i^+].$$

To complete the proof, we show that, for all sufficiently small $\delta > 0$, the solution $x(t, y_0)$ crosses the surfaces $t = \tau_i(x)$ in the same order as the solution $x(t, x_0)$. If (t_*, t) is the interval without pulses for the solutions $x(t, x_0)$ and $x(t, y_0)$, and

$t = \tau_i(x(t, x_0))$, then the condition $NC < 1$ yields $t_* < \tau_i(x(t_*, x_0))$. According to estimate (2.21), we can choose a sufficiently small δ so that $t_* < \tau_i(x(t_*, y_0))$. Since the surfaces $\tau_i(x)$ are disjoint, the proper choice of $\delta > 0$ and the condition $NC < 1$ guarantee that $x(t, y_0)$ crosses $t = \tau_i(x)$ prior to crossing another surface. Then, by induction, we establish that the solution $x(t, y_0)$ crosses the surfaces $t = \tau_i(x)$ in the same order as the solution $x(t, x_0)$.

Theorem 7 is proved. □

Remark 4. The statement of Theorem 7 remains true if we assume that the functions $\tau_i(x)$ satisfy the Lipschitz condition with constant N, $NC < 1$, and condition (2.18) is satisfied.

Consider the problem of continuous dependence of solutions of impulsive differential inclusions upon the right-hand sides, impulsive surfaces, and initial data.

Lemma 1 ([114]). *Let*

$$\delta_{i+1}^{+} \le a_1 \delta_{i+1}^{-} + a_2,$$

$$\delta_{i+1}^{-} \le a_3 \delta_i^{+} + a_4,$$

where $a_j \ge 0$, $j = \overline{1,4}$, *and* $i = 1, 2, \ldots$. *Then*

$$\delta_{i+1}^{-} \le (a_2 a_3 + a_4)\frac{(a_1 a_3)^i - 1}{a_1 a_3 - 1} + (a_1 a_3)^i \delta_0^{+} \quad \text{if } a_1 a_3 \ne 1, \tag{2.22}$$

$$\delta_{i+1}^{-} \le (a_2 a_3 + a_4) i + \delta_0^{+} \quad \text{if } a_1 a_3 = 1. \tag{2.23}$$

Proof. Indeed,

$$\begin{aligned} \delta_{i+1}^{-} &\le a_3 a_1 \delta_i^{-} + a_3 a_2 + a_4 \le a_1 a_3^2 \delta_{i-1}^{+} + a_3 a_1 a_4 + a_3 a_2 + a_4 \\ &\le \cdots \le [1 + a_1 a_3 + \cdots + (a_1 a_3)^{i-1}](a_2 a_3 + a_4) + (a_1 a_3)^i \delta_0^{+}. \end{aligned} \tag{2.24}$$

If $a_1 a_3 \ne 1$, then (2.24) yields (2.22), and if $a_1 a_3 = 1$, then we obtain (2.23). Lemma 1 is proved. □

Consider the impulsive differential inclusions

$$\dot{x} \in F_1(t, x), \quad t \ne \tau_1^i(x), \ x(t_0) = x_0 \in X_0, \tag{2.25}$$

$$\Delta x|_{t=\tau_1^i(x)} \in I_1^i(x), \quad i = 1, 2, \ldots, \tag{2.26}$$

$$\dot{y} \in F_2(t, y), \quad t \ne \tau_2^i(y), \ y(t_0) = y_0 \in Y_0, \tag{2.27}$$

$$\Delta y|_{t=\tau_2^i(y)} \in I_2^i(y), \quad i = 1, 2, \ldots. \tag{2.28}$$

Let τ_i^- (τ_i^+) denote the times when the left-continuous solution $x(t)$ of inclusion (2.25), (2.26) (solution $y(t)$ of inclusion (2.27), (2.28)) hits the surfaces $t = \tau_1^i(x)$ ($t = \tau_2^i(x)$).

Theorem 8 ([114]). *Suppose that, in the domain Q, the conditions of Theorem 6 are satisfied for inclusions (2.25), (2.26) and (2.27), (2.28), and the following inequalities are true:*

$$\mu C < 1, \quad \mu = \max\{\lambda, N\},$$
$$h(F_1(t,x), F_2(t,x)) \le \eta, \quad h(I_1^i(x), I_2^i(x)) \le \eta,$$
$$|\tau_1^i(x) - \tau_2^i(x)| \le \eta, \quad h(X_0, Y_0) \le \delta.$$

Then, for any $\xi > 0$ and every solution $x(t)$ of inclusion (2.25), (2.26), there exist $\eta > 0$, $\delta > 0$, and a solution $y(t)$ of inclusion (2.27), (2.28) such that

$$\|x(t) - y(t)\| \le \xi, \quad t \in [t_0, T] \setminus \bigcup_i [\tau_i^-, \tau_i^+]. \tag{2.29}$$

Proof. Without loss of generality, we can assume that $\tau_i^- \le \tau_i^+$. Let $x(t)$ be an arbitrary solution of the impulsive differential inclusion (2.25), (2.26) with initial condition $x(t_0) = x_0 \in X_0$. According to the Filippov theorem [51], there exist a solution $y(t)$ of the differential inclusion (2.27) and constants C_1 and C_2 such that

$$\|y(t) - x(t)\| \le C_1\delta_i^+ + C_2\eta,$$
$$t \in (\tau_i^+, \tau_{i+1}^-], \quad i = 0, 1, \dots, \quad \tau_0^+ = t_0,$$

where $\delta_i^- = \|x(\tau_i^-) - y(\tau_i^-)\|$ and $\delta_i^+ = \|x(\tau_i^+ + 0) - y(\tau_i^+ + 0)\|$ ($\delta_0^+ = \delta$).

Note that

$$\delta_{i+1}^- = \|x(\tau_{i+1}^-) - y(\tau_{i+1}^-)\| \le C_1\delta_i^+ + C_2\eta. \tag{2.30}$$

The following estimates are true:

$$\begin{aligned}
\|x(\tau_i^-) - y(\tau_i^+)\| &\le \|x(\tau_i^-) - y(\tau_i^-)\| + \|y(\tau_i^-) - y(\tau_i^+)\| \\
&\le \delta_i^- + C|\tau_i^- - \tau_i^+|, \\
|\tau_i^- - \tau_i^+| &= |\tau_1^i(x(\tau_i^-)) - \tau_2^i(y(\tau_i^+))| \\
&\le |\tau_1^i(x(\tau_i^-)) - \tau_2^i(x(\tau_i^-))| + |\tau_2^i(x(\tau_i^-)) - \tau_2^i(y(\tau_i^+))| \\
&\le \eta + \mu\|x(\tau_i^-) - y(\tau_i^+)\| \le \eta + \mu(\delta_i^- + C|\tau_i^- - \tau_i^+|),
\end{aligned}$$

i.e.,

$$|\tau_i^- - \tau_i^+| \le \frac{\eta + \mu\delta_i^-}{1 - \mu C}. \tag{2.31}$$

We now determine $y(\tau_i^+ + 0)$ from the following problem of minimization:

$$\delta_i^+ = \|x(\tau_i^+ + 0) - y(\tau_i^+ + 0)\| = \min_{z \in I_2^i(y(\tau_i^+))} \|x(\tau_i^+ + 0) - y(\tau_i^+) - z\|.$$

Then

$$\begin{aligned}
\delta_i^+ &\le h\Big(x(\tau_i^-) + I_1^i(x(\tau_i^-)) + \int_{\tau_i^-}^{\tau_i^+} u(s)\,ds\,,\ y(\tau_i^-) + \int_{\tau_i^-}^{\tau_i^+} v(s)\,ds + I_2^i(y(\tau_i^+))\Big)\\
&\le \delta_i^- + \left\| \int_{\tau_i^-}^{\tau_i^+} (u(s) - v(s))\,ds \right\|\\
&\quad + h(I_1^i(x(\tau_i^-)), I_2^i(x(\tau_i^-))) + h(I_2^i(x(\tau_i^-)), I_2^i(y(\tau_i^+)))\\
&\le \delta_i^- + 2C|\tau_i^- - \tau_i^+| + \eta + \mu(\delta_i^- + C|\tau_i^- - \tau_i^+|)\\
&\le \frac{1+\mu+\mu C}{1-\mu C}\delta_i^- + \frac{2C+1}{1-\mu C}\eta.
\end{aligned}$$

Taking inequality (2.30) into account, we obtain

$$\delta_{i+1}^- = \|x(\tau_{i+1}^-) - y(\tau_{i+1}^-)\| \le C_1\delta_i^+ + C_2\eta,$$

$$\delta_i^+ \le \frac{1+\mu+\mu C}{1-\mu C}\delta_i^- + \frac{2C+1}{1-\mu C}\eta.$$

By virtue of Lemma 1, the above inequalities yield

$$\delta_i^- \le K_1\eta + K_2\delta, \quad i = 1, 2, \ldots,$$

where K_1 and K_2 are constants independent of η and δ. Properly choosing $\eta > 0$ and $\delta > 0$, we establish that the solution $y(t)$ crosses the surfaces $t = \tau_2^i(x)$ in the same order as the solution $x(t)$ crosses the surfaces $t = \tau_1^i(x)$. Thus, inequality (2.29) is true.

Theorem 8 is proved. □

2.3 Examples

In this section, we consider several examples illustrating various cases of beating of solutions (strong and weak beating) and various types of extendability (strong and weak extendability), as well as the property of continuous dependence of solutions, depending on whether the conditions of the theorems from the previous section are satisfied.

Example 2. This example shows that the condition $NC < 1$ in the theorems on continuous dependence cannot be omitted.

Consider the system

$$\dot{x} = u(t), \quad x(0) = \alpha > 0,\ |u(t)| \le C,\ C \ge \frac{1}{2},\ t \ne x^2 \equiv \tau(x),$$

$$\Delta x|_{t=x^2} = -x \equiv I(x), \quad 0 \le t \le 1.$$

If $u(t) = C$ and $\alpha C = \frac{1}{4}$, then the solution $x(t) = \alpha + Ct$ crosses the curve $t = x^2$ only at the point $t = 4, x = 2$. If $\alpha C > \frac{1}{4}$, then the solution does not cross the curve $t = x^2$ because the system $t = x^2$, $x = Ct + \alpha$ has a unique solution for $\alpha C = \frac{1}{4}$ and does not have solutions for $\alpha C > \frac{1}{4}$.

Let $\alpha C = \frac{1}{4}$. By varying the initial data α or the right-hand sides, we establish that the continuous dependence is violated because the solutions are not subject to pulse action for $\alpha C > \frac{1}{4}$ and have a discontinuity for $\alpha C = \frac{1}{4}$. It remains to verify that, for $\alpha C = \frac{1}{4}$, we have $NC \geq 1$. Indeed,

$$\begin{aligned} NC &= C \sup_{t\in[0,1]} \left| \frac{\partial \tau(x(t))}{\partial x} \right| = C \sup_{t\in[0,1]} 2(\alpha + Ct) \\ &= 2C(\alpha + C) = \frac{1}{2} + 2C^2 \geq 1. \end{aligned}$$

Example 3. Consider one more example in which the condition $\tau_i(x + z_i) \leq \tau_i(x)$, $z_i \in I_i(x)$ is violated and the phenomenon of beating is observed (see Figure 1):

$$\begin{aligned} \dot{x} &= 0, \quad x(0) = \alpha > 0, \; t \neq x \equiv \tau(x), \\ \Delta x|_{t=x} &= x - 1 \equiv I(x). \end{aligned}$$

For $\alpha < 1$, the solutions undergo a single pulse action. If $\alpha = 1$, then the solution is $x(t) \equiv 1$. Finally, if $\alpha > 1$, then the solution undergoes pulse action infinitely many times (in this case, the condition $\tau(x + z) = x + x - 1 \leq x = \tau(x)$ is not satisfied because the inequality $x > 1$ holds for all t).

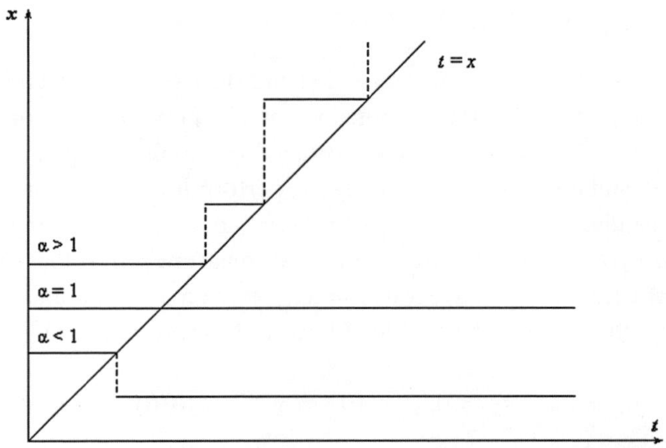

Figure 1.

Example 4. Consider the following differential equation with multivalued pulses:

$$\ddot{x} + x = 0 \quad \text{for } \dot{x} \neq 0,\ x(0) = x_0,\ \dot{x}(0) = \dot{x}_0, \tag{2.32}$$

$$\Delta x|_{\dot{x}=0} = 0, \quad \Delta \dot{x}|_{\dot{x}=0} \in W = [w^-, w^+],\ w^-, w^+ \in \mathbb{R}.$$

Equation (2.32) can be interpreted as the equation of oscillation of a swing: at the uppermost position, the swing gets a push, which increases the swinging amplitude.

We introduce the auxiliary variables $x = x_1$ and $x_2 = \dot{x}_1$ and rewrite (2.32) in the form of a system:

$$\dot{x}_1 = x_2, \quad \dot{x}_2 = -x_1 \quad \text{for } x_2 \neq 0, \tag{2.33}$$

$$\Delta x_1|_{x_2=0} = 0, \quad \Delta x_2|_{x_2=0} \in W, \tag{2.34}$$

$$x_1(0) = x_0, \quad x_2(0) = \dot{x}_0. \tag{2.35}$$

We introduce new variables by the relations

$$x_1 = y_1 \cos(t + y_2), \quad x_2 = -y_1 \sin(t + y_2), \quad y_1 \geq 0. \tag{2.36}$$

Substituting them in (2.33), we get

$$\dot{y}_1 \cos(t + y_2) - y_1 \sin(t + y_2)(1 + \dot{y}_2) = -y_1 \sin(t + y_2),$$

$$-\dot{y}_1 \sin(t + y_2) - y_1 \cos(t + y_2)(1 + \dot{y}_2) = -y_1 \cos(t + y_2).$$

Hence,

$$\dot{y}_1 \cos(t + y_2) - y_1 \dot{y}_2 \sin(t + y_2) = 0, \tag{2.37}$$

$$\dot{y}_1 \sin(t + y_2) + y_1 \dot{y}_2 \cos(t + y_2) = 0. \tag{2.38}$$

Multiplying (2.37) and (2.38) by $\cos(t + y_2)$ and $\sin(t + y_2)$, respectively, performing summation, and reducing similar terms, we obtain $\dot{y}_1 = 0$. Thus, on the intervals between pulses, we have $y_1(t) \equiv \text{const}$. By analogy, multiplying (2.37) and (2.38) by $-\sin(t + y_2)$ and $\cos(t + y_2)$, respectively, performing summation, and reducing similar terms, we obtain $y_1 \dot{y}_2 = 0$, whence $\dot{y}_2 = 0$ because, according to the nature of the problem, $y_1(t)$ may vanish only on a set of measure zero (otherwise, by virtue of the change of variables, we have $x_2(t) = \dot{x}(t) \equiv 0$ on a certain interval I).

Consequently, the change of variables (2.36) reduces system (2.33), (2.35) to the form

$$\dot{y}_1 = 0, \quad \dot{y}_2 = 0, \quad y_1(0) = y_{10}, \quad y_2(0) = y_{20}, \tag{2.39}$$

where the initial data y_{10} and y_{20} satisfy the relations

$$y_{10} \cos y_{20} = x_0, \quad y_{10} \sin y_{20} = -\dot{x}_0,$$

i.e., $y_{10} = \sqrt{x_0^2 + \dot{x}_0^2}$, and y_{20} is not uniquely determined (it is determined to within a value multiple of 2π). We assume that $y_{20} \in (-\pi, \pi]$.

Let us determine the values of the pulses Δy_1 and Δy_2 and the times at which they appear. To this end, we denote the values of y_1 and y_2 before and after a pulse by y_1^-, y_2^- and y_1^+, y_2^+, respectively.

At the time of pulse action, we have

$$x_2 = 0 \Longrightarrow \sin(t_i + y_2^-) = 0 \Longrightarrow t_i = -y_2^- + i\pi$$

and $\cos(t_i + y_2^-) = \pm 1$. By assumption, there is no pulse in the variable x_1 at the time when $x_2 = 0$, i.e.,

$$\begin{aligned}\Delta x_1 = 0 &\Longrightarrow y_1^+ \cos(t_i + y_2^+) - y_1^- \cos(t_i + y_2^-) = 0 \\ &\Longrightarrow y_1^+ \cos(t_i + y_2^- + \Delta y_2) = y_1^- \cos(t_i + y_2^-) \\ &\Longrightarrow y_1^+[\cos(t_i + y_2^-)\cos\Delta y_2 - \sin(t_i + y_2^-)\sin\Delta y_2] \\ &= y_1^- \cos(t_i + y_2^-) \Longrightarrow y_1^+ \cos(t_i + y_2^-)\cos\Delta y_2 = y_1^- \cos(t_i + y_2^-).\end{aligned}$$

Since $\cos(t_i + y_2^-) \neq 0$, we have

$$y_1^+ \cos\Delta y_2 = y_1^-. \tag{2.40}$$

For the value of the pulse in the variable x_2 at the time when $x_2 = 0$, we have

$$\begin{aligned}&\Delta x_2 = -y_1^+ \sin(t_i + y_2^+) + y_1^- \sin(t_i + y_2^-) = w_i \in W \\ &\Longrightarrow y_1^+ \sin(t_i + y_2^+) = -w_i \Longrightarrow y_1^+ \sin(t_i + y_2^- + \Delta y_2) = -w_i \\ &\Longrightarrow y_1^+[\sin(t_i + y_2^-)\cos\Delta y_2 + \cos(t_i + y_2^-)\sin\Delta y_2] = -w_i \\ &\Longrightarrow y_1^+ \cos(t_i + y_2^-)\sin\Delta y_2 = -w_i \\ &\Longrightarrow y_1^+ \operatorname{sign}(\cos(t_i + y_2^-))\sin\Delta y_2 = -w_i.\end{aligned} \tag{2.41}$$

Raising (2.40) and (2.41) to the second power and performing summation, we obtain

$$\begin{aligned}(y_1^+)^2 = (y_1^-)^2 + w_i^2 &\Longrightarrow y_1^+ = \sqrt{(y_1^-)^2 + w_i^2} \quad \text{(so that } y_1 \geq 0\text{)} \\ &\Longrightarrow \Delta y_1 = y_1^+ - y_1^- = \sqrt{(y_1^-)^2 + w_i^2} - y_1^-.\end{aligned} \tag{2.42}$$

Let us determine Δy_2. Using (2.40) and (2.42), we obtain

$$\cos\Delta y_2 = \frac{y_1^-}{y_1^+} = \frac{y_1^-}{y_1^- + \Delta y_1} = \frac{y_1^-}{\sqrt{(y_1^-)^2 + w_i^2}} > 0. \tag{2.43}$$

Relations (2.41) and (2.42) yield

$$\sin \Delta y_2 = \frac{-w_i}{\sqrt{(y_1^-)^2 + w_i^2}} \operatorname{sign}(\cos(t + y_2^-)). \tag{2.44}$$

Note that Δy_2 is not uniquely determined (it is determined to within a value multiple of 2π). Moreover, by virtue of (2.36), different choices of Δy_2 lead the same variation in x_1 and x_2. Therefore, we assume that $\Delta y_2 \in (-\pi, \pi]$. Then

$$\Delta y_2 = -\arctan\left(\frac{w_i}{y_1^-} \operatorname{sign}(\cos(t + y_2^-))\right). \tag{2.45}$$

In this case, system (2.33)–(2.35) takes the form

$$\begin{gathered} \dot{y}_1 = 0, \quad \dot{y}_2 = 0 \quad \text{at } t_i \neq -y_2 + i\pi,\ i = 0, 1, \dots, \\ \Delta y_1|_{t_i = -y_2 + i\pi} \in \sqrt{(y_1^-)^2 + W^2} - y_1^-, \\ \Delta y_2|_{t_i = -y_2 + i\pi} \in -\arctan\left(\frac{W}{y_1^-} \operatorname{sign}(\cos(t + y_2^-))\right), W = [w^-, w^+], \\ y_1(0) = y_{10}, \quad y_2(0) = y_{20}. \end{gathered} \tag{2.46}$$

We now study the behavior of solutions of system (2.46), depending on the restrictions imposed on w^- and w^+. The following three cases are possible:

a) $w^- < w^+ \leq 0$;

b) $w^- < 0 < w^+$;

c) $0 \leq w^- < w^+$.

All conditions of Lemma 1 except condition 3) are obviously satisfied. Since $\frac{\partial \tau_i}{\partial y} = (0, -1)^T$, condition 3) is satisfied if

$$w \operatorname{sign}(\cos(t + y_2^-)) \leq 0 \quad \text{for any } w \in W. \tag{2.47}$$

a$_1$) $w^- < w^+ < 0$.

First, assume that $y_{10} > 0$ and $0 < y_{20} \leq \pi$. The solution of system (2.46) before the first pulse is given by $y_1(t) \equiv y_{10}$ and $y_2(t) \equiv y_{20}$. The solution $y_2(t)$ hits the surface $\tau_1 : t = -y_2 + \pi$. In this case, we have $\cos(t + y_2^-) = -1 < 0$, and, hence, $\Delta y_2 < 0$ by virtue of (2.45). Thus, "beating" takes place, and the solution cannot cross the surface $\tau_1 : t = -y_2 + \pi$ (Figure 2).

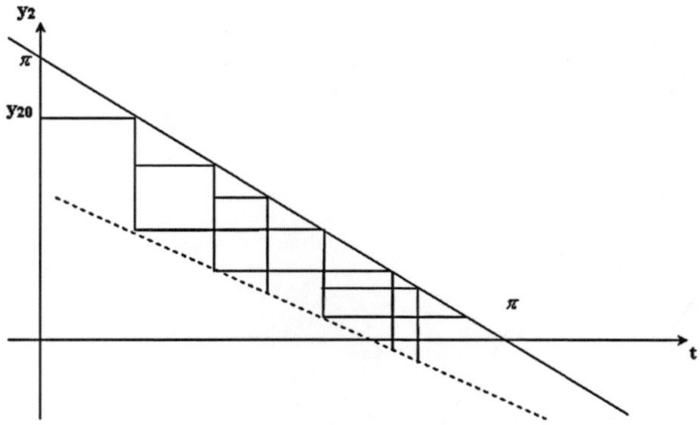

Figure 2.

We now show that $y_1(t) \to \infty$ as $t \to \infty$. Assume the contrary, i.e., let there exist a constant $C > 0$ such that $y_1(t) \le C$ for any t. Then, for any i, we get

$$\Delta y_1^i = \sqrt{(y_1^-)^2 + w_i^2} - y_1^- = \frac{w_i^2}{\sqrt{(y_1^-)^2 + w_i^2} + y_1^-}$$

$$\ge \frac{w_i^2}{\sqrt{C^2 + w_i^2} + C} \ge c_1 > 0.$$

Since the "beating" of a solution occurs, we have

$$y_1(t) \to y_{10} + \sum_{i=1}^{\infty} \Delta y_1^i = \infty \quad \text{as } t \to 0.$$

In other words, we arrive at a contradiction with the assumption that $y_1(t)$ is bounded.

It follows from (2.42) and (2.45) that $\Delta y_1 \to 0$ and $\Delta y_2 \to 0$ as $t \to \infty$.

In the case where $y_{10} > 0$ and $-\pi < y_{20} \le 0$, the solution $y_2(t)$ hits the surface $\tau_0 \colon t = -y_2$, and no beating is observed because inequality (2.47) is satisfied. The subsequent behavior is analogous to that in the previous case.

a$_2$) $w^- < w^+ \le 0$.

In this case, there exist solutions with "beating" and without "beating" (corresponding to $w_i = 0$), i.e., the weak extendability of solutions takes place (Figure 3).

b) $0 \le w^- < w^+$.

This case is analogous to case a). The solutions cross the first impulsive surface, and "beating" occurs on the second surface (Figure 4).

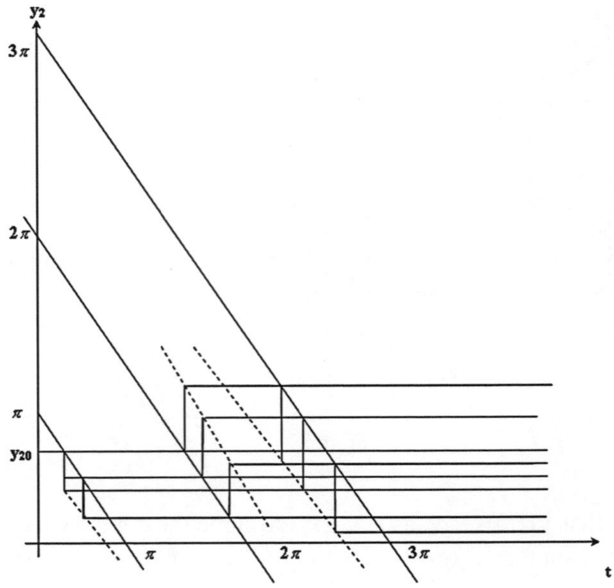

Figure 3.

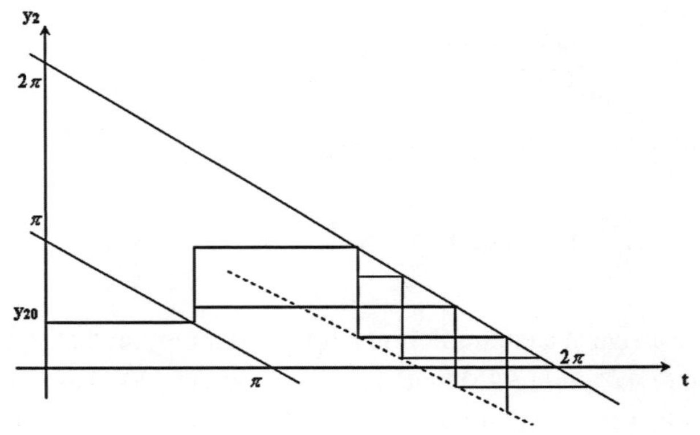

Figure 4.

c) $w^- < 0 < w^+$.

It is obvious that Δy_1 is always positive for $w \neq 0$. If $y_{10} \geq 0$, then $y_1 > 0$ for all t. Since, in this case, the set W contains both positive and negative values of $w \in W$, we conclude that, for each impulsive surface, there exist solutions that cross it and solutions with "beating" (Figure 5).

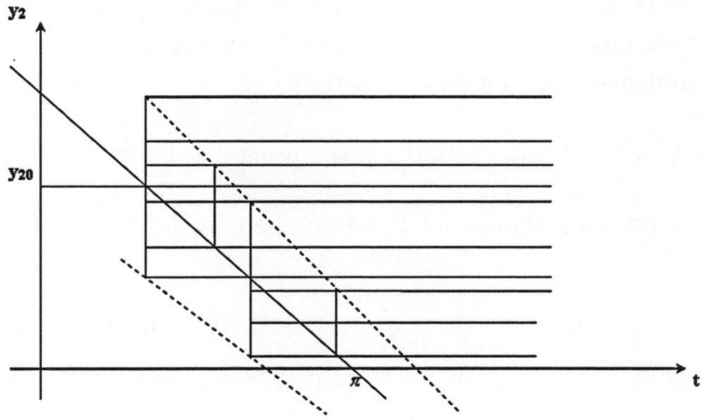

Figure 5.

Example 5. Consider the following system with multivalued pulses:

$$\ddot{x} + x = 0 \quad \text{for } x \neq 0,\ x(0) = x_0,\ \dot{x}(0) = \dot{x}_0, \tag{2.48}$$

$$\Delta x|_{x=0} = 0, \quad \Delta \dot{x}|_{x=0} \in W,\ W = [w^-, w^+],\ w^-, w^+ \in \mathbb{R}. \tag{2.49}$$

By analogy with the previous example, we reduce this inclusion to a system and change variables according to (2.36). As a result, we obtain

$$\dot{y}_1 = 0, \quad \dot{y}_2 = 0, \quad y_1(0) = y_{10}, \quad y_2(0) = y_{20}, \tag{2.50}$$

where the initial data y_{10} and y_{20} satisfy the conditions

$$y_{10} \cos y_{20} = x_0, \quad y_{10} \sin y_{20} = -\dot{x}_0,$$

i.e., $y_{10} = \sqrt{x_0^2 + \dot{x}_0^2}$ and y_{20} is chosen from the interval $(-\pi, \pi]$.

We now determine the values of the pulses Δy_1 and Δy_2 and the times at which they occur. To this end, we denote the values of y_1 and y_2 before and after a pulse by y_1^-, y_2^- and y_1^+, y_2^+, respectively.

By assumption, there is no pulse in the variable x_1, i.e.,

$$\Delta x_1 = 0 \Longrightarrow y_1^+ \cos(t_i + y_2^+) = y_1^- \cos(t_i + y_2^-). \tag{2.51}$$

The relation $\Delta x_2 = w_i \in W$ yields

$$-y_1^+ \sin(t_i + y_2^+) + y_1^- \sin(t_i + y_2^-) = w_i. \tag{2.52}$$

At the time of pulse action, we have $x_1 = 0$. Since $y_1 > 0$, this means that $\cos(t_i + y_2^-) = 0$, i.e., $t_i + y_2^- = \frac{\pi}{2} + i\pi$, $i = 0, 1, 2, \ldots$. Furthermore, $\sin(t_i + y_2^-) = \pm 1$. By virtue of (2.51), we conclude that $\cos(t_i + y_2^+) = 0$ ($\sin(t_i + y_2^+) = \pm 1$).

We also exclude the cases where the velocity changes its direction at the time of pulse action (these cases are improbable from the viewpoint of the physical interpretation of the problem as the motion of a swing):

$$\mathrm{a}_1)\ x_2^- > 0,\ w_i \le 0,\ x_2^- \le |w_i| \quad \text{and} \quad \mathrm{a}_2)\ x_2^- < 0,\ w_i > 0,\ |x_2^-| \le w_i.$$

If the velocity does not change its direction, then one of the following situations is realized:

$$\mathrm{b}_1)\ x_2^- > 0,\ w_i \ge 0; \qquad \mathrm{b}_2)\ x_2^- > 0,\ w_i < 0,\ x_2^- > |w_i|;$$
$$\mathrm{b}_3)\ x_2^- < 0,\ w_i \le 0; \qquad \mathrm{b}_4)\ x_2^- < 0,\ w_i > 0,\ |x_2^-| > w_i.$$

For example, consider the case b_1). Since $x_2^- > 0$, we have

$$-y_1^- \sin(t_i + y_2^-) > 0 \Longrightarrow \sin(t_i + y_2^-) < 0 \Longrightarrow \sin(t_i + y_2^-) = -1. \tag{2.53}$$

By virtue of (2.52), we obtain

$$\begin{aligned} &y_1^+ \sin(t_i + y_2^+) = y_1^- \sin(t_i + y_2^-) - w_i < 0 \\ &\Longrightarrow \sin(t_i + y_2^+) < 0 \Longrightarrow \sin(t_i + y_2^+) = -1. \end{aligned} \tag{2.54}$$

Relations (2.53) and (2.54) yield $y_2^- = y_2^+ \Longrightarrow \Delta y_2 = 0$.
In cases b_2), b_3), and b_4), we establish by analogy that

$$y_2^- = y_2^+ \Longrightarrow \Delta y_2 = 0, \tag{2.55}$$

i.e., there is no pulse in the variable y_2. It follows from relations (2.52) and (2.55) that $(-y_1^+ + y_1^-)\sin(t_i + y_2^\pm) = w_i$. Since $\sin(t_i + y_2^\pm) = \pm 1$ (depending on which of cases b_1)–b_4) is considered), we obtain

$$\Delta y_1 = y_1^+ - y_1^- = -w_i\ \mathrm{sign}(\sin(t_i + y_2^\pm)). \tag{2.56}$$

Thus, the original system is reduced to the form

$$\begin{aligned} &\dot y_1 = 0, \quad \dot y_2 = 0 \quad \text{for } t_i \ne -y_2 + \frac{\pi}{2} + i\pi, \\ &\Delta y_1|_{t_i = -y_2 + \frac{\pi}{2} + i\pi} \in -W\ \mathrm{sign}(\sin(t_i + y_2)), \\ &\Delta y_2|_{t_i = -y_2 + \frac{\pi}{2} + i\pi} = 0, \\ &y_1(0) = y_{10}, \quad y_2(0) = y_{20}. \end{aligned} \tag{2.57}$$

Pulses appear at the times $t_i = \tau_i(y_2) = -y_2 + \frac{\pi}{2} + i\pi$, $i = 0, 1, \ldots$. Since $y_2(t) \equiv y_{20} = \mathrm{const}$ by virtue of system (2.57), the times of pulses in y_1 are fixed (the points of intersection of the curve $y_2 = y_{20}$ and the surfaces (curves) $t_i = -y_{20} + \frac{\pi}{2} + i\pi$). In other words, the system

$$\dot{y}_1 = 0 \quad \text{for } t_i \neq -y_{20} + \frac{\pi}{2} + i\pi, \; y_1(0) = y_{10},$$

$$\Delta y_1|_{t_i = -y_{20} + \frac{\pi}{2} + i\pi} \in -W \,\text{sign}(\sin(t_i + y_{20}))$$

is a system with pulse action at fixed times. Thus, there is no "beating" in this system, and its solutions are extendable and depend continuously on the initial data and right-hand sides.

Chapter 3
Linear Impulsive Differential Inclusions

3.1 Statement of the Problem. Theorem on Existence and Uniqueness

Consider the linear impulsive differential inclusion

$$\dot{x} \in \mathcal{A}(t)x + F(t), \quad t \neq \tau_i, \tag{3.1}$$

$$\Delta x|_{t=\tau_i} \in \mathcal{B}_i x + P_i, \tag{3.2}$$

where $x \in \mathbb{R}^n$ is a phase vector, $t \in I = [t_0, T]$ is time, $T \leq +\infty$, $\mathcal{A}: I \to \operatorname{comp}(\mathbb{R}^{n \times n})$ is a measurable set-valued mapping, $|\mathcal{A}(t)| \leq \alpha(t)$, $F: I \to \operatorname{comp}(\mathbb{R}^n)$ is a measurable set-valued mapping, $|F(t)| \leq \mu(t)$, the functions $\alpha(t)$ and $\mu(t)$ are measurable and summable on any finite segment $I^* \subset I$, $\tau_i \in I$, $i = 1, 2, \ldots$, are fixed times numbered in the increasing order ($\tau_i < \tau_{i+1}$), the set $\{\tau_i\}$ does not have points of accumulation, $\mathcal{B}_i$ are compact sets of $n \times n$ matrices, and $P_i \in \operatorname{comp}(\mathbb{R}^n)$.

Definition 1. A function $x: I \to \mathbb{R}^n$ is called a solution of inclusion (3.1), (3.2) if it is absolutely continuous, satisfies (3.1) almost everywhere on intervals that do not contain τ_i, and has discontinuities of the first kind at points $t = \tau_i$ with jumps $\Delta x(\tau_i)$ satisfying inclusion (3.2).

Definition 2. A set-valued function $R: I \to \operatorname{comp}(\mathbb{R}^n)$ is called an R-solution generated by the impulsive differential inclusion (3.1), (3.2) if $R(t)$ is absolutely continuous on intervals that do not contain τ_i, for almost all $t \neq \tau_i$ one has

$$\lim_{\Delta \downarrow 0} \frac{1}{\Delta} h\left(R(t+\Delta), \bigcup_{x \in R(t)} \left\{ x + \int_t^{t+\Delta} (\mathcal{A}(s)x + F(s))ds \right\} \right) = 0,$$

and, for $t = \tau_i$, the function $R(t)$ satisfies the condition of jump

$$R(\tau_i + 0) = \bigcup_{x \in R(\tau_i)} \{x + \mathcal{B}_i x + P_i\}. \tag{3.3}$$

Let $A(t) \in \mathcal{A}(t)$, let $B_i \in \mathcal{B}_i$, and let $\Phi_{AB_i}(t, t_0)$ be the matrizant of the system

$$\dot{x} = A(t)x, \quad t \neq \tau_i,$$

$$\Delta x|_{t=\tau_i} = B_i x.$$

Using the formula for a solution of an impulsive differential equation obtained in Chapter 1, we represent an arbitrary solution of inclusion (3.1), (3.2) with the initial condition $x(t_0) = x_0$ for $t \in I$ in the form

$$x(t, x_0) = \Phi_{AB_i}(t, t_0)x_0 + \int_{t_0}^{t} \Phi_{AB_i}(t, \tau) f(\tau) d\tau + \sum_{t_0 \le \tau_i < t} \Phi_{AB_i}(t, \tau_i) p_i,$$

where $f(t)$ is a measurable branch of the mapping $F(t)$ and $p_i \in P_i$.

The integral funnel $X(t, X_0)$, $X(t_0, X_0) = X_0$, of inclusion (3.1), (3.2) (bundle of solutions) is determined by the relation

$$X(t, X_0) = \bigcup_{\substack{A(t)\in\mathcal{A}(t)\\ B_i\in\mathcal{B}_i}} \left\{ \Phi_{AB_i}(t, t_0)X_0 + \int_{t_0}^{t} \Phi_{AB_i}(t, \tau) F(\tau)\, d\tau + \sum_{t_0 \le \tau_i < t} \Phi_{AB_i}(t, \tau_i) P_i \right\}. \tag{3.4}$$

Lemma 1. *The bundle of solutions*

$$X(t, X_0) = \bigcup_{A(t)\in\mathcal{A}(t)} \left\{ \Phi_A(t, t_0)X_0 + \int_{t_0}^{t} \Phi_A(t, s) F(s)\, ds \right\} \tag{3.5}$$

of the linear inhomogeneous differential inclusion

$$\dot{x} \in \mathcal{A}(t)x + F(t) \tag{3.6}$$

is a nonempty compact set for any fixed $t \in I$.

Proof. The set $X(t, X_0)$ is nonempty because, by virtue of the Filippov theorem [49], there exist summable branches of the set-valued mappings $\mathcal{A}(t)$ and $F(t)$, and, furthermore, the set X_0 is nonempty. Any element $x \in X(t, x_0)$ can be represented in the form

$$x = x(t, x_0) = \Phi_A(t, t_0)x_0 + \int_{t_0}^{t} \Phi_A(t, s) f(s) ds,$$

where $x_0 \in X_0$, $A(s) \in \mathcal{A}(s)$, and $f(s) \in F(s)$. Therefore,

$$\|x\| \le e^{\int_{t_0}^{t} \alpha(s) ds} |X_0| + \int_{t_0}^{t} \mu(s) e^{\int_s^t \alpha(\tau) d\tau} ds = K < \infty, \tag{3.7}$$

i.e., the set $X(t, x_0)$ is bounded. We now prove its closedness, i.e., we prove that the limit of any convergent sequence of points $x_k \in X(t, X_0)$ also belongs to the set $X(t, X_0)$.

Since $x_k \in X(t, X_0)$, there exist $x_0^k \in X_0$, $A_k(s) \in \mathcal{A}(s)$, and $f_k(s) \in F(s)$ such that x_k is the value of a solution of the Cauchy problem

$$\dot{x} = A_k(s)x + f_k(s), \quad x(t_0) = x_0^k$$

at time t. By virtue of the equivalence of a differential equation and a Volterra integral equation, the following representation is true:

$$x_k = x_k(t) = x_0^k + \int_{t_0}^{t} A_k(s)x_k(s)ds + \int_{t_0}^{t} f_k(s)ds. \tag{3.8}$$

By analogy with (3.7), we get $\|x_k(s)\| \leq K$. Since

$$\begin{aligned}\|x_k(s_2) - x_k(s_1)\| &= \left\| \int_{s_1}^{s_2} [A_k(s)x_k(s) + f_k(s)]ds \right\| \\ &\leq \left| \int_{s_1}^{s_2} [k\alpha(s) + \mu(s)]ds \right| = |\varphi(s_2) - \varphi(s_1)|,\end{aligned}$$

where

$$\varphi(\tau) = \int_{t_0}^{\tau} [k\alpha(s) + \mu(s)]ds$$

is a function absolutely continuous and, hence, uniformly continuous on $[t_0, t]$, for any $\varepsilon > 0$ there exists $\delta(\varepsilon) > 0$ such that, for any $s_1, s_2 \in I\colon |s_2 - s_1| < \delta$, one has

$$|\varphi(s_2) - \varphi(s_1)| < \frac{\varepsilon}{K},$$

whence

$$\|x_k(s_2) - x_k(s_1)\| < \varepsilon.$$

Thus, the sequence of functions $x_k(s)$ is uniformly bounded and equicontinuous on $[t_0, t]$. Therefore, by virtue of the Arzelà theorem, one can choose its subsequence that converges uniformly to a continuous function $x_*(s)$.

This means that, for any $\varepsilon > 0$, there exists k_0 such that the following inequality holds for all $k > k_0$ and $s \in [t_0, t]$:

$$\|x_k(s) - x_*(s)\| < \frac{\varepsilon}{\gamma(t, t_0)},$$

where

$$\gamma(t, s) = \int_{s}^{t} \alpha(\sigma)d\sigma.$$

Since

$$\int_{t_0}^{t} A_k(s)x_k(s)ds = \int_{t_0}^{t} A_k(s)[x_k(s) - x_*(s)]ds + \int_{t_0}^{t} A_k(s)x_*(s)ds,$$

where

$$\left\| \int_{t_0}^{t} A_k(s)[x_k(s) - x_*(s)]ds \right\| \le \int_{t_0}^{t} \alpha(s)\|x_k(s) - x_*(s)\|ds < \varepsilon,$$

and, by virtue of the Lyapunov theorem [78], there exists a subsequence $A_{k_1}(s)$ of the sequence $A_k(s)$ that converges weakly to a matrix $A_*(s) \in \mathcal{A}(s)$ on $[t_0, t]$, we conclude that

$$\int_{t_0}^{t} A_{k_1}(s)x_{k_1}(s)ds \to \int_{t_0}^{t} A_*(s)x_*(s)ds \quad \text{as } k_1 \to \infty.$$

Furthermore, $x_0^{k_1} \in X_0 \in \text{comp}(\mathbb{R}^n)$. Hence, there exists a subsequence $\{x_0^{k_2}\}$ that converges to a certain vector $x_0^* \in X_0$. Moreover, by virtue of the Lyapunov theorem, there exists a subsequence of the sequence $f_{k_2}(s)$ that converges weakly to $f_*(s) \in F(s)$ on $[t_0, t]$.

Passing to the limit in (3.8), we obtain

$$x_* = x_*(t) = x_0^* + \int_{t_0}^{t} A_*(s)x_*(s)ds + \int_{t_0}^{t} f_*(s)ds.$$

In other words, x_* is the value of a solution of the differential inclusion (3.6) at time t, i.e., $x_* \in X(t, X_0)$. Thus, the compactness of the set $X(t, X_0)$ is proved. □

By virtue of Lemma 1 and the compactness of the sets $\mathcal{B}_i$ and P_i, the integral funnel $X(t, X_0)$ of inclusion (3.1), (3.2) is a compact set for every fixed $t \in I$.

Moreover, since the right-hand side of inclusion (3.1) possesses the Lipschitz property with respect to x, with regard for [101] and the condition of jump (3.3) we conclude that the set $X(t, X_0)$ is the unique R-solution of inclusion (3.1), (3.2).

Any finite interval $[t_0, T^*]$, $T^* \le T$, contains finitely many points τ_i. By virtue of Theorem 6 in Chapter 1, the solutions $x(t, x_0)$, $x(t_0, x_0) = x_0$, of inclusion (3.1), (3.2) exist for all $t \in [t_0, T^*]$ and any $x_0 \in \mathbb{R}^n$. Therefore, for any $X_0 \in \text{comp}(\mathbb{R}^n)$, the R-solution $X(t, X_0)$, $X(t_0, X_0) = X_0$, exists for $t \in [t_0, T^*]$. It is obvious that, in the case $T = +\infty$, the solution $x(t, x_0)$ and the R-solution $X(t, x_0)$ can be extended to the infinite interval.

Note that, in this case, R-solutions that start from different points can coincide at a certain moment of time.

Example 1. Consider the impulsive differential inclusion

$$\dot{x} \in [0, 1], \quad t \neq 1,$$

$$\Delta x|_{t=1} \in \{-2, 0\}x, \quad x(0) = x_0.$$

Then $X(1, x_0) = [x_0, x_0 + 1]$, and, for $t = 1 + 0$, we have

$$X(1 + 0, x_0) = [x_0, x_0 + 1] \cup [-1 - x_0, -x_0].$$

Hence,

$$X(1+0,0) = X(1+0,-1) = [-1,1].$$

We now consider the question of the possibility of coincidence of R-solutions that start from different points for linear impulsive differential inclusions in the special case

$$\dot{x} \in A(t)x + F(t), \quad t \neq \tau_i, \tag{3.9}$$

$$\Delta x|_{t=\tau_i} \in B_i x + P_i, \tag{3.10}$$

where $A(t)$ is an $n \times n$ matrix measurable on the interval I, and B_i are constant $n \times n$ matrices.

Theorem 1 ([117]). *If the matrices $E + B_i$ are nondegenerate for all i such that $\tau_i \in I$, then $X(t, x_0) \neq X(t, y_0)$ for all $t \in I$ whenever $x_0 \neq y_0$, where $X(t, x_0)$ and $X(t, y_0)$ are R-solutions of inclusion* (3.9), (3.10).

Proof. The sets $X(t, x_0)$ and $X(t, y_0)$ are nonempty compact sets. For $t \in (\tau_i, \tau_{i+1}]$, the following representation is true:

$$X(t, x_0) = \Phi_{AB_i}(t, \tau_i + 0)X(\tau_i + 0, x_0) + \int_{\tau_i+0}^{t} \Phi_{AB_i}(t, \tau)F(\tau)d\tau. \tag{3.11}$$

Therefore, by virtue of the nondegeneracy of the matrizant, we have

$$\operatorname{co} X(t, x_0) \neq \operatorname{co} X(t, y_0) \quad \text{for all } \tau_i < t \leq \tau_{i+1},\ i = 1, 2, \dots,$$

provided that $\operatorname{co} X(\tau_i + 0, x_0) \neq \operatorname{co} X(\tau_i + 0, y_0)$.

Indeed, assume the contrary, i.e., assume that the following equality holds for some $t \in (\tau_i, \tau_{i+1}]$:

$$\operatorname{co} X(t, x_0) = \operatorname{co} X(t, y_0).$$

By virtue of a property of support functions [23], for all $\psi \in \mathbb{R}^n$ we have

$$\begin{aligned}
& c(\operatorname{co} X(t, x_0), \psi) = c(\operatorname{co} X(t, y_0), \psi) \\
& \Rightarrow c(X(t, x_0), \psi) = c(X(t, y_0), \psi) \\
& \Rightarrow c\left(\Phi_{AB_i}(t, \tau_i + 0)X(\tau_i + 0, x_0) + \int_{\tau_i+0}^{t} \Phi_{AB_i}(t, \tau)F(\tau)d\tau, \psi\right) \\
& \quad = c\left(\Phi_{AB_i}(t, \tau_i + 0)X(\tau_i + 0, y_0) + \int_{\tau_i+0}^{t} \Phi_{AB_i}(t, \tau)F(\tau)d\tau, \psi\right) \\
& \Rightarrow c(\Phi_{AB_i}(t, \tau_i + 0)X(\tau_i + 0, x_0), \psi) \\
& \quad = c(\Phi_{AB_i}(t, \tau_i + 0)X(\tau_i + 0, y_0), \psi) \\
& \Rightarrow c(X(\tau_i + 0, x_0), \Phi^T_{AB_i}(t, \tau_i + 0)\psi) \\
& \quad = c(X(\tau_i + 0, y_0), \Phi^T_{AB_i}(t, \tau_i + 0)\psi).
\end{aligned}$$

Since the matrices $\Phi_{AB_i}(t, \tau_i + 0)$ are nondegenerate, the set of vectors $\{\Phi^T_{AB_i}(t, \tau_i+0)\psi, \ \psi \in \mathbb{R}^n\}$ coincides with $\mathbb{R}^n$, whence $\operatorname{co} X(\tau_i+0, x_0) = \operatorname{co} X(\tau_i+0, y_0)$, which contradicts the condition.

We now show that the equality

$$\operatorname{co} X(\tau_i + 0, x_0) = \operatorname{co} X(\tau_i + 0, y_0)$$

yields $\operatorname{co} X(\tau_i, x_0) = \operatorname{co} X(\tau_i, y_0)$. Indeed, by virtue of a property of support functions [23], for all $\psi \in \mathbb{R}^n$ we get

$$\begin{aligned} &c(\operatorname{co} X(\tau_i + 0, x_0), \psi) = c(\operatorname{co} X(\tau_i + 0, y_0), \psi) \\ \Rightarrow\ &c(X(\tau_i + 0, x_0), \psi) = c(X(\tau_i + 0, y_0), \psi), \end{aligned}$$

i.e.,

$$\begin{aligned} &c((E + B_i)X(\tau_i, x_0) + P_i, \psi) = c((E + B_i)X(\tau_i, y_0) + P_i, \psi) \\ \Rightarrow\ &c((E + B_i)X(\tau_i, x_0), \psi) = c((E + B_i)X(\tau_i, y_0), \psi) \\ \Rightarrow\ &c(X(\tau_i, x_0), (E + B_i)^T \psi) = c(X(\tau_i, y_0), (E + B_i)^T \psi). \end{aligned}$$

Since the matrices $E + B_i$ are nondegenerate, the set of vectors $\{(E + B_i)^T \psi, \ \psi \in \mathbb{R}^n\}$ coincides with $\mathbb{R}^n$, whence

$$\operatorname{co} X(\tau_i, x_0) = \operatorname{co} X(\tau_i, y_0).$$

Thus, we get $\operatorname{co} X(t_0, x_0) = \operatorname{co} X(t_0, y_0)$, i.e., $x_0 = y_0$, whenever $X(t, x_0) = X(t, y_0)$ for some $t \in I$. □

Remark 1. If the initial condition takes the form

$$X(t_0) = X_0 \in \operatorname{comp}(\mathbb{R}^n),$$

then we can only prove that the equality $X(t, X_0) = X(t, Y_0)$ for some $t \in I$ yields $\operatorname{co} X_0 = \operatorname{co} Y_0$, i.e., the uniqueness may be violated.

Example 2. Consider the differential inclusion

$$\begin{aligned} &\dot{x} = 0, \quad t \neq 1, \\ &\Delta x|_{t=1} \in S_1(0). \end{aligned}$$

Then the solutions $X(t, S_1(0))$ and $X(t, B_1(0))$ that start from the sets $S_1(0)$ and $B_1(0)$, respectively, coincide at $t = 1+0$ because $S_1(0) + S_1(0) = B_1(0) + S_1(0) = B_2(0)$.

Remark 2. Separate solutions $x(t, x_0)$ and $x(t, y_0)$ for $x_0 \neq y_0$ may coincide for $t \geq t_1$, where $t_1 \in I$.

Remark 3. In what follows, we assume that the matrices $E + B_i$ are nondegenerate for all $B_i \in \mathcal{B}_i$, $i = 1, 2, \ldots$.

Remark 4. If $P_i \in \operatorname{conv}(\mathbb{R}^n)$ in (3.10), then $X(t, x_0) \in \operatorname{conv}(\mathbb{R}^n)$. It should be noted that, in this case, the set of R-solutions of the impulsive differential inclusion (3.9), (3.10) is not a linear n-dimensional space because the space $\operatorname{conv}(\mathbb{R}^n)$ is not linear.

Of interest is the following special case: Let $x_0 \in X_0$. Assume that the set X_0 has a finite number m of corner points (i.e., it is a polyhedron). Then any point $x_0 \in X_0$ can be represented as a convex combination of corner points x_i, $i = \overline{1, m}$.

Consequently, any bundle of solutions $X(t, x_0)$ can be represented as a convex combination of "basic bundles" $X(t, x_i)$, i.e.,

$$X(t, x_0) = \sum_{i=1}^{m} \lambda_i X(t, x_i), \quad \text{where } \lambda_i \geq 0, \ \sum_{i=1}^{m} \lambda_i = 1, \ \sum_{i=1}^{m} \lambda_i x_i = x_0.$$

3.2 Stability of Solutions of Linear Impulsive Differential Inclusions

By analogy with various notions of stability of solutions and R-solutions for differential inclusions (see [23] and [107]), we introduce analogous notions for impulsive differential inclusions.

Definition 3. An R-solution $R(t)$ $(t_0 \leq t < +\infty)$ of the impulsive differential inclusion (3.1), (3.2) is called stable in the sense of Lyapunov if, for any $\varepsilon > 0$, one can find $\delta(\varepsilon) > 0$ such that the following conditions are satisfied:

(1) all R-solutions $X(t)$ of inclusion (3.1), (3.2) that satisfy the condition

$$h(X(t_0), R(t_0)) < \delta \tag{3.12}$$

are defined for all $t \geq t_0$;

(2) for all solutions satisfying inequality (3.12), the following relation is true:

$$h(X(t), R(t)) < \varepsilon.$$

Definition 4. An R-solution $R(t)$ $(t_0 \leq t < +\infty)$ of the impulsive differential inclusion (3.1), (3.2) is called asymptotically stable if the following conditions are satisfied:

(1) it is stable in the sense of Lyapunov;

(2) it satisfies the following condition:

$$\lim_{t\to\infty} h(X(t), R(t)) = 0.$$

Definition 5. A solution $\psi(t)$ ($t_0 \le t < +\infty$) of the impulsive differential inclusion (3.1), (3.2) is called stable if, for every $\varepsilon > 0$, there exists $\delta(\varepsilon) > 0$ such that, for every $\tilde{x}_0$ such that $\|\tilde{x}_0 - \psi(t_0)\| < \delta$, every solution $\tilde{x}(t)$ with the initial condition $\tilde{x}(t_0) = \tilde{x}_0$ exists and satisfies the following inequality for $t_0 \le t < +\infty$:

$$\|\tilde{x}(t) - \psi(t)\| < \varepsilon.$$

Definition 6. A solution $\psi(t)$ ($t_0 \le t < +\infty$) of the impulsive differential inclusion (3.1), (3.2) is called weakly stable if, for every $\varepsilon > 0$, there exists $\delta(\varepsilon) > 0$ such that, for every $\tilde{x}_0$ such that $\|\tilde{x}_0 - \psi(t_0)\| < \sigma$, some solution $\tilde{x}(t)$ with the initial condition $\tilde{x}(t_0) = \tilde{x}_0$ exists and satisfies the following inequality for $t_0 \le t < +\infty$:

$$\|\tilde{x}(t) - \psi(t)\| < \varepsilon.$$

Definition 7. A solution $\psi(t)$ ($t_0 \le t < +\infty$) of the impulsive differential inclusion (3.1), (3.2) is called asymptotically stable if the following conditions are satisfied:

(1) it is stable;

(2) it satisfies the following condition:

$$\lim_{t\to\infty} \|\tilde{x}(t) - \psi(t)\| = 0.$$

Definition 8. A solution $\psi(t)$ ($t_0 \le t < +\infty$) of the impulsive differential inclusion (3.1), (3.2) is called weakly asymptotically stable if the following conditions are satisfied:

(1) it is weakly stable;

(2) it satisfies the following condition:

$$\lim_{t\to\infty} \|\tilde{x}(t) - \psi(t)\| = 0.$$

We now study the problem of stability of solutions of linear homogeneous impulsive inclusions of the form

$$\dot{x} \in \mathcal{A}(t)x, \quad t \ne \tau_i, \quad \Delta x|_{t=\tau_i} \in \mathcal{B}_i x, \tag{3.13}$$

where $\mathcal{A}(t)$ is a compact set of $n \times n$ matrices measurable on $[t_0, +\infty)$, $\mathcal{B}_i$ are compact sets of $n \times n$ matrices, and the times of pulse action $\tau_i \to +\infty$ as $i \to \infty$.

Theorem 2 ([117]). *For the linear homogeneous impulsive differential inclusion* (3.13), *the following assertions are true:*

(a) *for the asymptotic stability of a solution $x(t, x_0)$, it is necessary and sufficient that the matrizants $\Phi_{AB_i}(t, t_0)$ satisfy the condition*

$$\lim_{t\to\infty} \Phi_{AB_i}(t, t_0) = 0 \tag{3.14}$$

uniformly in all $A(t) \in \mathcal{A}(t)$ and $B_i \in \mathcal{B}_i$, $i = 1, 2, \ldots$;

(b) *for the stability of the trivial solution, it is necessary and sufficient that the matrizants $\Phi_{AB_i}(t, t_0)$ be uniformly bounded for $t \geq t_0$ (i.e., there should exist a constant $M > 0$ such that $\|\Phi_{AB_i}(t, t_0)\| \leq M$ for $t \geq t_0$ and all $A(t) \in \mathcal{A}(t)$ and $B_i \in \mathcal{B}_i$, $i = 1, 2, \ldots$);*

(c) *for the weak stability of a nontrivial solution $x(t, x_0)$, it is sufficient that this solution be bounded for $t \geq t_0$;*

(d) *for the weak stability of the trivial solution, it is sufficient that there exist at least one matrizant $\Phi_{AB_i}(t, t_0)$ bounded for $t \geq t_0$;*

(e) *for the weak asymptotic stability of a nontrivial solution $x(t, x_0)$, it is sufficient that*

$$\lim_{t\to\infty} x(t, x_0) = 0;$$

(f) *for the weak asymptotic stability of the trivial solution, it is sufficient that there exist at least one matrizant $\Phi_{AB_i}(t, t_0) \to 0$ as $t \to \infty$.*

Proof. (a) If condition (3.14) is satisfied, then

$$\begin{aligned}&\lim_{t\to\infty} \|x(t, x_0) - x(t, y_0)\| \\ &= \lim_{t\to\infty} \|\Phi_{A^1 B_i^1}(t, t_0)x_0 - \Phi_{A^2 B_i^2}(t, t_0)y_0\| = 0,\end{aligned} \tag{3.15}$$

i.e., the solution is asymptotically stable. If condition (3.15) is satisfied for any y_0, then relation (3.14) is true.

(c) Assume that a nontrivial solution $x(t, x_0)$ is bounded for $t \geq t_0$ by a constant M_0 and corresponds to the matrices $A^1(t) \in \mathcal{A}(t)$ and $B_i^1 \in \mathcal{B}_i$.

Choosing the matrices $A(t) = A^1(t)$ and $B_i = B_i^1$ for the solution that starts at the point $y_0 = \alpha x_0$, $|\alpha - 1| < \delta$, we obtain

$$\begin{aligned}&\|x(t, x_0) - x(t, y_0)\| \\ &= \|\Phi_{A^1 B_i^1}(t, t_0)x_0 - \Phi_{A^1 B_i^1}(t, t_0)\alpha x_0\| < \delta M_0 = \varepsilon\end{aligned}$$

for all $t \geq t_0$ whenever $\delta = \frac{\varepsilon}{M_0}$. This means that the solution $x(t, x_0)$ is weakly stable. Moreover, it can be unstable.

Indeed, the difference of solutions corresponding to different $A(t)$ and B_i may not be small even for $y_0 = x_0$:

$$\begin{aligned}\|x^1(t,x_0) - x^2(t,x_0)\| &= \|\Phi_{A^1 B_i^1}(t,t_0)x_0 - \Phi_{A^2 B_i^2}(t,t_0)x_0\| \\ &= \|(\Phi_{A^1 B_i^1}(t,t_0) - \Phi_{A^2 B_i^2}(t,t_0))x_0\|.\end{aligned}$$

For example, if we consider the differential inclusion with zero pulses

$$\dot{x} \in [-1, 0]x, \quad x(0) = x_0 \neq 0$$

and take $A^1(t) \equiv 0$ and $A^2(t) \equiv -1$, then we obtain

$$\Phi_{A^1,0}(t,0) \equiv 1, \quad \Phi_{A^2,0}(t,0) = e^{-t}.$$

Then

$$\|x^1(t,x_0) - x^2(t,x_0)\| = \|x_0\|(1 - e^{-t})$$

is an increasing function tending to $\|x_0\|$ as $t \to \infty$.

(e) Let

$$\lim_{t\to\infty} x(t,x_0) = \lim_{t\to\infty} \Phi_{AB_i}(t,t_0)x_0 = 0.$$

Consider the solutions

$$x(t,y_0) = \Phi_{AB_i}(t,t_0)y_0, \quad \text{where } y_0 = \delta x_0.$$

Then

$$\lim_{t\to\infty}[x(t,x_0) - x(t,y_0)] = \lim_{t\to\infty}[\Phi_{AB_i}(t,t_0)x_0 - \Phi_{AB_i}(t,t_0)\delta x_0] = 0,$$

i.e., the solution $x(t,x_0)$ is weakly asymptotically stable.

Assertions (b), (d), and (f) are proved by analogy. □

Theorem 3 ([117]). *For the linear homogeneous impulsive differential inclusion* (3.13), *the following assertions are true:*

(a) *for the stability of an R-solution $X(t,x_0)$, it is necessary and sufficient that the matrizants $\Phi_{AB_i}(t,t_0)$ be uniformly bounded for $t \geq t_0$;*

(b) *for the asymptotic stability of an R-solution, it is necessary and sufficient that the matrizants $\Phi_{AB_i}(t,t_0)$ uniformly satisfy condition* (3.14)*;*

(c) *for the instability of an R-solution, it is necessary and sufficient that the matrizants $\Phi_{AB_i}(t,t_0)$ be not uniformly bounded for $t \geq t_0$.*

Proof. (a) We prove the sufficiency. Let $X(t, x_0)$ be an R-solution of inclusion (3.13) and let there exist a constant $M_0 > 0$ such that $\|\Phi_{A,B_i}(t, t_0)\| \le M_0$ for $t \ge t_0$ and all $A(t) \in \mathcal{A}(t)$ and $B_i \in \mathcal{B}_i$. Then

$$\begin{aligned} &h(X(t, x_0), X(t, y_0)) \\ &= h\Bigg(\bigcup_{\substack{A(t)\in\mathcal{A}(t)\\ B_i\in\mathcal{B}_i}} \Phi_{AB_i}(t, t_0)x_0, \bigcup_{\substack{A(t)\in\mathcal{A}(t)\\ B_i\in\mathcal{B}_i}} \Phi_{AB_i}(t, t_0)y_0 \Bigg) \\ &= \sup_{\substack{A(t)\in\mathcal{A}(t)\\ B_i\in\mathcal{B}_i}} \{d_1(t), d_2(t)\}, \end{aligned}$$

where

$$\begin{aligned} d_1(t) &= \rho\Bigg(\Phi_{AB_i}(t, t_0)x_0, \bigcup_{\substack{A(t)\in\mathcal{A}(t)\\ B_i\in\mathcal{B}_i}} \Phi_{AB_i}(t, t_0)y_0 \Bigg) \\ &\le \|\Phi_{AB_i}(t, t_0)x_0 - \Phi_{AB_i}(t, t_0)y_0\| \le M_0\|x_0 - y_0\|, \\ d_2(t) &= \rho\Bigg(\Phi_{AB_i}(t, t_0)y_0, \bigcup_{\substack{A(t)\in\mathcal{A}(t)\\ B_i\in\mathcal{B}_i}} \Phi_{AB_i}(t, t_0)x_0 \Bigg) \\ &\le \|\Phi_{AB_i}(t, t_0)x_0 - \Phi_{AB_i}(t, t_0)y_0\| \le M_0\|x_0 - y_0\|. \end{aligned}$$

Thus,

$$h(X(t, x_0), X(t, y_0)) \le M_0\|x_0 - y_0\| < \varepsilon$$

for $\|x_0 - y_0\| < \delta$, $\delta = \frac{\varepsilon}{M_0}$, and $t \ge t_0$. Hence, the R-solution $X(t, x_0)$ is stable.

Let us prove the necessity. Assume that the R-solution $X(t, x_0)$ is stable, i.e., for any $\varepsilon > 0$, there exists $\delta > 0$ such that, for $\|x_0 - y_0\| < \delta$, one has

$$\begin{aligned} &h(X(t, x_0), X(t, y_0)) \\ &= h\Bigg(\bigcup_{\substack{A(t)\in\mathcal{A}(t)\\ B_i\in\mathcal{B}_i}} \Phi_{AB_i}(t, t_0)x_0, \bigcup_{\substack{A(t)\in\mathcal{A}(t)\\ B_i\in\mathcal{B}_i}} \Phi_{AB_i}(t, t_0)y_0 \Bigg) < \varepsilon. \end{aligned}$$

We fix $\varepsilon > 0$, select the corresponding $\delta > 0$, and take $y_0 = \alpha x_0$, where $|1 - \alpha|\|x_0\| < \delta$. Then

$$\begin{aligned} &h\Bigg(\bigcup_{\substack{A(t)\in\mathcal{A}(t)\\ B_i\in\mathcal{B}_i}} \Phi_{AB_i}(t, t_0)x_0, \bigcup_{\substack{A(t)\in\mathcal{A}(t)\\ B_i\in\mathcal{B}_i}} \Phi_{AB_i}(t, t_0)y_0 \Bigg) \\ &= h\Bigg(\bigcup_{\substack{A(t)\in\mathcal{A}(t)\\ B_i\in\mathcal{B}_i}} \Phi_{AB_i}(t, t_0)x_0, \alpha \bigcup_{\substack{A(t)\in\mathcal{A}(t)\\ B_i\in\mathcal{B}_i}} \Phi_{AB_i}(t, t_0)x_0 \Bigg) \end{aligned}$$

$$= \max_{\psi \in S_1(0)} \left| c\Big(\bigcup_{\substack{A(t)\in\mathcal{A}(t)\\ B_i\in\mathcal{B}_i}} \Phi_{AB_i}(t,t_0)x_0, \psi \Big) - \alpha\, c\Big(\bigcup_{\substack{A(t)\in\mathcal{A}(t)\\ B_i\in\mathcal{B}_i}} \Phi_{AB_i}(t,t_0)x_0, \psi \Big) \right|$$

$$= |1-\alpha| \max_{\psi \in S_1(0)} \left| c\Big(\bigcup_{\substack{A(t)\in\mathcal{A}(t)\\ B_i\in\mathcal{B}_i}} \Phi_{AB_i}(t,t_0)x_0, \psi \Big) \right|$$

$$= |1-\alpha| \cdot h\Big(\bigcup_{\substack{A(t)\in\mathcal{A}(t)\\ B_i\in\mathcal{B}_i}} \Phi_{AB_i}(t,t_0)x_0, \{0\} \Big) < \varepsilon.$$

Hence,

$$h\Big(\bigcup_{\substack{A(t)\in\mathcal{A}(t)\\ B_i\in\mathcal{B}_i}} \Phi_{AB_i}(t,t_0)x_0, \{0\} \Big) < \frac{\varepsilon}{|1-\alpha|} = M.$$

Thus, the R-solution $X(t,x_0)$ is bounded.

Now assume that the matrizants $\Phi_{AB_i}(t,t_0)$ are not uniformly bounded. Then, for any sequence $M_k \to \infty$, $k \to \infty$, there exists a matrizant $\Phi_{A^k B_i^k}(t,t_0)$ such that

$$\|\Phi_{A^k B_i^k}(t^k,t_0)\| > M_k.$$

Then there exists at least one element $\varphi_{j^k l^k}(t^k,t_0)$ of the matrix $\Phi_{A^k B_i^k}(t^k,t_0)$ such that $|\varphi_{j^k l^k}(t^k,t_0)| > L_k$, where $L_k \to \infty$ as $k \to \infty$.

By definition, we have

$$h\Big(\bigcup_{\substack{A(t)\in\mathcal{A}(t)\\ B_i\in\mathcal{B}_i}} \Phi_{AB_i}(t,t_0)x_0, \bigcup_{\substack{A(t)\in\mathcal{A}(t)\\ B_i\in\mathcal{B}_i}} \Phi_{AB_i}(t,t_0)y_0 \Big)$$
$$= \sup_{\substack{A(t)\in\mathcal{A}(t)\\ B_i\in\mathcal{B}_i}} \{d_1(t), d_2(t)\},$$

where

$$d_2(t) = \rho\Big(\Phi_{AB_i}(t,t_0)y_0, \bigcup_{\substack{A(t)\in\mathcal{A}(t)\\ B_i\in\mathcal{B}_i}} \Phi_{AB_i}(t,t_0)x_0 \Big).$$

As the matrices $A(t)$ and B_i, we take the matrices $A^k(t)$ and B_i^k, respectively. We also choose the initial value y_0^k as follows:

$$y_{01}^k = x_{01}, \quad \dots, \quad y_{0l^k}^k = x_{0l^k} + \frac{\delta}{2}, \quad \dots, \quad y_{0n}^k = x_{0n}.$$

In this case, we have

$$\begin{aligned}&\|(\Phi_{A^k B_i^k}(t^k, t_0) y_0^k)_j\| \\ &= \|(\Phi_{A^k B_i^k}(t^k, t_0) x_0)_j + \varphi_{j^k l^k}(t^k, t_0)(y_{0l^k}^k - x_{0l^k})\| \\ &\geq |\varphi_{j^k l^k}(t^k, t_0)| \cdot |y_{0l^k}^k - x_{0l^k}| - \|(\Phi_{A^k B_i^k}(t^k, t_0) x_0)_j\| > L_k \frac{\delta}{2} - M.\end{aligned}$$

Hence,

$$h(X(t, x_0), X(t, y_0))|_{t=t_k} \geq \sup_{\substack{A^k(t) \in \mathcal{A}(t) \\ B_i^k \in \mathcal{B}_i}} d_2(t^k) = \infty.$$

We arrive at a contradiction with the assertion that the R-solution $X(t, x_0)$ is stable. Thus, we have shown that the requirement of the uniform boundedness of the matrizants $\Phi_{AB_i}(t, t_0)$ is necessary and sufficient for the stability of the R-solution $X(t, x_0)$. Assertion (c) of the theorem follows automatically from (a). The criterion for the asymptotic stability of an R-solution is proved by analogy. □

We now illustrate the obtained results by several examples.

Example 3. Consider a linear impulsive differential inclusion of the form

$$\begin{aligned}\dot{x} &\in [\alpha, \beta] x, \quad t \neq i, \ i \in N, \ x(0) = x_0 \geq 0, \\ \Delta x|_{t=i} &\in [-0.1; 0] x,\end{aligned} \tag{3.16}$$

where $\alpha < \beta$ are arbitrary constants. The matrizant corresponding to a measurable function $a(t) \in [\alpha, \beta]$ and a sequence $b_i \in [-0.1; 0]$ takes the form

$$\Phi_{a(t), b_i}(t, t_0) = e^{\int_0^t a(s)\, ds} \prod_{0<i<t} (1 + b_i).$$

It is obvious that $0.9^{[t]} e^{\alpha t} \leq \Phi_{a(t), b_i}(t, t_0) \leq e^{\beta t}$. Also note that $0.9^t e^{\alpha t} \leq 0.9^{[t]} e^{\alpha t} < 0.9^{t-1} e^{\alpha t}$.

We now consider the problem of stability of the trivial solution $x(t, 0) \equiv 0$.

The trivial solution is asymptotically stable if and only if $\beta < 0$. The trivial solution is stable if and only if $\beta \leq 0$. In order that the trivial solution (3.16) be weakly stable, it is sufficient that $0.9 e^{\alpha} \leq 1$, i.e., $\alpha \leq \ln \frac{10}{9}$. For the weak asymptotic stability of the trivial solution, it is sufficient that $\alpha < \ln \frac{10}{9}$.

The trivial R-solution is stable if and only if $\beta \leq 0$, asymptotically stable if and only if $\beta < 0$, and unstable if and only if $\beta > 0$.

Example 4. Consider the impulsive differential inclusion

$$\dot{x} \in \mathcal{A}x, \quad t \neq i, \; i \in N,$$

$$\Delta x|_{t=i} \in \mathcal{B}_i x, \quad \text{where } x \in R^2, \; x(0) = x_0 \in R^2,$$

$$\mathcal{A} = \left\{ \begin{pmatrix} 0 & a \\ -a & 0 \end{pmatrix}, \; a \in [a_1, a_2] \right\},$$

$$\mathcal{B}_i = \left\{ \begin{pmatrix} 0 & 0 \\ 0 & b_i \end{pmatrix}, \; b_i \in [-1, 0] \right\}.$$

The matrizant corresponding to the matrices A and B_i has the form

$$\Phi_{AB_i}(t,0) = e^{A(t-k)} \prod_{i=1}^{k} (E + B_i) e^A.$$

For $0 \le t \le 1$, we have

$$\Phi_{AB_i}(t,0) = e^{At} = \begin{pmatrix} \cos at & \sin at \\ -\sin at & \cos at \end{pmatrix} \quad \text{and} \quad \|\Phi_{AB_i}(t,0)\|_1 = \sqrt{2},$$

where

$$\|M\|_1 = \sqrt{\sum_{i,j=1}^{n} |m_{ij}|^2}.$$

We now show that the function $\|\Phi_{AB_i}(t,0)\|_1$ is nonincreasing for $t \ge 0$. Assume that $k < t \le k+1$. Let

$$\Phi_{AB_i}(k+0,0) = \begin{pmatrix} x_{11} & x_{12} \\ x_{21} & x_{22} \end{pmatrix}.$$

Then

$$\Phi_{AB_i}(t,0) = \begin{pmatrix} \cos a(t-k) & \sin a(t-k) \\ -\sin a(t-k) & \cos a(t-k) \end{pmatrix} \begin{pmatrix} x_{11} & x_{12} \\ x_{21} & x_{22} \end{pmatrix}$$

$$= \begin{pmatrix} x_{11} \cos a(t-k) + x_{21} \sin a(t-k) & x_{12} \cos a(t-k) + x_{22} \sin a(t-k) \\ -x_{11} \sin a(t-k) + x_{21} \cos a(t-k) & -x_{12} \sin a(t-k) + x_{22} \cos a(t-k) \end{pmatrix},$$

$$\begin{aligned} \|\Phi_{AB_i}(t,0)\|_1^2 &= x_{11}^2 \cos^2 a(t-k) + 2x_{11}x_{21} \sin a(t-k) \cos a(t-k) \\ &\quad + x_{21}^2 \sin^2 a(t-k) + x_{12}^2 \cos^2 a(t-k) \\ &\quad + 2x_{12}x_{22} \sin a(t-k) \cos a(t-k) + x_{22}^2 \sin^2 a(t-k) \\ &\quad + x_{11}^2 \sin^2 a(t-k) - 2x_{11}x_{21} \sin a(t-k) \cos a(t-k) \\ &\quad + x_{21}^2 \cos^2 a(t-k) + x_{12}^2 \sin^2 a(t-k) \\ &\quad - 2x_{12}x_{22} \sin a(t-k) \cos a(t-k) + x_{22}^2 \cos^2 a(t-k) \\ &= x_{11}^2 + x_{12}^2 + x_{21}^2 + x_{22}^2 = \|\Phi_{AB_i}(k+0,0)\|_1^2. \end{aligned}$$

Thus, the function $\|\Phi_{AB_i}(t,0)\|_1$ is constant on the semi-segments $(k, k+1]$. We determine its jump at the time of pulse action $t = k$. Let

$$\Phi_{AB_i}(k,0) = \begin{pmatrix} x_{11}^- & x_{12}^- \\ x_{21}^- & x_{22}^- \end{pmatrix}.$$

Then

$$\begin{aligned}
\Phi_{AB_i}(k+0,0) &= \begin{pmatrix} 1 & 0 \\ 0 & 1+b_k \end{pmatrix} \begin{pmatrix} x_{11}^- & x_{12}^- \\ x_{21}^- & x_{22}^- \end{pmatrix} \\
&= \begin{pmatrix} x_{11}^- & x_{12}^- \\ (1+b_k)x_{21}^- & (1+b_k)x_{22}^- \end{pmatrix}, \\
\|\Phi_{AB_i}(k+0,0)\|_1^2 &= {x_{11}^-}^2 + {x_{12}^-}^2 + (1+b_k)^2 {x_{21}^-}^2 + (1+b_k)^2 {x_{22}^-}^2 \\
&= {x_{11}^-}^2 + {x_{12}^-}^2 + {x_{21}^-}^2 + {x_{22}^-}^2 + b_k(2+b_k)({x_{21}^-}^2 + {x_{22}^-}^2) \\
&\le \|\Phi_{AB_i}(k-0,0)\|_1^2
\end{aligned}$$

because $b_k(2+b_k) \le 0$.

Thus, the function $\|\Phi_{AB_i}(t,0)\|_1$ is nonincreasing. Hence, $\|\Phi_{AB_i}(t,0)\|_1 \le \sqrt{2}$ for all matrices $A \in \mathcal{A}$ and $B_i \in \mathcal{B}_i$, $i = 1, 2, \ldots$, and all $t \ge 0$.

By virtue of the equivalence of the matrix norms $\|\cdot\|$ and $\|\cdot\|_1$, there exist constants $0 < \alpha < \beta < \infty$ such that

$$\alpha\|M\|_1 \le \|M\| \le \beta\|M\|_1$$

for any matrix M (Theorem 6.1.2 [77]).

Thus, $\|\Phi_{AB_i}(t,0)\| \le \beta\sqrt{2}$. Hence, all solutions of the original impulsive differential inclusion are weakly stable because they are bounded by virtue of the equality $x(t,0) = \Phi_{AB_i}(t,0)x_0$.

The jump of the function $\|\Phi_{AB_i}(t,0)\|_1^2$ at $t = k$ is equal to

$$\Delta_k = -b_k(2+b_k)({x_{21}^-}^2 + {x_{22}^-}^2).$$

Thus, if there exist a and b_k such that $\sum_{k=1}^{\infty} \Delta_k = 2$, then the solution corresponding to these A and B_k is weakly asymptotically stable because $\|\Phi_{AB_i}(t,0)\|_1 \to 0$ as $t \to \infty$. Consequently, $x(t,0) \to 0$ as $t \to \infty$.

Assume that, in the differential inclusion (3.13), we have

$$\mathcal{A}(t) = A + \mathcal{A}^0(t), \quad \mathcal{B}_i = B_i + \mathcal{B}_i^0,$$

where A and B_i are constant matrices.

Then the differential inclusion (3.13) can be rewritten in the form

$$\dot{x} \in Ax + \mathcal{A}^0(t)x \quad t \neq \tau_i, \quad \Delta x|_{t=\tau_i} \in B_i x + \mathcal{B}_i^0 x. \tag{3.17}$$

Consider the system of linear impulsive differential equations

$$\dot{x} = Ax, \quad t \neq \tau_i, \quad \Delta x|_{t=\tau_i} = B_i x. \tag{3.18}$$

Theorem 4 ([117]). *If the solutions of the system of equations* (3.18) *are stable and there exist* $A^0(t) \in \mathcal{A}^0(t)$ *and* $B_i^0 \in \mathcal{B}_i^0$ *such that*

$$\int_{t_0}^{\infty} \|A^0(t)\| dt < \infty, \quad \prod_{\tau_i \geq t_0} (1 + \|B_i^0\|) < \infty, \tag{3.19}$$

then the solutions of the differential inclusion (3.17) *that correspond to the matrices* $A^0(t)$ *and* B_i^0 *are weakly stable.*

Proof. In Theorem 11 of Chapter 1, we have considered the problem of stability of solutions of an impulsive differential equation of the type (3.17) under the condition of stability of solutions of an impulsive differential equation of the type (3.18), where $B_i \equiv B$. It follows from the proof of this theorem that the last requirement can be omitted. Thus, the stability of solutions of the system of equations (3.18) under conditions (3.19) yields the stability of solutions of the system

$$\dot{x} = Ax + A^0(t)x \quad t \neq \tau_i, \quad \Delta x|_{t=\tau_i} = B_i x + B_i^0 x.$$

Thus, these solutions are weakly stable solutions of inclusion (3.17). □

Remark 5. The solutions of inclusion (3.17) can be unstable even if condition (3.19) is satisfied for all matrices $A^0(t) \in \mathcal{A}^0(t)$ and $B_i^0 \in \mathcal{B}_i^0$.

Example 5. Consider the linear homogeneous differential inclusion with zero pulses

$$\dot{x} \in \mathcal{A}^0(t)x, \quad \text{where } \mathcal{A}^0(t) = \left\{\frac{a}{t^2}, a \in [1,2]\right\}, \ x(1) = x_0 \neq 0. \tag{3.20}$$

The solution of the truncated system $x(t, x_0) \equiv x_0$ is stable in the sense of Lyapunov in the positive direction. Condition (3.19) is satisfied for all $A^0(t) \in \mathcal{A}^0(t)$. In this case, solution (3.20) that corresponds to the matrix $A_1^0(t) = \frac{1}{t^2}$ is not stable because, by choosing $y_0 = x_0$ and $A_2^0(t) = \frac{2}{t^2}$, we obtain

$$\begin{aligned}\|x^1(t, x_0) - x^2(t, x_0)\| &= \|\Phi_{A_1^0,0}(t, t_0)x_0 - \Phi_{A_2^0,0}(t, t_0)x_0\| \\ &= |e^{1-\frac{1}{t}} - e^{2-\frac{2}{t}}| \cdot \|x_0\| \to e(e-1)\|x_0\| \neq 0 \quad \text{as } t \to \infty.\end{aligned}$$

Theorem 5 ([117]). *If the solutions of the system of equations* (3.18) *are stable, then an R-solutions of inclusion* (3.17) *are also stable under the condition that*

$$\int_{t_0}^{\infty} |\mathcal{A}^0(t)|\,dt < \infty, \qquad \prod_{\tau_i \ge t_0} (1 + |\mathcal{B}_i^0|) < \infty, \tag{3.21}$$

where

$$|\mathcal{A}^0(t)| = \max_{A^0(t)\in\mathcal{A}^0(t)} \|A^0(t)\| \quad \text{and} \quad |\mathcal{B}_i^0| = \max_{B_i^0\in\mathcal{B}_i^0} \|B_i^0\|.$$

Proof. Let $X(t, x_0)$ be an R-solution of (3.17). Then

$$X(t,x_0) = \Big\{ x(t,x_0) = \Phi_{AB_i}(t,t_0)x_0 + \int_{t_0}^{t} \Phi_{AB_i}(t,\tau)A^0(\tau)x(\tau,x_0)\,d\tau + \sum_{t_0\le\tau_i<t} \Phi_{AB_i}(t,\tau_i)B_i^0 x(\tau_i,x_0),\ A^0(t)\in\mathcal{A}^0(t),\ B_i^0\in\mathcal{B}_i^0 \Big\}.$$

Using the estimates presented in Theorem 11 of Chapter 1, we get

$$\begin{aligned}
d_1(t) &= \rho\Big(\Phi_{AB_i}(t,t_0)x_0 + \int_{t_0}^{t} \Phi_{AB_i}(t,\tau)A^0(\tau)x(\tau,x_0)d\tau \\
&\quad + \sum_{t_0\le\tau_i<t} \Phi_{AB_i}(t,\tau_i)B_i^0 x(\tau_i,x_0), X(t,y_0)\Big) \\
&\le \Big\| \Phi_{AB_i}(t,t_0)x_0 + \int_{t_0}^{t} \Phi_{AB_i}(t,\tau)A^0(\tau)x(\tau,x_0)\,d\tau \\
&\quad + \sum_{t_0\le\tau_i<t} \Phi_{AB_i}(t,\tau_i)B_i^0 x(\tau_i,x_0) \\
&\quad - \Big[\Phi_{AB_i}(t,t_0)y_0 + \int_{t_0}^{t} \Phi_{AB_i}(t,\tau)A^0(\tau)x(\tau,y_0)\,d\tau \\
&\quad + \sum_{t_0\le\tau_i<t} \Phi_{AB_i}(t,\tau_i)B_i^0 x(\tau_i,y_0)\Big]\Big\| \le K\|x_0-y_0\|, \\
d_2(t) &= \rho\Big(\Phi_{AB_i}(t,t_0)y_0 + \int_{t_0}^{t} \Phi_{AB_i}(t,\tau)A^0(\tau)x(\tau,y_0)\,d\tau \\
&\quad + \sum_{t_0\le\tau_i<t} \Phi_{AB_i}(t,\tau_i)B_i^0 x(\tau_i,y_0), X(t,x_0)\Big)
\end{aligned}$$

$$\leq \Bigg\| \Phi_{AB_i}(t,t_0)y_0 + \int_{t_0}^{t} \Phi_{AB_i}(t,\tau)A^0(\tau)x(\tau,y_0)\,d\tau$$
$$+ \sum_{t_0 \leq \tau_i < t} \Phi_{AB_i}(t,\tau_i)B_i^0 x(\tau_i,y_0)$$
$$- \Bigg[\Phi_{AB_i}(t,t_0)x_0 + \int_{t_0}^{t} \Phi_{AB_i}(t,\tau)A^0(\tau)x(\tau,x_0)\,d\tau$$
$$+ \sum_{t_0 \leq \tau_i < t} \Phi_{AB_i}(t,\tau_i)B_i^0 x(\tau_i,x_0) \Bigg] \Bigg\| \leq K\|x_0 - y_0\|,$$

where

$$K = \left(\prod_{\tau_i \geq t_0} (1 + |\mathcal{B}_i^0|) \right)^M e^{M \int_{t_0}^{\infty} |\mathcal{A}^0(t)|\,dt}$$

and M is a positive constant such that $\|\Phi_{AB_i}(t,t_0)\| \leq M$.

Thus,

$$h(X(t,x_0), X(t,y_0))$$
$$= \sup_{A^0(t) \in \mathcal{A}^0(t),\, B_i^0 \in \mathcal{B}_i^0} \{d_1(t), d_2(t)\} \leq K\|x_0 - y_0\| < \varepsilon$$

for $\|x_0 - y_0\| < \delta$, $\delta = \frac{\varepsilon}{K}$. Hence, the R-solution $X(t,x_0)$ of inclusion (3.17) is stable. □

Remark 6. Note that condition (3.21) implies that condition (3.19) is satisfied for all $A^0(t)$ and B_i^0.

We now examine the relationship between the stability of solutions of the linear homogeneous impulsive inclusion (3.13) and the stability of solutions of the linear inhomogeneous impulsive differential inclusion

$$\dot{x} \in \mathcal{A}(t)x + F(t), \quad t \neq \tau_i, \tag{3.22}$$
$$\Delta x|_{t=\tau_i} \in \mathcal{B}_i x + P_i, \tag{3.23}$$

where $F\colon [t_0, +\infty) \to \mathrm{comp}(R^n)$ is a measurable set-valued mapping and $P_i \in \mathrm{comp}(R^n)$.

Theorem 6. *If the solutions of* (3.13) *are stable, then the solutions of* (3.22), (3.23) *are weakly stable.*

Proof. Let

$$x(t,x_0) = \Phi_{AB_i}(t,t_0)x_0$$
$$+ \int_{t_0}^{t} \Phi_{AB_i}(t,\tau)f(\tau)\,d\tau + \sum_{t_0 \leq \tau_i < t} \Phi_{AB_i}(t,\tau_i)p_i$$

be a solution of (3.22), (3.23). By virtue of a criterion for the stability of solutions of (3.13), there exists the constant

$$M = \sup_{t \geq t_0} \|\Phi_{AB_i}(t, t_0)\|.$$

We set

$$\|y_0 - x_0\| < \delta = \frac{\varepsilon}{M}.$$

Consider

$$x(t, y_0) = \Phi_{AB_i}(t, t_0) y_0 + \int_{t_0}^{t} \Phi_{AB_i}(t, \tau) f(\tau)\, d\tau + \sum_{t_0 \leq \tau_i < t} \Phi_{AB_i}(t, \tau_i) p_i.$$

Then

$$\begin{aligned} \|x(t, x_0) - x(t, y_0)\| &= \|\Phi_{AB_i}(t, t_0)(x_0 - y_0)\| \\ &\leq \|\Phi_{AB_i}(t, t_0)\| \, \|x_0 - y_0\| \leq M \|x_0 - y_0\| < \varepsilon, \end{aligned}$$

i.e., $x(t, t_0)$ is weakly stable.

Note that the stability of solutions of inclusion (3.13) does not yield the stability of solutions of (3.22), (3.23) in the general case. Consider the inclusion with zero pulses

$$\dot{x} \in [0, 1]. \tag{3.24}$$

A solution of the homogeneous system $x(t, x_0) \equiv x_0$ is stable. In this case, the solution of (3.24) that corresponds to $f(t) \equiv 0 \in [0, 1]$ is not stable because if we choose, say, $y_0 = x_0$ and $f_1(t) \equiv 1 \in [0, 1]$, then

$$\begin{aligned} \|x^1(t, x_0) - x^2(t, x_0)\| &= \|x_0 - (t - t_0 + x_0)\| \\ &= |t - t_0| \to \infty \quad \text{as } t \to \infty. \end{aligned}$$ □

Theorem 7. *If the R-solutions of* (3.13) *are stable, then the R-solutions of* (3.22), (3.23) *are also stable.*

Proof. Let $X(t, x_0)$ be an R-solution of inclusion (3.22), (3.23) with initial condition $X(t_0, x_0) = x_0$. Then

$$X(t, x_0) = \bigcup_{\substack{A(t) \in \mathcal{A}(t) \\ B_i \in \mathcal{B}_i}} \Big\{ \Phi_{AB_i}(t, t_0) x_0 + \int_{t_0}^{t} \Phi_{AB_i}(t, \tau) F(\tau)\, d\tau + \sum_{t_0 \leq \tau_i < t} \Phi_{AB_i}(t, \tau_i) P_i \Big\}.$$

It is necessary to prove that, for any $\varepsilon > 0$, there exists $\delta = \delta(\varepsilon) > 0$ such that, for all y_0 such that $\|x_0 - y_0\| < \delta$, one has $h(X(t, x_0), X(t, y_0)) < \varepsilon$.

We now estimate

$$h(X(t, x_0), X(t, y_0)) = \sup_{A(t) \in \mathcal{A}(t), B_i \in \mathcal{B}_i} \{d_1(t), d_2(t)\},$$

where

$$\begin{aligned} d_1(t) &= \rho\Big(\Phi_{AB_i}(t, t_0)x_0 + \int_{t_0}^{t} \Phi_{AB_i}(t, \tau) f(\tau)\, d\tau \\ &\quad + \sum_{t_0 \le \tau_i < t} \Phi_{AB_i}(t, \tau_i) p_i, X(t, y_0)\Big) \\ &\le \Big\| \Phi_{AB_i}(t, t_0)x_0 + \int_{t_0}^{t} \Phi_{AB_i}(t, \tau) f(\tau)\, d\tau + \sum_{t_0 \le \tau_i < t} \Phi_{AB_i}(t, \tau_i) p_i \\ &\quad - \Big[\Phi_{AB_i}(t, t_0)y_0 + \int_{t_0}^{t} \Phi_{AB_i}(t, \tau) f(\tau)\, d\tau + \sum_{t_0 \le \tau_i < t} \Phi_{AB_i}(t, \tau_i) p_i\Big]\Big\| \\ &= \|\Phi_{AB_i}(t, t_0)(x_0 - y_0)\| \le \|\Phi_{AB_i}(t, t_0)\|\, \|x_0 - y_0\| \le M\|x_0 - y_0\|, \end{aligned}$$

the constant $M > 0$ exists by virtue of the criterion for stability of R-solutions for inclusion (3.13), and

$$\begin{aligned} d_2(t) &= \rho\Big(\Phi_{AB_i}(t, t_0)y_0 + \int_{t_0}^{t} \Phi_{AB_i}(t, \tau) f(\tau)\, d\tau \\ &\quad + \sum_{t_0 \le \tau_i < t} \Phi_{AB_i}(t, \tau_i) p_i, X(t, x_0)\Big) \\ &\le \Big\| \Phi_{AB_i}(t, t_0)y_0 + \int_{t_0}^{t} \Phi_{AB_i}(t, \tau) f(\tau)\, d\tau + \sum_{t_0 \le \tau_i < t} \Phi_{AB_i}(t, \tau_i) p_i \\ &\quad - \Big[\Phi_{AB_i}(t, t_0)x_0 + \int_{t_0}^{t} \Phi_{AB_i}(t, \tau) f(\tau)\, d\tau + \sum_{t_0 \le \tau_i < t} \Phi_{AB_i}(t, \tau_i) p_i\Big]\Big\| \\ &= \|\Phi_{AB_i}(t, t_0)(x_0 - y_0)\| \le M\|x_0 - y_0\|. \end{aligned}$$

We set $\delta = \frac{\varepsilon}{M}$. Then, for $\|x_0 - y_0\| < \delta$, we have

$$h(X(t, x_0), X(t, y_0)) < \varepsilon,$$

which was to be proved. □

We prove the converse statement in a special case.

Consider linear inhomogeneous inclusions of the form

$$\dot{x} \in A(t)x + F(t), \quad t \neq \tau_i, \tag{3.25}$$

$$\Delta x|_{t=\tau_i} \in B_i x + P_i, \tag{3.26}$$

where $A(t)$ is an $n \times n$ matrix measurable on the interval $I = [t_0, +\infty)$ and B_i are constant $n \times n$ matrices. The following statement is true:

Theorem 8. *The R-solutions of* (3.25), (3.26) *are stable if and only if the solutions of*

$$\dot{x} = A(t)x, \quad t \neq \tau_i, \tag{3.27}$$

$$\Delta x|_{t=\tau_i} = B_i x \tag{3.28}$$

are stable.

Proof. The sufficiency follows from Theorem 7. Let us prove the necessity. Assume that the R-solution of (3.25), (3.26)

$$X(t, x_0) = \Phi_{AB_i}(t, t_0)x_0 + \int_{t_0}^{t} \Phi_{AB_i}(t, \tau)F(\tau)\, d\tau + \sum_{t_0 \le \tau_i < t} \Phi_{AB_i}(t, \tau_i)P_i$$

is stable. Then, for any $\varepsilon > 0$, there exists $\delta = \delta(\varepsilon) > 0$ such that, for all y_0: $\|x_0 - y_0\| < \delta$, one has

$$h(X(t, x_0), X(t, y_0)) = h(\Phi_{AB_i}(t, t_0)x_0, \Phi_{AB_i}(t, t_0)y_0) < \varepsilon,$$

i.e., the solutions of (3.27), (3.28) are stable. □

Stability in the First Approximation. Consider a nonlinear impulsive differential inclusion of the form

$$\dot{x} \in \mathcal{A}(t)x + F(t, x), \quad t \neq \tau_i, \tag{3.29}$$

$$\Delta x|_{t=\tau_i} \in \mathcal{B}_i x + P_i(x), \tag{3.30}$$

where $x \in \mathbb{R}^n$ is a phase vector, $t \in I = [t_0, +\infty]$ is time, $\mathcal{A}(t)$ is a compact set of $n \times n$ matrices measurable on I, $F: I \times \mathbb{R}^n \to \operatorname{conv}(\mathbb{R}^n)$ is a set-valued mapping continuous in x for every fixed t and measurable in t for every fixed x, $|F(t, x)| \le \mu(t)$, the function $\mu(t)$ is measurable and summable on any finite segment $I^* \subset I$, $\tau_i \in I$, $i = 1, 2, \ldots$, are fixed times enumerated in the increasing order ($\tau_i < \tau_{i+1}$), the set $\{\tau_i\}$ does not have points of accumulation, $\mathcal{B}_i$ are compact sets of $n \times n$ matrices, and $P_i: \mathbb{R}^n \to \operatorname{conv}(\mathbb{R}^n)$ is a continuous set-valued mapping.

Assume that the set-valued mappings $F(t,x)$ and $P_i(x)$ satisfy the following conditions:

(i) $0 \in F(t,0)$ for all $t \in I$;

(ii) $0 \in P_i(0)$ for all $i \in N$.

It is obvious that inclusion (3.29), (3.30) has the trivial solution. Let us study its stability.

Together with inclusion (3.29), (3.30), we consider the linear inclusion

$$\dot{x} \in \mathcal{A}(t)x, \quad t \neq \tau_i, \tag{3.31}$$

$$\Delta x|_{t=\tau_i} \in \mathcal{B}_i x \tag{3.32}$$

and call it the inclusion of the first approximation with respect to inclusion (3.29), (3.30).

Theorem 9. *Suppose that the matrizants* $\Phi_{AB_i}(t,\tau)$ *of inclusion* (3.31), (3.32) *admit the following estimate for all* t *and* τ, $t_0 \le \tau \le t$:

$$\|\Phi_{AB_i}(t,\tau)\| \le Ke^{-\gamma(t-\tau)}, \quad K \ge 1,\ \gamma > 0, \tag{3.33}$$

where the constants K *and* γ *do not depend on the choice of matrices* $A(t) \in \mathcal{A}(t)$ *and* $B_i \in \mathcal{B}_i$, *and the functions* $F(t,x)$ *and* $P_i(x)$ *satisfy the inequalities*

$$|F(t,x)| \le a\|x\|, \quad |P_i(x)| \le a\|x\| \tag{3.34}$$

for all $t \ge t_0$, $i \in N$, $\|x\| \le h$, *and* $h > 0$. *Then, for sufficiently small* a, *the trivial solution of inclusion* (3.29), (3.30) *is asymptotically stable.*

Proof. Every solution of inclusion (3.29), (3.30) can be represented in the form

$$x(t,x_0) = \Phi_{AB_i}(t,t_0)x_0 + \int_{t_0}^{t} \Phi_{AB_i}(t,\tau) f(\tau, x(\tau,x_0))d\tau + \sum_{t_0 \le \tau_i < t} \Phi_{AB_i}(t,\tau_i) p_i(x(\tau_i,x_0)),$$

where $A(t) \in \mathcal{A}(t)$, $B_i \in \mathcal{B}_i$, $f(t,x)$ is the branch of the mapping $F(t,x)$ measurable in t and continuous in x, and $p_i(x)$ is the continuous branch of the mapping $P_i(x)$ [52]. Then, by analogy with Theorem 12 of Chapter 1, we obtain the estimate

$$\|x(t,x_0)\| \le Ke^{-(\gamma - Ka - \frac{1}{\theta}\ln(1+Ka))(t-t_0)}\|x_0\|. \tag{3.35}$$

Thus, if a is so small that

$$\gamma - Ka - \frac{1}{\theta}\ln(1+Ka) > 0,$$

then any solution $x(t, x_0)$, $\|x_0\| \leq \frac{h}{K}$, of inclusion (3.29), (3.30) is defined for all $t \leq t_0$ and

$$\lim_{t\to\infty} \|x(t, x_0)\| = 0,$$

i.e., the trivial solution of inclusion (3.29), (3.30) is asymptotically stable. □

Corollary 1. *If inequality* (3.33) *in the conditions of Theorem* 9 *is satisfied only for some* $A(t) \in \mathcal{A}(t)$ *and* $B_i \in \mathcal{B}_i$ *and condition* (3.34) *is replaced by*

$$\|f(t, x)\| \leq a\|x\|, \quad \|p_i(x)\| \leq a\|x\| \tag{3.36}$$

for some branch $f(t, x)$ *of the mapping* $F(t, x)$ *measurable in* t *and continuous in* x *and the continuous branch* $p_i(x)$ *of the mapping* $P_i(x)$*, then we can prove only the weak asymptotic stability of the trivial solution of inclusion* (3.29), (3.30).

Remark 7. By virtue of estimate (3.35), inclusion (3.29), (3.30) has a unique solution with initial condition $x(0) = 0$. Hence, inclusion (3.29), (3.30) has the unique R-solution ($R(t) \equiv 0$) with initial condition $R(0) = 0$.

Corollary 2. *If the conditions of the theorem are satisfied, then the trivial* R*-solution of inclusion* (3.29), (3.30) *is asymptotically stable.*

3.3 Periodic Solutions of Linear Impulsive Differential Inclusions

Consider the problem of existence of periodic R-solutions and ordinary solutions of linear inhomogeneous periodic impulsive differential inclusions of the form

$$\begin{aligned} \dot{x} &\in A(t)x + F(t), \quad t \neq \tau_i, \\ \Delta x|_{t=\tau_i} &\in B_i x + P_i, \end{aligned} \tag{3.37}$$

where $A(t)$ is a continuous T-periodic matrix function, $F: \mathbb{R} \to \operatorname{comp}(\mathbb{R}^n)$ is a continuous T-periodic mapping, and the constant matrices B_i, sets $P_i \in \operatorname{conv}(\mathbb{R}^n)$, and times τ_i are such that, for a certain natural number r, one has

$$B_{i+r} = B_i, \quad P_{i+r} = P_i, \quad \tau_{i+r} = \tau_i + T$$

for all $i \in \mathbb{Z}$. Also assume that $0 \leq \tau_1 < \cdots < \tau_r < T$ and $\det(E + B_i) \neq 0$ for all $i = \overline{1, r}$.

Let $\Phi(t, s) = \Phi_{AB_i}(t, s)$ be the matrizant of the homogeneous system corresponding to (3.37). A solution $x(t, x_0)$, $x(0, x_0) = x_0$, of the linear impulsive differential inclusion (3.37) can be represented in the form

$$x(t, x_0) = \Phi(t, 0)x_0 + \int_0^t \Phi(t, \tau) f(\tau) d\tau + \sum_{0 \leq \tau_i < t} \Phi(t, \tau_i) p_i,$$

where $f(t) \in F(t)$ and $p_i \in P_i$.

By analogy with (3.4), we write an R-solution $X(t, R_0)$, $X(t_0, R_0) = R_0$, of the impulsive differential inclusion (3.37) in the form

$$X(t, R_0) = \Phi(t,0)R_0 + \int_0^t \Phi(t,\tau)F(\tau)d\tau + \sum_{0\le\tau_i<t} \Phi(t,\tau_i)P_i.$$

If $R_0 \in \operatorname{conv}(\mathbb{R}^n)$, then, using the Lyapunov theorem, we establish that $X(t, R_0) \in \operatorname{conv}(\mathbb{R}^n)$ for any t. Thus, the existence of T-periodic R-solutions and ordinary solutions is directly related to the existence of solutions of the following equations in the space $\operatorname{conv}(\mathbb{R}^n)$:

$$R_0 = \Phi(T,0)R_0 + G, \tag{3.38}$$

$$G = \int_0^T \Phi(T,\tau)F(\tau)d\tau + \sum_{0\le\tau_i<T} \Phi(T,\tau_i)P_i,$$

$$x_0 = \Phi(T,0)x_0 + g, \tag{3.39}$$

$$g = \int_0^T \Phi(T,\tau)f(\tau)d\tau + \sum_{0\le\tau_i<T} \Phi(T,\tau_i)p_i \in G.$$

In Eq. (3.39), different $f(t) \in F(t)$ and $p_i \in P_i$ are associated with points $g \in G \in \operatorname{conv}(\mathbb{R}^n)$. If the matrix $E-\Phi(T,0)$ is nondegenerate (i.e., the eigenvalues of the matrix $\Phi(T,0)$ do not contain 1), then ordinary periodic solutions exist for all $g \in G$, i.e., for all $f(t) \in F(t)$ and $p_i \in P_i$. If the matrix $E - \Phi(T,0)$ is degenerate, then ordinary periodic solutions exist for $g \in G$ for which the rank of the extended matrix $(E - \Phi(T,0), g)$ is equal to the rank of the matrix $E - \Phi(T,0)$ [142].

Thus, if Eq. (3.39) has a solution for at least one $g \in G$, then there exists an ordinary T-periodic solution of problem (3.37).

In the general case, this solution may not be unique, i.e., a single point $g \in G$ (and, hence, a single initial point x_0) may correspond to several T-periodic solutions.

Example 6. Consider the impulsive differential inclusion

$$\dot{x} \in [0,1], \quad t \neq 2n-1,\ n \in N,$$
$$\Delta x|_{t=2n-1} = -2x.$$

In this case, we have

$$\Phi(2,s) = \begin{cases} -1, & s \in [0,1], \\ 1, & s \in (1,2], \end{cases} \quad \text{and} \quad G = \int_0^2 \Phi(2,s)[0,1]ds = [-1,1].$$

Thus, the initial points corresponding to 2-periodic solutions can be found from the equations

$$x_0 = -x_0 + g, \quad |g| \le 1.$$

We take $g_0 = 0$. Then $x_0 = 0$. This initial condition corresponds to infinitely many 2-periodic solutions, e.g.,

$$x_k(t) = \begin{cases} k(t-2m), & t \in [2m, 2m+1], \\ -2k + k(t-2m), & t \in (2m+1, 2(m+1)], \end{cases}$$

where $k \in [0,1]$ and $m \in \mathbb{N}$.

The initial points of ordinary periodic solutions (if periodic solutions exist) correspond to different points $g \in G$. Let X_0 denote the set of initial points x_0 of ordinary periodic solutions, i.e.,

$$X_0 = \{x_0 \in \mathbb{R}^n \colon x_0 = \Phi(T,0)x_0 + g,\ g \in G\}.$$

Consider Eq. (3.38). For the matrix $\Phi(T,0)$, there exists a real nondegenerate matrix M such that

$$\Phi(T,0) = M^{-1}JM, \quad \text{where } J = \begin{pmatrix} J_1 & 0 & \cdots & 0 \\ 0 & J_2 & \cdots & 0 \\ \vdots & \vdots & \ddots & \vdots \\ 0 & 0 & \cdots & J_s \end{pmatrix} \tag{3.40}$$

is the real canonical form [32]. Thus, Eq. (3.38) takes the form

$$\begin{aligned} &R_0 = M^{-1}JMR_0 + G \\ \Leftrightarrow\ &MR_0 = JMR_0 + MG \\ \Leftrightarrow\ &Z_0 = JZ_0 + D, \end{aligned} \tag{3.41}$$

where $Z_0 = MR_0$ and $D = MG \in \operatorname{conv}(\mathbb{R}^n)$.

Let Λ denote the spectral radius of the matrix $\Phi(T,0)$. Consider the problem of existence of solutions of Eq. (3.41) in dependence on the value of Λ.

Case $\Lambda < 1$ ([118]). For the matrix J, there exists a diagonal matrix $L = \operatorname{diag}\{1, \varepsilon, \ldots, \varepsilon^{n-1}\}$ such that $J = L^{-1}J(\varepsilon)L$, where $J(\varepsilon)$ is a modified form [131], i.e., $J_i(\varepsilon) = J_i$ if λ_i is a simple root and

$$J_i(\varepsilon) = \begin{pmatrix} \lambda_i & 0 & \cdots & 0 \\ \varepsilon_i^1 & \lambda_i & \cdots & 0 \\ \vdots & \ddots & \ddots & \vdots \\ 0 & \cdots & \varepsilon_i^{k_i-1} & \lambda_i \end{pmatrix}$$

or

$$J_i(\varepsilon) = \begin{pmatrix} \begin{array}{|cc|} \operatorname{Re}\lambda_i & -\operatorname{Im}\lambda_i \\ \operatorname{Im}\lambda_i & \operatorname{Re}\lambda_i \end{array} & \mathbf{0} & \cdots & \mathbf{0} \\ \varepsilon_i^1 E_2 & \begin{array}{|cc|} \operatorname{Re}\lambda_i & -\operatorname{Im}\lambda_i \\ \operatorname{Im}\lambda_i & \operatorname{Re}\lambda_i \end{array} & \cdots & \mathbf{0} \\ \mathbf{0} & \ddots & \ddots & \vdots \\ \mathbf{0} & \cdots & \varepsilon_i^{k_i-1} E_2 & \begin{array}{|cc} \operatorname{Re}\lambda_i & -\operatorname{Im}\lambda_i \\ \operatorname{Im}\lambda_i & \operatorname{Re}\lambda_i \end{array} \end{pmatrix}$$

in the case of a multiple (respectively, real or complex) root, $\varepsilon_i^j \in \{0, \varepsilon\}$, ε is an arbitrary positive number, and E_2 is the 2×2 identity matrix.

In this case, Eq. (3.41) can be transformed as follows:

$$Z_0(\varepsilon) = J(\varepsilon)Z_0(\varepsilon) + D(\varepsilon), \tag{3.42}$$

where $Z_0(\varepsilon) = LZ_0$ and $D(\varepsilon) = LD \in \operatorname{conv}(\mathbb{R}^n)$.

The spectral radius ρ of the matrix $J^T(\varepsilon)J(\varepsilon)$ is a continuous function of ε at the point $\varepsilon = 0$. Therefore,

$$\rho(J^T(\varepsilon)J(\varepsilon)) = \rho(J^T(0)J(0)) + \sigma(\varepsilon) = \Lambda^2 + \sigma(\varepsilon),$$

where $\sigma(\varepsilon) \to 0$ as $\varepsilon \to 0$.

We choose the constant $\varepsilon > 0$ so that $\|J(\varepsilon)\| = \sqrt{\Lambda^2 + \sigma(\varepsilon)} < 1$.

The space $\operatorname{conv}(\mathbb{R}^n)$ is complete [23,28]. The existence of a T-periodic R-solution of Eq. (3.37), i.e., the existence of an initial set $Z_0(\varepsilon)$ that satisfies Eq. (3.42), is related to the existence of a fixed point of the mapping

$$\chi(Z) = J(\varepsilon)Z + D(\varepsilon). \tag{3.43}$$

Using the properties of the Hausdorff distance and support function, we obtain

$$\begin{aligned} h(\chi(Z_1), \chi(Z_2)) &= h(J(\varepsilon)Z_1 + D(\varepsilon), J(\varepsilon)Z_2 + D(\varepsilon)) \\ &= h(J(\varepsilon)Z_1, J(\varepsilon)Z_2) = \sup_{\psi \in S} |c(J(\varepsilon)Z_1, \psi) - c(J(\varepsilon)Z_2, \psi)| \\ &= |c(Z_1, J^T(\varepsilon)\psi^*) - c(Z_2, J^T(\varepsilon)\psi^*)| \le \|J^T(\varepsilon)\psi^*\|\, h(Z_1, Z_2) \\ &\le \|J(\varepsilon)\|\, h(Z_1, Z_2) < h(Z_1, Z_2), \end{aligned}$$

i.e., the mapping $\chi(Z)$ is contracting. Hence, all conditions of the Banach theorem [14] are satisfied, and, therefore, mapping (3.43) has a unique fixed point in $\operatorname{conv}(\mathbb{R}^n)$.

Thus, if the spectral radius of the matrix $\Phi(T, 0)$ is less than 1, then, for any T-periodic set-valued mapping $F(t)$ and periodic sequence P_i, the linear impulsive differential inclusion (3.37) has a unique T-periodic R-solution $R(t)$ whose initial value is determined by the formula $R_0 = M^{-1}L^{-1}Z_0(\varepsilon)$, where $Z_0(\varepsilon)$ is determined as a fixed point of mapping (3.43).

Example 7. Consider the impulsive differential inclusion

$$\begin{pmatrix}\dot{x}_1\\ \dot{x}_2\end{pmatrix} \in \begin{pmatrix}-1 & 0\\ 0 & -1\end{pmatrix}\begin{pmatrix}x_1\\ x_2\end{pmatrix} + \alpha(1+\sin t)B_1(0), \quad t \neq (2i-1)\pi,$$

$$\left.\begin{pmatrix}\Delta x_1\\ \Delta x_2\end{pmatrix}\right|_{t=(2i-1)\pi} \in \begin{pmatrix}\frac{1}{2}e^{2\pi}-1 & 0\\ 0 & -\frac{1}{2}e^{2\pi}-1\end{pmatrix}\begin{pmatrix}x_1\\ x_2\end{pmatrix} + pB_1(0),$$

where

$$\alpha = \frac{4}{3e^{\pi}+1} \quad \text{and} \quad p = \frac{12e^{-2\pi}}{3e^{\pi}+1}.$$

The initial set R_0 corresponding to a 2π-periodic R-solution satisfies Eq. (3.38). In this example, we have $T = 2\pi$,

$$\Phi(2\pi, s) = \begin{cases} \frac{1}{2}e^{s}\begin{pmatrix}1 & 0\\ 0 & -1\end{pmatrix}, & s \le \pi,\\ e^{s-2\pi}E, & s > \pi,\end{cases}$$

$$\begin{aligned} G &= \int_0^{2\pi} \Phi(T,s)\alpha(1+\sin s)B_1(0)ds + \Phi(2\pi,\pi)pB_1(0)\\ &= G_1 + G_2 + G_3, \end{aligned}$$

where

$$\begin{aligned} G_1 &= \frac{1}{2}\alpha\int_0^{\pi} e^{s}(1+\sin s)\begin{pmatrix}1 & 0\\ 0 & -1\end{pmatrix}B_1(0)ds\\ &= \frac{1}{2}\alpha\int_0^{\pi} e^{s}(1+\sin s)B_1(0)ds\\ &= \frac{1}{2}\alpha\int_0^{\pi} e^{s}(1+\sin s)ds\ B_1(0) = \frac{1}{2}\alpha\left(\frac{3}{2}e^{\pi} - \frac{1}{2}\right)B_1(0),\\ G_2 &= \alpha e^{-2\pi}\int_{\pi}^{2\pi} e^{s}(1+\sin s)B_1(0)ds = \alpha\left(\frac{1}{2} - \frac{3}{2}e^{-\pi}\right)B_1(0),\\ G_3 &= \frac{1}{2}pe^{\pi}\begin{pmatrix}1 & 0\\ 0 & -1\end{pmatrix}B_1(0) = \frac{1}{2}pe^{\pi}B_1(0). \end{aligned}$$

Thus,

$$G = \left[\frac{1}{2}\alpha\left(\frac{3}{2}e^{\pi} - \frac{1}{2}\right) + \alpha\left(\frac{1}{2} - \frac{3}{2}e^{-\pi}\right) + \frac{1}{2}pe^{\pi}\right]B_1(0) = B_1(0).$$

Hence, Eq. (3.38) takes the form

$$R_0 = \begin{pmatrix}\frac{1}{2} & 0\\ 0 & -\frac{1}{2}\end{pmatrix}R_0 + B_1(0). \tag{3.44}$$

By virtue of the uniqueness of a solution of Eq. (3.44) in the space $\operatorname{conv}(\mathbb{R}^2)$, we get $R_0 = B_2(0)$.

For the determination of the set X_0, we obtain the system

$$\begin{cases} x_1^0 = \frac{1}{2}x_1^0 + g_1, \\ x_2^0 = -\frac{1}{2}x_2^0 + g_2, \quad g_1^2 + g_2^2 \le 1, \end{cases}$$

which implies that

$$\frac{(x_1^0)^2}{4} + 9\frac{(x_2^0)^2}{4} \le 1.$$

Thus, X_0 is an ellipse and $X_0 \subset R_0$.

Case $\Lambda > 1$ ([118]). Since

$$\Lambda = \max_{i=\overline{1,n}} |\lambda_i| > 1,$$

where λ_i are the eigenvalues of the matrix $\Phi(T,0)$, there exists $\lambda_{i_0}\colon |\lambda_{i_0}| > 1$ (in what follows, we assume, for simplicity, that $i_0 = 1$).

In Eq. (3.41), we separately consider the cases where λ_1 is real (simple or multiple) and complex (simple or multiple).

(1) Let λ_1 be simple and real. Then the Jordan cell has the form

$$J_1 = (\lambda_1).$$

Let $\mathrm{pr}_{x_1} A$ be the projection of the set A to the axis x_1. Then

$$\begin{aligned} \operatorname{mes}(\mathrm{pr}_{x_1} Z_0) &= \operatorname{mes}(\mathrm{pr}_{x_1}(JZ_0 + D)) = \operatorname{mes}(\mathrm{pr}_{x_1} JZ_0 + \mathrm{pr}_{x_1} D) \\ &= \operatorname{mes}(\mathrm{pr}_{x_1} JZ_0) + \operatorname{mes}(\mathrm{pr}_{x_1} D) = \operatorname{mes}(\lambda_1 \mathrm{pr}_{x_1} Z_0) + \operatorname{mes}(\mathrm{pr}_{x_1} D) \\ &= |\lambda_1| \operatorname{mes}(\mathrm{pr}_{x_1} Z_0) + \operatorname{mes}(\mathrm{pr}_{x_1} D). \end{aligned}$$

Consequently, Eq. (3.41) has solutions if and only if $\operatorname{mes}(\mathrm{pr}_{x_1} D) = 0$ and $\operatorname{mes}(\mathrm{pr}_{x_1} Z_0) = 0$. In other words, the set D lies in the hyperplane $x_1 = d_1^*$, and the set Z_0 lies in the hyperplane $x_1 = x_1^*$, where x_1^* is determined from the equation $x_1^* = \lambda_1 x_1^* + d_1^*$, i.e.,

$$x_1^* = \frac{d_1^*}{1 - \lambda_1}.$$

In this case, Z_0 and D can be considered as sets of dimension $n-1$, and Eq. (3.41) reduces to the equation

$$\tilde{Z}_0 = \tilde{J}\tilde{Z}_0 + \tilde{D}, \quad \text{where } \tilde{J} = \left(\begin{array}{c|c} J_2 & \mathbf{0} \\ \hline \mathbf{0} & \ddots \end{array}\right),$$

$$\tilde{Z}_0 = \{(x_2, \dots, x_n)\colon (x_1^*, x_2, \dots, x_n) \in Z_0\},$$

$$\tilde{D} = \{(d_2, \dots, d_n)\colon (d_1^*, d_2, \dots, d_n) \in D\}.$$

Thus, we have obtained a problem whose dimension is less by 1, which can be studied by analogy with the original problem.

Example 8. Consider a linear impulsive differential inclusion of the form

$$\begin{pmatrix} \dot{x}_1 \\ \dot{x}_2 \\ \dot{x}_3 \end{pmatrix} \in \begin{pmatrix} \frac{1}{2\pi} & 0 & 0 \\ 0 & 1 & 0 \\ 0 & 0 & 1 \end{pmatrix} \begin{pmatrix} x_1 \\ x_2 \\ x_3 \end{pmatrix} + \alpha(1+\cos t)\tilde{B}_1(0),$$

$$\left.\begin{pmatrix} \Delta x_1 \\ \Delta x_2 \\ \Delta x_3 \end{pmatrix}\right|_{t=(2i-1)\pi} \in \begin{pmatrix} 2e^{-1}-1 & 0 & 0 \\ 0 & \frac{1}{2}e^{-2\pi}-1 & 0 \\ 0 & 0 & \frac{1}{2}e^{-2\pi}-1 \end{pmatrix} \begin{pmatrix} x_1 \\ x_2 \\ x_3 \end{pmatrix} + p\tilde{B}_1(0),$$

where $\tilde{B}_1(0) = \{(x_1, x_2, x_3)^T : x_1 = 0,\ (x_2, x_3)^T \in B_1(0)\}$, $\alpha = \frac{2}{2e^{\pi}-e^{-\pi}-3}$, and $p = e^{\pi}$.

In this case, the fundamental matrix has the form

$$\Phi(2\pi, s) = \begin{cases} \begin{pmatrix} 2e^{\frac{-s}{2\pi}} & 0 & 0 \\ 0 & \frac{1}{2}e^{-s} & 0 \\ 0 & 0 & \frac{1}{2}e^{-s} \end{pmatrix}, & s \le \pi \\ \begin{pmatrix} e^{\frac{1}{2\pi}(2\pi-s)} & 0 & 0 \\ 0 & e^{2\pi-s} & 0 \\ 0 & 0 & e^{2\pi-s} \end{pmatrix}, & s > \pi \end{cases}$$

We now determine the set G. We have

$$\begin{aligned} G &= \int_0^{2\pi} \Phi(T,s)\alpha(1+\sin s)\tilde{B}_1(0)ds + X(2\pi,\pi)p\tilde{B}_1(0) \\ &= G_1 + G_2 + G_3, \end{aligned}$$

where

$$\begin{aligned} G_1 &= \alpha \int_0^{\pi} \begin{pmatrix} 2e^{-\frac{s}{2\pi}} & 0 & 0 \\ 0 & \frac{1}{2}e^{-s} & 0 \\ 0 & 0 & \frac{1}{2}e^{-s} \end{pmatrix} (1+\cos s)\tilde{B}_1(0)ds \\ &= \frac{\alpha}{2}\int_0^{\pi} e^{-s}(1+\cos s)\tilde{B}_1(0)ds \\ &= \frac{\alpha}{2}\int_0^{\pi} e^{-s}(1+\cos s)ds\,\tilde{B}_1(0) = \frac{\alpha}{4}(3-e^{-\pi})\tilde{B}_1(0), \\ G_2 &= \alpha \int_{\pi}^{2\pi} \begin{pmatrix} e^{\frac{2\pi-s}{2\pi}} & 0 & 0 \\ 0 & e^{2\pi-s} & 0 \\ 0 & 0 & e^{2\pi-s} \end{pmatrix} (1+\cos s)\tilde{B}_1(0)ds \\ &= \alpha \int_{\pi}^{2\pi} e^{2\pi-s}(1+\cos s)\tilde{B}_1(0)ds \end{aligned}$$

$$= \alpha \int_{\pi}^{2\pi} e^{2\pi - s}(1 + \cos s) ds \tilde{B}_1(0) = \frac{1}{2}\alpha e^{2\pi}(e^{-\pi} - 3e^{-2\pi})\tilde{B}_1(0),$$

$$G_3 = p \begin{pmatrix} 2e^{-\frac{1}{2}} & 0 & 0 \\ 0 & \frac{1}{2}e^{-\pi} & 0 \\ 0 & 0 & \frac{1}{2}e^{-\pi} \end{pmatrix} \tilde{B}_1(0) = \frac{p}{2}e^{-\pi}\tilde{B}_1(0).$$

Hence,

$$G = \left[\alpha\left(\frac{1}{4}(3 - e^{-\pi}) + \frac{1}{2}e^{2\pi}(e^{-\pi} - 3e^{-2\pi})\right) + \frac{p}{2}e^{-\pi}\right]\tilde{B}_1(0)$$
$$= \tilde{B}_1(0).$$

Thus, Eq. (3.38) takes the form

$$R_0 = \begin{pmatrix} 2 & 0 & 0 \\ 0 & \frac{1}{2} & 0 \\ 0 & 0 & \frac{1}{2} \end{pmatrix} R_0 + \tilde{B}_1(0).$$

Since $\lambda_1 = 2$ and $\tilde{B}_1(0)$ belongs to the hyperplane $x_1 = 0$, we conclude that R_0 must belong to the hyperplane $x_1 = 0$, and Eq. (3.38) reduces to the equation

$$\tilde{R}_0 = \begin{pmatrix} \frac{1}{2} & 0 \\ 0 & \frac{1}{2} \end{pmatrix} \tilde{R}_0 + B_1(0),$$

whose solution is $\tilde{R}_0 = B_2(0)$. In other words,

$$R_0 = \{(x_1, x_2, x_3)^T : x_1 = 0,\ (x_2, x_3)^T \in B_2(0)\} = \tilde{B}_2(0).$$

Ordinary periodic solutions exist for all $g \in \tilde{B}_1(0)$, and the corresponding initial conditions are determined from the system

$$x_0 = \begin{pmatrix} 2 & 0 & 0 \\ 0 & \frac{1}{2} & 0 \\ 0 & 0 & \frac{1}{2} \end{pmatrix} x_0 + g.$$

Thus, the set of initial values for ordinary periodic solutions X_0 has the form

$$X_0 = \begin{pmatrix} -1 & 0 & 0 \\ 0 & \frac{1}{2} & 0 \\ 0 & 0 & \frac{1}{2} \end{pmatrix}^{-1} \tilde{B}_1(0) = 2\tilde{B}_1(0) = R_0.$$

Example 9. Consider the linear impulsive differential inclusion from Example 8 with the mapping $F(t)$ and set P chosen as follows: $F(t) = \alpha(1 + \cos t)B_1(0)$ and $P = pB_1(0)$. In this case, we have $G = B_1(0)$. Therefore, a 2π-periodic R-solution does not exist because $\lambda_1 > 1$ and G does not belong to the hyperplane $x_1 = g_1^*$. Ordinary T-periodic solutions exist for any $g \in G$ because the eigenvalues of the matrix $\Phi(T, 0)$ do not contain 1. In this case, the set X_0 is an ellipsoid.

(2) Let λ_1 be of multiplicity k. Then the cell J_1 takes the form

$$J_1 = \begin{pmatrix} \lambda_1 & 0 & \cdots & 0 \\ e_1^1 & \lambda_1 & \cdots & 0 \\ \vdots & \ddots & \ddots & \vdots \\ 0 & \cdots & e_1^{k-1} & \lambda_1 \end{pmatrix},$$

where $e_1^i \in \{0, 1\}$.

In this case, by analogy with the previous one, we establish that Eq. (3.41) has solutions if and only if $\operatorname{mes}(\operatorname{pr}_{x_1} D) = 0$ and $\operatorname{mes}(\operatorname{pr}_{x_1} Z_0) = 0$. In other words, the set D lies in the hyperplane $x_1 = d_1^*$, and the set Z_0 lies in the hyperplane $x_1 = x_1^*$. Since

$$\begin{aligned} \operatorname{mes}(\operatorname{pr}_{x_2} Z_0) &= \operatorname{mes}(\operatorname{pr}_{x_2}(JZ_0 + D)) = \operatorname{mes}(\operatorname{pr}_{x_2} JZ_0 + \operatorname{pr}_{x_2} D) \\ &= \operatorname{mes}(\operatorname{pr}_{x_2} JZ_0) + \operatorname{mes}(\operatorname{pr}_{x_2} D) = \operatorname{mes}(\lambda_1 \operatorname{pr}_{x_2} Z_0 + x_1^*) \\ &\quad + \operatorname{mes}(\operatorname{pr}_{x_2} D) = |\lambda_1| \operatorname{mes}(\operatorname{pr}_{x_2} Z_0) + \operatorname{mes}(\operatorname{pr}_{x_2} D), \end{aligned}$$

we conclude that Eq. (3.41) has solutions if and only if $\operatorname{mes}(\operatorname{pr}_{x_2} D) = 0$ and $\operatorname{mes}(\operatorname{pr}_{x_2} Z_0) = 0$, i.e., the set D lies in the hyperplane $x_2 = d_2^*$, and the set Z_0 lies in the hyperplane $x_2 = x_2^*$, etc.

Thus, Eq. (3.41) has solutions if and only if the set D lies in the hyperplanes $x_i = d_i^*$, $i = \overline{1,k}$, and the set Z_0 lies in the hyperplanes $x_i = x_i^*$, $i = \overline{1,k}$. In this case, the constants x_i^* are determined from the system

$$\begin{pmatrix} x_1^* \\ \vdots \\ x_k^* \end{pmatrix} = J_1 \begin{pmatrix} x_1^* \\ \vdots \\ x_k^* \end{pmatrix} + \begin{pmatrix} d_1^* \\ \vdots \\ d_k^* \end{pmatrix}, \tag{3.45}$$

which is solved successively from the top downward and has a unique solution because the matrix $E - J_1$ is nondegenerate.

(3) Let λ_1 be simple and complex. Then the Jordan cell has the form

$$J_1 = \begin{pmatrix} \operatorname{Re}\lambda_1 & -\operatorname{Im}\lambda_1 \\ \operatorname{Im}\lambda_1 & \operatorname{Re}\lambda_1 \end{pmatrix}.$$

Let $\operatorname{pr}_{(x_1,x_2)} A$ be the projection of a set A to the hyperplane (x_1, x_2). Then

$$\operatorname{pr}_{(x_1,x_2)} Z_0 = \operatorname{pr}_{(x_1,x_2)}(JZ_0 + D).$$

The projection of the set Z_0 to the hyperplane (x_1, x_2) is a convex compact set. Therefore, either $\mathrm{pr}_{(x_1,x_2)} Z_0$ is a segment, or we can choose a disk with maximum area among the disks contained in $\mathrm{pr}_{(x_1,x_2)} Z_0$ (let this be $S_r(a)$, $r > 0$).

In the first case, if $\mathrm{mes}(\mathrm{pr}_{(x_1,x_2)} D) > 0$, then Eq. (3.41) does not have solutions because

$$\begin{aligned}\mathrm{mes}(\mathrm{pr}_{(x_1,x_2)}(JZ_0 + D)) \geq \mathrm{mes}(\mathrm{pr}_{(x_1,x_2)} D) &> 0\\ &= \mathrm{mes}(\mathrm{pr}_{(x_1,x_2)} Z_0).\end{aligned}$$

If $\mathrm{mes}(\mathrm{pr}_{(x_1,x_2)} D) = 0$, i.e., $\mathrm{pr}_{(x_1,x_2)} D$ is a segment, then it is obvious that the segment $\mathrm{pr}_{(x_1,x_2)} JZ_0$ must be parallel to the given one, otherwise $\mathrm{pr}_{(x_1,x_2)}(JZ_0 + D)$ is a parallelogram by virtue of the definition of the sum of two sets and cannot coincide with the set $\mathrm{pr}_{(x_1,x_2)} Z_0$. By (x_1^1, x_2^1) and (x_1^2, x_2^2) we denote the endpoints of the segment $\mathrm{pr}_{(x_1,x_2)} Z_0$. Then

$$\begin{aligned}
&\mathrm{length}\,\mathrm{pr}_{(x_1,x_2)}(JZ_0 + D)\\
&= \mathrm{length}\,\mathrm{pr}_{(x_1,x_2)}(JZ_0) + \mathrm{length}\,\mathrm{pr}_{(x_1,x_2)} D\\
&= [(\mathrm{Re}\,\lambda_1 x_1^1 - \mathrm{Im}\,\lambda_1 x_2^1 - (\mathrm{Re}\,\lambda_1 x_1^2 - \mathrm{Im}\,\lambda_1 x_2^2))^2\\
&\quad + (\mathrm{Im}\,\lambda_1 x_1^1 + \mathrm{Re}\,\lambda_1 x_2^1 - (\mathrm{Im}\,\lambda_1 x_1^2 + \mathrm{Re}\,\lambda_1 x_2^2))^2]^{\frac{1}{2}}\\
&\quad + \mathrm{length}\,\mathrm{pr}_{(x_1,x_2)} D\\
&= [(\mathrm{Re}\,\lambda_1^2 + \mathrm{Im}\,\lambda_1^2)((x_1^1 - x_1^2)^2 + (x_2^1 - x_2^2)^2)]^{\frac{1}{2}}\\
&\quad + \mathrm{length}\,\mathrm{pr}_{(x_1,x_2)} D = |\lambda_1|\mathrm{length}\,\mathrm{pr}_{(x_1,x_2)} Z_0 + \mathrm{length}\,\mathrm{pr}_{(x_1,x_2)} D.
\end{aligned}$$

Thus, in this case, Eq. (3.41) has a solution if and only if $\mathrm{length}\,\mathrm{pr}_{(x_1,x_2)} Z_0 = 0$ and $\mathrm{length}\,\mathrm{pr}_{(x_1,x_2)} D = 0$, i.e., $\mathrm{pr}_{(x_1,x_2)} Z_0$ and $\mathrm{pr}_{(x_1,x_2)} D$ are points.

In the second case, since $S_r(a) \subset \mathrm{pr}_{(x_1,x_2)} Z_0$, we have

$$J_1 S_r(a) + \mathrm{pr}_{(x_1,x_2)} D \subset \mathrm{pr}_{(x_1,x_2)}(JZ_0 + D),$$

and the set $J_1 S_r(a) + \mathrm{pr}_{(x_1,x_2)} D$ contains the disk

$$J_1 S_r(a) + d = |\lambda_1| \begin{pmatrix} \cos\varphi & -\sin\varphi \\ \sin\varphi & \cos\varphi \end{pmatrix} S_r(a) + d,$$

$$\cos\varphi = \frac{\mathrm{Re}\,\lambda_1}{|\lambda_1|}, \quad \sin\varphi = \frac{\mathrm{Im}\,\lambda_1}{|\lambda_1|}, \quad d \in \mathrm{pr}_{(x_1,x_2)} D,$$

i.e., a disk of radius $r_1 = |\lambda_1| r > r$, which, by virtue of Eq. (3.41), contradicts the choice of the disk $S_r(a)$.

Thus, a T-periodic R-solution may exist only in the case where the sets $\mathrm{pr}_{(x_1,x_2)} Z_0$ and $\mathrm{pr}_{(x_1,x_2)} D$ are singletons. For the determination of $\mathrm{pr}_{(x_1,x_2)} Z_0$, we obtain the system

$$\begin{pmatrix} x_1^* \\ x_2^* \end{pmatrix} = \begin{pmatrix} \operatorname{Re}\lambda_1 & -\operatorname{Im}\lambda_1 \\ \operatorname{Im}\lambda_1 & \operatorname{Re}\lambda_1 \end{pmatrix} \begin{pmatrix} x_1^* \\ x_2^* \end{pmatrix} + \begin{pmatrix} d_1^* \\ d_2^* \end{pmatrix}, \tag{3.46}$$

or

$$\begin{pmatrix} 1-\operatorname{Re}\lambda_1 & \operatorname{Im}\lambda_1 \\ -\operatorname{Im}\lambda_1 & 1-\operatorname{Re}\lambda_1 \end{pmatrix} \begin{pmatrix} x_1^* \\ x_2^* \end{pmatrix} = \begin{pmatrix} d_1^* \\ d_2^* \end{pmatrix},$$

whose determinant is $(1-\operatorname{Re}\lambda_1)^2 + \operatorname{Im}\lambda_1^2 > 0$. Hence, this system has a unique solution.

The sets Z_0 and D can be considered as sets in the space $\operatorname{conv}(\mathbb{R}^{n-2})$. Thus, the dimension of the problem is decreased by 2, and the obtained problem

$$\tilde{Z}_0 = \tilde{J}\tilde{Z}_0 + \tilde{D},$$

where

$$\tilde{J} = \left(\begin{array}{c|c} J_2 & \mathbf{0} \\ \hline \mathbf{0} & \ddots \end{array}\right),$$

$\tilde{Z}_0 = \{(x_3,\dots,x_n)\colon (x_1^*, x_2^*, x_3,\dots,x_n) \in Z_0\}$, and $\tilde{D} = \{(d_3,\dots,d_n)\colon (d_1^*, d_2^*, d_3,\dots,d_n) \in D\}$, can be studied by analogy with the original one.

Example 10. Consider the linear impulsive differential inclusion

$$\begin{aligned} \dot{x}_1 &= x_2, \\ \dot{x}_2 &= -x_1, \\ \dot{x}_3 &\in x_3 + \alpha[-1,1], \quad t \neq (2i-1)\pi, \\ \Delta x_1|_{t=(2i-1)\pi} &= x_2, \\ \Delta x_2|_{t=(2i-1)\pi} &= -x_1, \\ \Delta x_3|_{t=(2i-1)\pi} &\in \left(\frac{1}{2}e^{-2\pi} - 1\right) x_3 + p[-1,1], \end{aligned}$$

where $i \in N$, $\alpha = \frac{1}{2e^{\pi} - e^{-\pi} - 1}$, and $p = e^{\pi}$.

The fundamental matrix of this system has the form

$$\Phi(2\pi, s) = \begin{cases} \begin{pmatrix} \cos s + \sin s & \cos s - \sin s & 0 \\ \sin s - \cos s & \cos s + \sin s & 0 \\ 0 & 0 & \frac{1}{2}e^{-s} \end{pmatrix}, & s \le \pi, \\ \begin{pmatrix} \cos s & -\sin s & 0 \\ \sin s & \cos s & 0 \\ 0 & 0 & e^{2\pi - s} \end{pmatrix}, & s > \pi. \end{cases}$$

In this case, we have

$$G_1 = \alpha \int_0^{\pi} \begin{pmatrix} \cos s + \sin s & \cos s - \sin s & 0 \\ \sin s - \cos s & \cos s + \sin s & 0 \\ 0 & 0 & \frac{1}{2}e^{-s} \end{pmatrix} \overline{B}_1(0) ds$$

$$= \frac{\alpha}{2} \int_0^{\pi} e^{-s} ds\, \overline{B}_1(0) = \frac{\alpha}{2}(1 - e^{-\pi})\overline{B}_1(0),$$

$$G_2 = \alpha \int_{\pi}^{2\pi} \begin{pmatrix} \cos s & -\sin s & 0 \\ \sin s & \cos s & 0 \\ 0 & 0 & e^{2\pi - s} \end{pmatrix} \overline{B}_1(0) ds$$

$$= \frac{\alpha}{2} \int_{\pi}^{2\pi} e^{2\pi - s} ds\, \overline{B}_1(0) = \alpha(e^{\pi} - 1)\overline{B}_1(0),$$

$$G_3 = \Phi(2\pi, \pi)P = p \begin{pmatrix} -1 & -1 & 0 \\ 1 & -1 & 0 \\ 0 & 0 & \frac{1}{2}e^{-\pi} \end{pmatrix} \overline{B}_1(0) = \frac{p}{2} e^{-\pi} \overline{B}_1(0),$$

where $\overline{B}_1(0) = \{(x_1, x_2, x_3)^T : x_1 = 0,\ x_2 = 0,\ |x_3| \le 1\}$.

Thus,

$$G = \left[\alpha\left(\frac{1}{2}(1 - e^{-\pi}) + (e^{\pi} - 1)\right) + \frac{p}{2}e^{-\pi}\right] \overline{B}_1(0) = \overline{B}_1(0).$$

Then Eq. (3.38) takes the form

$$R_0 = \begin{pmatrix} 1 & 1 & 0 \\ -1 & 1 & 0 \\ 0 & 0 & \frac{1}{2} \end{pmatrix} R_0 + \overline{B}_1(0).$$

By virtue of the arguments presented above, R_0 is a point in the plane (x_1, x_2), which is determined from the system

$$\begin{pmatrix} 0 & -1 \\ 1 & 0 \end{pmatrix} \begin{pmatrix} x_1 \\ x_2 \end{pmatrix} = \begin{pmatrix} 0 \\ 0 \end{pmatrix}.$$

Thus, $x_1 = 0$ and $x_2 = 0$. Equation (3.38) reduces to the following one:

$$\tilde{R}_0 = \frac{1}{2}\tilde{R}_0 + [-1, 1].$$

Therefore, $\tilde{R}_0 = [-2, 2]$, whence $R_0 = \overline{B}_2(0)$.

Considering the problem of existence of ordinary periodic solutions, we conclude that the matrix $E - \Phi(T, 0)$ is nondegenerate, periodic solutions exist for all $d \in \overline{B}_1(0)$, and

$$X_0 = \begin{pmatrix} 0 & -1 & 0 \\ 1 & 0 & 0 \\ 0 & 0 & \frac{1}{2} \end{pmatrix}^{-1} \overline{B}_1(0) = \overline{B}_2(0) = R_0.$$

(4) Let λ_1 be a complex root of multiplicity k. Then the cell J_1 takes the form

$$\begin{pmatrix} \begin{array}{cc} \operatorname{Re}\lambda_1 & -\operatorname{Im}\lambda_1 \\ \operatorname{Im}\lambda_1 & \operatorname{Re}\lambda_1 \end{array} & \mathbf{0} & \cdots & \mathbf{0} \\ e_1^1 E_2 & \begin{array}{cc} \operatorname{Re}\lambda_1 & -\operatorname{Im}\lambda_1 \\ \operatorname{Im}\lambda_1 & \operatorname{Re}\lambda_1 \end{array} & \cdots & \mathbf{0} \\ \mathbf{0} & \ddots & \ddots & \vdots \\ \mathbf{0} & \cdots & e_1^{k-1} E_2 & \begin{array}{cc} \operatorname{Re}\lambda_1 & -\operatorname{Im}\lambda_1 \\ \operatorname{Im}\lambda_1 & \operatorname{Re}\lambda_1 \end{array} \end{pmatrix}.$$

Consequently, similarly to the case of a simple complex root, the sets Z_0 and D must lie in the hyperplanes $x_1 = x_1^*$ and $x_2 = x_2^*$, as well as in the hyperplanes $x_{2i-1} = x_{2i-1}^*, x_{2i} = x_{2i}^*, i = \overline{2,k}$ (which is established by successively decreasing the dimension of the problem by analogy with the case of a multiple real root). Thus, Z_0 and D are sets of dimension $n - 2k$. For the determination of x_i^*, $i = \overline{1,2k}$, we obtain the system

$$\begin{pmatrix} x_1^* \\ \vdots \\ x_{2k}^* \end{pmatrix} = J_1 \begin{pmatrix} x_1^* \\ \vdots \\ x_{2k}^* \end{pmatrix} + \begin{pmatrix} d_1^* \\ \vdots \\ d_{2k}^* \end{pmatrix}. \tag{3.47}$$

Case $\Lambda = 1$ ([118]). Without loss of generality, we assume that

$$\max_{i=\overline{1,n}} |\lambda_i| = |\lambda_1|.$$

(1) Let λ_1 be simple and real, i.e., $\lambda_1 = \pm 1$. By virtue of (3.41), we have

$$\mathrm{pr}_{x_1} Z_0 = \mathrm{pr}_{x_1}(J Z_0 + D) = \mathrm{pr}_{x_1} J Z_0 + \mathrm{pr}_{x_1} D = \lambda_1 \mathrm{pr}_{x_1} Z_0 + \mathrm{pr}_{x_1} D.$$

If $\lambda_1 = 1$, then it is clear that a solution of Eq. (3.41) exists if and only if $\mathrm{pr}_{x_1} D = \{d_1^*\} = \{0\}$. In this case, $\mathrm{pr}_{x_1} Z_0$ is an arbitrary segment in $\mathbb{R}$.

Let $\lambda_1 = -1$. Denote $\mathrm{pr}_{x_1} Z_0 = [x_1^-, x_1^+]$ and $\mathrm{pr}_{x_1} D = [d_1^-, d_1^+]$. Then

$$x_1^- = -x_1^+ + d_1^-, \quad x_1^+ = -x_1^- + d_1^+,$$

whence $d_1^- = d_1^+ = d_1^*$, i.e., D can be considered as a set of dimension $n - 1$.

In this case, we have $\mathrm{pr}_{x_1} Z_0 = [x_1^*, d_1^* - x_1^*]$, where $x_1^* \le \frac{d_1^*}{2}$ is an arbitrary constant.

(2) Let $\lambda_1 = \pm 1$ be a root of multiplicity k. Then, by analogy with the previous case, we consider projections to the axis x_1 and establish that, for $\lambda_1 = 1$, the set D must lie in the hyperplane $x_1 = 0$, and $\mathrm{pr}_{x_1} Z_0$ is an arbitrary segment. For $\lambda_1 = -1$, the set D must lie in the hyperplane $x_1 = d_1^*$, and $\mathrm{pr}_{x_1} Z_0 = [x_1^*, d_1^* - x_1^*]$, where $x_1^* \le \frac{d_1^*}{2}$ is an arbitrary constant.

Let $e_1^1 = 0$. Then

$$\mathrm{pr}_{x_2} Z_0 = \mathrm{pr}_{x_2}(JZ_0 + D) = \mathrm{pr}_{x_2} JZ_0 + \mathrm{pr}_{x_2} D = \lambda_1 \mathrm{pr}_{x_2} Z_0 + \mathrm{pr}_{x_2} D.$$

Similarly to the previous case, for $\lambda_1 = 1$ the set D must lie in the hyperplane $x_2 = 0$, and $\mathrm{pr}_{x_2} Z_0$ is an arbitrary segment. For $\lambda_1 = -1$, the set D must lie in the hyperplane $x_2 = d_2^*$, and $\mathrm{pr}_{x_2} Z_0 = [x_2^*, d_2^* - x_2^*]$, where $x_2^* \le \frac{d_2^*}{2}$ is an arbitrary constant.

Now assume that $e_1^1 \neq 0$.

Let

$$e_{\nu,\mu} = \left(\frac{\nu}{\sqrt{\nu^2+\mu^2}}, \frac{\mu}{\sqrt{\nu^2+\mu^2}}, 0, \ldots, 0\right)^T \in \mathbb{R}^n$$

and let $x_{\nu,\mu}$ be the straight line in the hyperplane (x_1, x_2) that passes through the point $(0, 0)$ and is determined by the vector $e_{\nu,\mu}$. Then

$$\begin{aligned}
&\mathrm{mes}(\mathrm{pr}_{x_2} Z_0) = \mathrm{mes}(\mathrm{pr}_{x_2}(JZ_0 + D))\\
&= \mathrm{mes}(\mathrm{pr}_{x_2} JZ_0) + \mathrm{mes}(\mathrm{pr}_{x_2} D)\\
&= c(JZ_0, e_{0,1}) + c(JZ_0, -e_{0,1}) + \mathrm{mes}(\mathrm{pr}_{x_2} D)\\
&= c(Z_0, J^T e_{0,1}) + c(Z_0, -J^T e_{0,1}) + \mathrm{mes}(\mathrm{pr}_{x_2} D)\\
&= \left\{J^T e_{0,1} = (1, \lambda_1, 0, \ldots, 0)^T = \sqrt{1+\lambda_1^2}\, e_{1,\lambda_1} = \sqrt{2} e_{1,\lambda_1}\right\}\\
&= \sqrt{2}[c(Z_0, e_{1,\lambda_1}) + c(Z_0, -e_{1,\lambda_1})] + \mathrm{mes}(\mathrm{pr}_{x_2} D)\\
&= \sqrt{2}\,\mathrm{mes}(\mathrm{pr}_{x_{1,\lambda_1}} Z_0) + \mathrm{mes}(\mathrm{pr}_{x_2} D)\\
&= \sqrt{2}(\mathrm{mes}(\mathrm{pr}_{x_{1,\lambda_1}} JZ_0) + \mathrm{mes}(\mathrm{pr}_{x_{1,\lambda_1}} D)) + \mathrm{mes}(\mathrm{pr}_{x_2} D)\\
&= \sqrt{2}[c(Z_0, J^T e_{1,\lambda_1}) + c(z_0, -J^T e_{1,\lambda_1})] + \sqrt{2}\,\mathrm{mes}(\mathrm{pr}_{x_{1,\lambda_1}} D) + \mathrm{mes}(\mathrm{pr}_{x_2} D)\\
&= \Bigg\{J^T e_{1,\lambda_1} = \frac{1}{\sqrt{2}}(2\lambda_1, \lambda_1^2, 0, \ldots, 0)^T = \frac{\sqrt{5}}{\sqrt{2}} e_{2\lambda_1,\lambda_1^2};\\
&\qquad \mathrm{mes}(\mathrm{pr}_{x_{1,\lambda_1}} D) = \mathrm{mes}(\mathrm{pr}_{x_2} D)|\cos \widehat{e_{0,1}, e_{1,\lambda_1}}| = \frac{1}{\sqrt{2}}\mathrm{mes}(\mathrm{pr}_{x_2} D)\Bigg\}\\
&= \sqrt{5}\,\mathrm{mes}(\mathrm{pr}_{x_{2\lambda_1,\lambda_1^2}} Z_0) + 2\,\mathrm{mes}(\mathrm{pr}_{x_2} D) = \cdots\\
&= \sqrt{m^2+1}\,\mathrm{mes}(\mathrm{pr}_{x_{m\lambda_1^{m-1},\lambda_1^m}} Z_0) + m\,\mathrm{mes}(\mathrm{pr}_{x_2} D).
\end{aligned}$$

As $m \to \infty$, the straight line $x_{m\lambda_1^{m-1},\lambda_1^m}$ tends to the straight line x_1. Therefore, the limit

$$\lim_{m\to\infty} \sqrt{m^2+1}\ \mathrm{mes}(\mathrm{pr}_{x_{m\lambda_1^{m-1},\lambda_1^m}} Z_0)$$

exists if and only if $\mathrm{mes}(\mathrm{pr}_{x_1} Z_0) = 0$, i.e., in the case where $\mathrm{pr}_{x_1} Z_0 = \{x_1^*\}$. Thus, Z_0 can be considered as a set of dimension $n-1$. Furthermore, $\mathrm{mes}(\mathrm{pr}_{x_2} D) = 0$, whence $\mathrm{pr}_{x_2} D = \{d_2^*\}$.

Then

$$\mathrm{pr}_{x_2} Z_0 = \mathrm{pr}_{x_2}(JZ_0 + D) = \mathrm{pr}_{x_2} JZ_0 + d_2^* = \lambda_1 \mathrm{pr}_{x_2} Z_0 + x_1^* + d_2^*.$$

By analogy with the case of a simple root, we establish that, for $\lambda_1 = 1$, the set D must lie in the hyperplane $d_2^* = -x_1^*$, and $\mathrm{pr}_{x_2} Z_0$ is an arbitrary segment in $\mathbb{R}$. For $\lambda_1 = -1$, we have $\mathrm{pr}_{x_1} Z_0 = \{\frac{1}{2} d_1^*\}$ and $\mathrm{pr}_{x_2} Z_0 = [x_2^*, (d_2^* + x_1^*) - x_2^*]$, where $x_2^* \le \frac{d_2^*+x_1^*}{2}$ is an arbitrary constant.

Thus, $\mathrm{pr}_{x_i} D = \{d_i^*\}$ for all $i = \overline{1,k}$. Let $i_1, \ldots, i_{m-1}$ be the set of indices for which $e_1^i = 0$ and $i_m = k$. Then $\mathrm{pr}_{x_{i_j}} Z_0$ is an arbitrary segment that satisfies the equation $\mathrm{pr}_{x_{i_j}} Z_0 = \lambda_1 \mathrm{pr}_{x_{i_j}} Z_0 + x_{i_j-1}^* + d_{i_j}^*$ if $e_1^{i_j-1} = 1$ and $\mathrm{pr}_{x_{i_j}} Z_0 = \lambda_1 \mathrm{pr}_{x_{i_j}} Z_0 + d_{i_j}^*$ if $e_1^{i_j-1} = 0$ (we assume here that $e_1^0 = 0$). For the other indices, we have $\mathrm{pr}_{x_j} Z_0 = \{x_j^*\}$, where

$$x_j^* = \begin{cases} -d_{j+1}^* & \text{if } \lambda_1 = 1, \\ \frac{x_{j-1}^*+d_j^*}{2} & \text{if } \lambda_1 = -1 \text{ and } j \ne i_q + 1,\ q = \overline{1,m-1}, \\ \frac{d_j^*}{2} & \text{if } \lambda_1 = -1 \text{ and } j = i_q + 1,\ q = \overline{1,m-1}. \end{cases}$$

By virtue of Eq. (3.41), we get

$$c(Z_0, \psi) = c(JZ_0 + D, \psi) = c(Z_0, J^T \psi) + c(D, \psi) \tag{3.48}$$

for all $\psi = (\psi_1, \ldots, \psi_k, 0, \ldots)^T \in \mathbb{R}^n$.

Let $\lambda_1 = 1$. We rewrite equality (3.48) as follows:

$$c(Z_0, \psi) = \max_{x\in Z_0} \sum_{i=1}^{k} x_i \psi_i = \max(x_{i_1}\psi_{i_1} + \cdots + x_{i_m}\psi_{i_m}) + \sum_{\substack{i\in\overline{1,k}\\ i\ne i_q}} x_i^* \psi_i,$$

$$c(Z_0, J^T\psi) + c(D, \psi) = \max(x_{i_1}\psi_{i_1} + \cdots + x_{i_m}\psi_{i_m})$$
$$+ \sum_{\substack{i\in\overline{1,k}\\ i\ne i_q}} x_i^*(\psi_i + \psi_{i+1}) + \sum_{i=1}^{k} d_i^* \psi_i.$$

Since $d_1^* = 0$, $d_{i_q+1}^* = 0$, $q \in \overline{1, m-1}$, and $x_i^* = -d_{i+1}^*$ for the other indices, it follows from equality (3.48) that the projection of the set Z_0 in the space $x_{i_1}, \ldots, x_{i_m}$ is an arbitrary convex compact set.

Let $\lambda_1 = -1$ and let the set Z be such that

$$Z_0 = Z + (x_1^*, \ldots, x_k^*, 0, \ldots, 0)^T,$$

where $(x_1^*, \ldots, x_k^*)^T$ is a solution of system (3.45). Then Eq. (3.48) is equivalent to the equation

$$c(Z, \psi) = c(Z, J^T \psi),$$

or

$$\begin{aligned}&\max((x_{i_1} - x_{i_1}^*)\psi_{i_1} + \cdots + (x_{i_m} - x_{i_m}^*)\psi_{i_m})\\ &= \max\{-((x_{i_1} - x_{i_1}^*)\psi_{i_1} + \cdots + (x_{i_m} - x_{i_m}^*)\psi_{i_m})\}.\end{aligned}$$

Thus, the projection of the set Z_0 in the space $x_{i_1}, \ldots, x_{i_m}$ is an arbitrary convex compact set symmetric with respect to the point $(x_{i_1}^*, \ldots, x_{i_m}^*)$.

(3) Let λ_1 be a simple complex root. Then, by virtue of Eq. (3.41), we obtain

$$\mathrm{pr}_{(x_1,x_2)} Z_0 = \mathrm{pr}_{(x_1,x_2)}(J Z_0 + D) = J_1 \mathrm{pr}_{(x_1,x_2)} Z_0 + \mathrm{pr}_{(x_1,x_2)} D. \tag{3.49}$$

In this case, J_1 is the matrix of rotation by an angle φ: $\cos\varphi = \operatorname{Re}\lambda_1$, $\sin\varphi = \operatorname{Im}\lambda_1$. It is obvious that $\mathrm{mes}(\mathrm{pr}_{(x_1,x_2)} D) = 0$, i.e., $\mathrm{pr}_{(x_1,x_2)} D$ is a segment because

$$\begin{aligned}\mathrm{mes}(\mathrm{pr}_{(x_1,x_2)} Z_0) &= \mathrm{mes}(J_1 \mathrm{pr}_{(x_1,x_2)} Z_0 + \mathrm{pr}_{(x_1,x_2)} D)\\ &\geq \mathrm{mes}(J_1 \mathrm{pr}_{(x_1,x_2)} Z_0) + \mathrm{mes}(\mathrm{pr}_{(x_1,x_2)} D)\\ &= \mathrm{mes}(\mathrm{pr}_{(x_1,x_2)} Z_0) + \mathrm{mes}(\mathrm{pr}_{(x_1,x_2)} D).\end{aligned}$$

We have $Z_0 \in \mathrm{conv}(\mathbb{R}^n)$. Therefore, $\mathrm{pr}_{(x_1,x_2)} Z_0 \in \mathrm{conv}(\mathbb{R}^2)$, i.e., $\mathrm{pr}_{(x_1,x_2)} Z_0$ is either a segment or a solid set. Hence, this set contains a segment AB of maximum length r. If $\mathrm{pr}_{(x_1,x_2)} D$ is not a singleton, then the set $J_1 \mathrm{pr}_{(x_1,x_2)} Z_0 + \mathrm{pr}_{(x_1,x_2)} D$ contains a segment of length greater than r (the major diagonal of the parallelogram $AB + \mathrm{pr}_{(x_1,x_2)} D$), which is impossible by virtue of (3.49).

Thus, $\mathrm{pr}_{(x_1,x_2)} D = (d_1^*, d_2^*)^T$ and the set $\mathrm{pr}_{(x_1,x_2)} Z_0$ is such that

$$\mathrm{pr}_{(x_1,x_2)} Z_0 = J_1 \mathrm{pr}_{(x_1,x_2)} Z_0 + (d_1^*, d_2^*)^T. \tag{3.50}$$

Assume that a set $Z \in \mathrm{conv}(\mathbb{R}^2)$ is such that

$$\mathrm{pr}_{(x_1,x_2)} Z_0 = Z + (x_1^*, x_2^*)^T,$$

where $(x_1^*, x_2^*)^T$ is a solution of the system of equations

$$\begin{pmatrix} x_1 \\ x_2 \end{pmatrix} = J_1 \begin{pmatrix} x_1 \\ x_2 \end{pmatrix} + \begin{pmatrix} d_1^* \\ d_2^* \end{pmatrix}. \tag{3.51}$$

Then Eq. (3.50) takes the form $Z = J_1 Z$.

Let $x \in Z$ be such that $\|x\| = h(\{0\}, Z)$. The point $J_1 x$ obtained from the point x as a result of rotation by an angle φ about the point $(0,0)^T$ also belongs to the boundary of the set Z.

Thus, if $\varphi = 2\pi \frac{k}{m}$, where $\frac{k}{m}$ is an irreducible fraction, then the solutions of the equation are "pseudo-m-gons". The term "pseudo-m-gon" stands for a convex compact set in $\mathbb{R}^2$, which remains invariant under the rotations by the angle $\frac{2\pi}{m}$ about the origin. It is clear that a disk of any radius centered at the point $(0,0)^T$ is a special case of "pseudo-m-gons".

In view of the convexity of the set Z, it is clear that

$$Z_m \subset Z \subset B_{\|x\|}(0),$$

where Z_m is a regular m-gon with vertices at the points $x, J_1 x, \dots, J_1^{m-1} x$.

For $m = 6$, the corresponding example is presented by the set depicted in Figure 1.

If $\varphi \neq 2\pi \frac{k}{m}$, then the set of boundary points $\{J_1^i x,\ i \in \mathbb{N}\}$ is everywhere dense on a circle of radius $\|x\|$ centered at the point $(0,0)^T$. Hence, $Z = B_{\|x\|}(0)$.

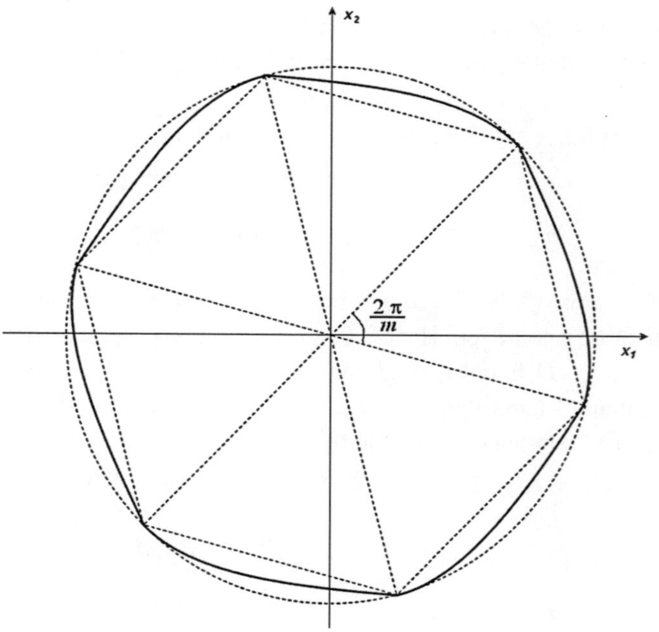

Figure 1.

(4) Let λ_1 be a complex root of multiplicity k.

By analogy with the previous case, we now consider the projections onto the hyperplane (x_1, x_2) and conclude that the set $\mathrm{pr}_{(x_1,x_2)} D$ must contain a single point. In this case, $\mathrm{pr}_{(x_1,x_2)} Z_0$ is an arbitrary convex compact set in $\mathbb{R}^2$ satisfying the equation

$$\mathrm{pr}_{(x_1,x_2)} Z_0 = \begin{pmatrix} \cos\varphi & -\sin\varphi \\ \sin\varphi & \cos\varphi \end{pmatrix} \mathrm{pr}_{(x_1,x_2)} Z_0 + (d_1^*, d_2^*)^T, \tag{3.52}$$

i.e., either a disk or a "pseudo-m-gon" whose center is located at the point $(x_1^*, x_2^*)^T$.

Consider the projections onto the hyperplane (x_3, x_4). By virtue of Eq. (3.41), we get

$$\mathrm{pr}_{(x_3,x_4)} Z_0 = \mathrm{pr}_{(x_3,x_4)}(J Z_0 + D) = \mathrm{pr}_{(x_3,x_4)} J Z_0 + \mathrm{pr}_{(x_3,x_4)} D.$$

If $e_1^1 = 0$, then, as in the previous case, we conclude that the set $\mathrm{pr}_{(x_3,x_4)} D$ must be one-point. In this case, $\mathrm{pr}_{(x_3,x_4)} Z_0$ is an arbitrary convex compact set in $\mathbb{R}^2$ satisfying an equation similar to (3.52), i.e., either a disk or a "pseudo-m-gon" centered at the point $(x_3^*, x_4^*)^T$.

Let $e_1^1 = 1$ and let $\mathrm{pr}_{x_1,\dots,x_4} A$ be the projection of the set $A \in \mathrm{conv}(\mathbb{R}^n)$ in the space $x_1, \dots, x_4$.

We choose a set Z such that

$$\mathrm{pr}_{x_1,\dots,x_4} Z_0 = Z + (x_1^*, x_2^*, 0, 0)^T,$$

where $(x_1^*, x_2^*)^T$ is a solution of system (3.51). Then

$$Z = J_* Z + D_1, \quad \text{where } D_1 \colon \mathrm{pr}_{x_1,\dots,x_4} D = D_1 + (d_1^*, d_2^*, 0, 0)^T,$$

$$J_* = \begin{pmatrix} \cos\varphi & -\sin\varphi & 0 & 0 \\ \sin\varphi & \cos\varphi & 0 & 0 \\ 1 & 0 & \cos\varphi & -\sin\varphi \\ 0 & 1 & \sin\varphi & \cos\varphi \end{pmatrix}.$$

If Z is a solution of this equation, then Z is also a solution of the equation

$$Z = J_*^m Z + (J_*^{m-1} D_1 + \cdots + D_1) \tag{3.53}$$

for all $m \in N$. The matrix J_*^i takes the form

$$J_*^i = \begin{pmatrix} \cos i\varphi & -\sin i\varphi & 0 & 0 \\ \sin i\varphi & \cos i\varphi & 0 & 0 \\ i\cos(i-1)\varphi & -i\sin(i-1)\varphi & \cos i\varphi & -\sin i\varphi \\ i\sin(i-1)\varphi & i\cos(i-1)\varphi & \sin i\varphi & \cos i\varphi \end{pmatrix}.$$

Furthermore, according to the properties of the operations in $\operatorname{conv}(\mathbb{R}^n)$, we conclude that

$$J_*^{m-1}D_1 + \cdots + D_1 \supset (J_*^{m-1} + \cdots + J + E)D_1$$
$$= \begin{pmatrix} \sum_{i=0}^{m-1}\cos i\varphi & -\sum_{i=1}^{m-1}\sin i\varphi & 0 & 0 \\ \sum_{i=1}^{m-1}\sin i\varphi & \sum_{i=0}^{m-1}\cos i\varphi & 0 & 0 \\ \sum_{i=1}^{m-1} i\cos(i-1)\varphi & -\sum_{i=1}^{m-1} i\sin(i-1)\varphi & \sum_{i=0}^{m-1}\cos i\varphi & -\sum_{i=1}^{m-1}\sin i\varphi \\ \sum_{i=1}^{m-1} i\sin(i-1)\varphi & \sum_{i=1}^{m-1} i\cos(i-1)\varphi & \sum_{i=1}^{m-1}\sin i\varphi & \sum_{i=0}^{m-1}\cos i\varphi \end{pmatrix} D_1.$$

Thus, for any $x = (x_1, x_2, x_3, x_4)^T \in Z$ and $d = (0, 0, d_3, d_4)^T \in D_1$, the point

$$y = (y_1, \ldots, y_4)^T = J_*^m x + (J_*^{m-1} + \cdots + E)d \in Z$$

for all $m \in \mathbb{N}$. In this case,

$$y_3 = mx_1\cos(m-1)\varphi - mx_2\sin(m-1)\varphi + x_3\cos m\varphi - x_4\sin m\varphi$$
$$+ d_3\sum_{i=0}^{m-1}\cos i\varphi - d_4\sum_{i=1}^{m-1}\sin i\varphi,$$
$$y_4 = mx_1\sin(m-1)\varphi + mx_2\cos(m-1)\varphi + x_3\sin m\varphi + x_4\cos m\varphi$$
$$+ d_3\sum_{i=1}^{m-1}\sin i\varphi + d_4\sum_{i=0}^{m-1}\cos i\varphi.$$

Moreover, if $x_1^2 + x_2^2 > 0$, then

$$y_3 = m\sqrt{x_1^2 + x_2^2}\cos((m-1)\varphi + \beta_{12}) + x_3\cos m\varphi - x_4\sin m\varphi$$
$$+ d_3\frac{\sin\frac{m+1}{2}\varphi\cos\frac{m}{2}\varphi}{\sin\frac{\varphi}{2}} - d_4\frac{\sin\frac{m+1}{2}\varphi\sin\frac{m}{2}\varphi}{\sin\frac{\varphi}{2}},$$
$$y_4 = m\sqrt{x_1^2 + x_2^2}\sin((m-1)\varphi + \beta_{12}) + x_3\sin m\varphi + x_4\cos m\varphi$$
$$+ d_3\frac{\sin\frac{m+1}{2}\varphi\sin\frac{m}{2}\varphi}{\sin\frac{\varphi}{2}} + d_4\frac{\sin\frac{m+1}{2}\varphi\cos\frac{m}{2}\varphi}{\sin\frac{\varphi}{2}},$$
$$\cos\beta_{12} = \frac{x_1}{\sqrt{x_1^2 + x_2^2}}, \quad \sin\beta_{12} = \frac{x_2}{\sqrt{x_1^2 + x_2^2}}.$$

In this case, $y_3^2 + y_4^2 \to \infty$ as $m \to \infty$, which is impossible because the set $Z \in \operatorname{conv}(\mathbb{R}^4)$.

Hence, equality (3.53) is possible only in the case $\mathrm{pr}_{(x_1,x_2)} Z = (0,0)^T$, i.e., $\mathrm{pr}_{(x_1,x_2)} Z_0 = (x_1^*, x_2^*)^T$.

Thus, Eq. (3.52) is reduced to

$$\mathrm{pr}_{(x_3,x_4)} Z_0 = (x_1^*, x_2^*)^T + \begin{pmatrix} \cos\varphi & -\sin\varphi \\ \sin\varphi & \cos\varphi \end{pmatrix} \mathrm{pr}_{(x_3,x_4)} Z_0 + \mathrm{pr}_{(x_3,x_4)} D.$$

As in the case of a simple complex root, we obtain $\mathrm{pr}_{(x_3,x_4)} D = (d_3^*, d_4^*)^T$, and the set $\mathrm{pr}_{(x_3,x_4)} Z_0$ is either a "pseudo-m-gon" or a disk centered at the point $(x_3^*, x_4^*)^T$.

Further, we consider the projections onto the hyperplanes (x_{2i-1}, x_{2i}), $i \le k$, etc. This yields $\mathrm{pr}_{(x_{2i-1},x_{2i})} D = (d_{2i-1}^*, d_{2i}^*)^T$ for all $i = \overline{1,k}$. Let $i_1, \ldots, i_{m-1}$ be the set of indices such that $e_1^i = 0$ and $i_m = k$.

Then $\mathrm{pr}_{(x_{2i_j-1},x_{2i_j})} Z_0$, $j \in \overline{1,m}$, is an arbitrary convex compact set in $\mathbb{R}^2$ satisfying an equation similar to (3.52), i.e., a disk or a "pseudo-m-gon" centered at the point $(x_{2i_j-1}^*, x_{2i_j}^*)^T$. For all remaining indices, we get $\mathrm{pr}_{(x_{2i-1},x_{2i})} Z_0 = (x_{2i-1}^*, x_{2i}^*)^T$.

Thus, we arrive at the following system for the determination of x_i^*, $i = \overline{1,2k}$:

$$\begin{pmatrix} x_1^* \\ \vdots \\ x_{2k}^* \end{pmatrix} = J_1 \begin{pmatrix} x_1^* \\ \vdots \\ x_{2k}^* \end{pmatrix} + \begin{pmatrix} d_1^* \\ \vdots \\ d_{2k}^* \end{pmatrix}.$$

Hence, by virtue of Eq. (3.41),

$$c(Z_0, \psi) = c(JZ_0 + D, \psi) = c(Z_0, J^T\psi) + c(D, \psi) \tag{3.54}$$

for all $\psi = (\psi_1, \ldots, \psi_k, 0, \ldots)^T \in \mathbb{R}^n$.

Assume that the set Z is such that

$$Z_0 = Z + (x_1^*, \ldots, x_{2k}^*, 0, \ldots, 0)^T.$$

Then Eq. (3.54) is equivalent to the equation

$$c(Z, \psi) = c(Z, J^T\psi) \tag{3.55}$$

for all $\psi = (\psi_1, \ldots, \psi_k, 0, \ldots)^T$.

We rewrite this equation in a more detailed form:

$$\begin{aligned} c(Z,\psi) &= \max_{x \in \tilde{Z}} \sum_{i=1}^{2k} x_i \psi_i \\ &= \max\{(x_{2i_1-1} - x_{2i_1-1}^*)\psi_{2i_1-1} + (x_{2i_1} - x_{2i_1}^*)\psi_{2i_1} + \cdots \\ &\quad + (x_{2i_m-1} - x_{2i_m-1}^*)\psi_{2i_m-1} + (x_{2i_m} - x_{2i_m}^*)\psi_{2i_m}\}, \end{aligned}$$

$$\begin{aligned}c(Z, J^T\psi) = \max\{&(\psi_{2i_1-1}\cos\varphi + \psi_{2i_1}\sin\varphi)(x_{2i_1-1} - x^*_{2i_1-1})\\ &+ (-\psi_{2i_1-1}\sin\varphi + \psi_{2i_1}\cos\varphi)(x_{2i_1} - x^*_{2i_1}) + \cdots\\ &+ (\psi_{2i_m-1}\cos\varphi + \psi_{2i_m}\sin\varphi)(x_{2i_m-1} - x^*_{2i_m-1})\\ &+ (-\psi_{2i_m-1}\sin\varphi + \psi_{2i_m}\cos\varphi)(x_{2i_m} - x^*_{2i_m})\}.\end{aligned}$$

Thus, equality (3.55) means that the projection of the set Z_0 in the space

$$x_{2i_1-1}, x_{2i_1}, \ldots, x_{2i_s-1}, x_{2i_s}$$

is an arbitrary convex compact set invariant under rotations by an angle φ in the hyperplanes (x_{2i_j-1}, x_{2i_j}), $j = \overline{1,m}$, about the points $(x^*_{2i_j-1}, x^*_{2i_j})^T$.

Consider the general case. Let $\lambda_1, \ldots, \lambda_m$ be different eigenvalues of the matrix $\Phi(T, 0)$ whose moduli are equal to 1 and let k_j, $j = \overline{1,m}$, be the dimensions of the corresponding Jordan cells J_j.

We now find the projection Z_1 of the set Z_0 in the space $x_1, \ldots, x_{m_0}$, where $m_0 = \sum_{j=1}^{m} k_j$. As shown above, the set Z_1 "degenerates" into points in a part of variables. In the variables corresponding to $\lambda = -1$, the set Z_1 is symmetric about the given point and, in the couples of variables corresponding to roots of the form $\lambda = \cos\varphi + i\sin\varphi$, the set Z_1 is invariant under rotations by the angle φ about the given point.

We now choose an arbitrary point $x^*_1, \ldots, x^*_{m_0}$ from the set Z_1. Consider a section of the set Z_0 by the hyperplanes $x_1 = x^*_1, \ldots, x_{m_0} = x^*_{m_0}$. By using the structure of the matrix J and the set D, we arrive at the following equation for the given section:

$$Z_2 = \tilde{J} Z_2 + \tilde{D}, \tag{3.56}$$

where $\tilde{D}$ is a section of the set D [as shown above, the set $\tilde{D}$ does not depend on the choice of the point $(x^*_1, \ldots, x^*_{m_0})$] and the matrix $\tilde{J}$ contains solely the Jordan cells of the matrix J corresponding to eigenvalues whose moduli are smaller than 1. As indicated above, since $\rho(\tilde{J}) < 1$, Eq. (3.56) possesses a unique solution Z_2 independent of the choice of the point $x^*_1, \ldots, x^*_{m_0}$. Thus, $Z_0 = Z_1 \times Z_2$.

Example 11. Consider a linear impulsive differential inclusion of the form

$$\begin{pmatrix}\dot{x}_1\\ \dot{x}_2\\ \dot{x}_3\end{pmatrix} \in \begin{pmatrix}1 & 0 & 0\\ 0 & 1 & 0\\ 0 & 0 & \frac{1}{2\pi}\end{pmatrix}\begin{pmatrix}x_1\\ x_2\\ x_3\end{pmatrix} + \alpha(1+\cos t)\overline{B}_1(0),$$

$$\left.\begin{pmatrix}\Delta x_1\\ \Delta x_2\\ \Delta x_3\end{pmatrix}\right|_{t=(2i-1)\pi} \in \begin{pmatrix}e^{-2\pi}-1 & 0 & 0\\ 0 & e^{-2\pi}-1 & 0\\ 0 & 0 & \frac{1}{2}e^{-1}-1\end{pmatrix}\begin{pmatrix}x_1\\ x_2\\ x_3\end{pmatrix} + p\overline{B}_1(0),$$

where

$$\alpha = \frac{1+4\pi^2}{16\pi^3(e^{\frac{3}{2}} - e^{-\frac{1}{2}})}, \qquad p = e^{\frac{1}{2}}.$$

The fundamental matrix takes the form

$$\Phi(2\pi,s)=\begin{cases}\begin{pmatrix} e^{-s} & 0 & 0\\ 0 & e^{-s} & 0\\ 0 & 0 & \frac{1}{2}e^{\frac{-s}{2\pi}}\end{pmatrix}, & s\le\pi,\\ \begin{pmatrix} e^{2\pi-s} & 0 & 0\\ 0 & e^{2\pi-s} & 0\\ 0 & 0 & e^{\frac{1}{2\pi}(2\pi-s)}\end{pmatrix}, & s>\pi.\end{cases}$$

In this case,

$$\begin{aligned}
G_1 &= \alpha\int_0^{\pi}\begin{pmatrix} e^{-s} & 0 & 0\\ 0 & e^{-s} & 0\\ 0 & 0 & \frac{1}{2}e^{-\frac{s}{2\pi}}\end{pmatrix}(1+\cos s)\overline{B}_1(0)ds\\
&= \frac{\alpha}{2}\int_0^{\pi} e^{-\frac{s}{2\pi}}(1+\cos s)ds\overline{B}_1(0)\\
&= \alpha\frac{4\pi}{1+4\pi^2}(2\pi^2(1-e^{-\frac{1}{2}})+1)\overline{B}_1(0),\\
G_2 &= \alpha\int_{\pi}^{2\pi}\begin{pmatrix} e^{2\pi-s} & 0 & 0\\ 0 & e^{2\pi-s} & 0\\ 0 & 0 & e^{\frac{2\pi-s}{2\pi}}\end{pmatrix}(1+\cos s)\overline{B}_1(0)ds\\
&= \alpha\int_{\pi}^{2\pi} e^{-\frac{2\pi-s}{2\pi}}(1+\cos s)ds\overline{B}_1(0)\\
&= \alpha e\frac{4\pi}{1+4\pi^2}(2\pi^2e^{\frac{1}{2}}-e^{-1}(1+2\pi^2))\overline{B}_1(0),\\
G_3 &= \Phi(2\pi,\pi)P = p\begin{pmatrix} e^{-2\pi}e^{\pi} & 0 & 0\\ 0 & e^{-2\pi}e^{\pi} & 0\\ 0 & 0 & \frac{1}{2e}e^{\frac{1}{2}}\end{pmatrix}\overline{B}_1(0)\\
&= \frac{p}{2}e^{-\frac{1}{2}}\overline{B}_1(0).
\end{aligned}$$

Hence,

$$\begin{aligned}
G = \Big[\alpha\Big(&\frac{4\pi}{1+4\pi^2}(2\pi^2(1-e^{-\frac{1}{2}})+1)\\
&+e\frac{4\pi}{1+4\pi^2}(2\pi^2e^{\frac{1}{2}}-e^{-1}(1+2\pi^2))\Big)+\frac{p}{2}e^{-\frac{1}{2}}\Big]\overline{B}_1(0)=\overline{B}_1(0).
\end{aligned}$$

Therefore, Eq. (3.38) takes the form

$$R_0=\begin{pmatrix}1&0&0\\0&1&0\\0&0&\frac{1}{2}\end{pmatrix}R_0+\overline{B}_1(0).$$

The eigenvalues $\lambda_{1,2} = 1$ and the set $\overline{B}_1(0)$ belongs to the hyperplanes $x_1 = 0$ and $x_2 = 0$. Hence, $\mathrm{pr}_{(x_1,x_2)} R_0$ is an arbitrary convex compact set in R^2, and the analyzed equation is reduced to

$$\overline{R}_0 = \frac{1}{2}\overline{R}_0 + [-1, 1], \quad \overline{R}_0 \in \mathbb{R}.$$

The solution of the last equation is $\overline{R}_0 = [-2, 2]$ and, thus,

$$R_0 = \mathrm{pr}_{(x_1,x_2)} R_0 \times [-2, 2].$$

As for the existence of ordinary periodic solutions, we note that the matrix $E - \Phi(T, 0)$ is degenerate but periodic solutions exist for all $d \in \overline{B}_1(0)$ because

$$\mathrm{rang}\begin{pmatrix} 0 & 0 & 0 & \vline & 0 \\ 0 & 0 & 0 & \vline & 0 \\ 0 & 0 & \frac{1}{2} & \vline & d_3 \end{pmatrix} = \mathrm{rang}\begin{pmatrix} 0 & 0 & 0 \\ 0 & 0 & 0 \\ 0 & 0 & \frac{1}{2} \end{pmatrix} = 1.$$

In this case, the system for x_0 degenerates into the equation

$$\frac{1}{2}x_3 = d_3, \quad d_3 \in [-1, 1]$$

and, therefore,

$$X_0 = \{(x_1, x_2, x_3)^T : x_1, x_2 \in R,\ x_3 \in [-2, 2]\}.$$

Remark 8. If $\Lambda \geq 1$ and there exists $\lambda_i = 1$, then ordinary periodic solutions exist for all $d \in D$ provided that a periodic R-solution exists.

Indeed, the rank of the extended matrix

$$(E - J, d) = \left(\begin{array}{cccc|c|c} 0 & 0 & \cdots & 0 & \mathbf{0} & d_1 \\ -e_i^1 & 0 & \cdots & 0 & \mathbf{0} & d_2 \\ \vdots & \ddots & \ddots & \vdots & \vdots & \vdots \\ 0 & \cdots & e_i^{k-1} & 0 & \mathbf{0} & d_k \\ \hline \mathbf{0} & \mathbf{0} & \cdots & \mathbf{0} & \ddots & \vdots \end{array}\right)$$

is equal to the rank of the matrix $E - J$ in view of the fact that $d_1 = 0$ and $d_{j+1} = 0$ for all $j \in \{1, \ldots, k-1\}$ such that $e_i^j = 0$. Otherwise, $d_j = d_j^* = \mathrm{const}$.

Relationship Between the Sets X_0 and R_0. As shown above, it is clear that Eq. (3.38) may have more than one solution. Therefore, we consider a set R^* defined as the union of the sets R_0 specified by Eq. (3.38).

We now show that $X_0 \subset R^*$ provided that R^* exists.

By using representation (3.40) for the matrix $\Phi(T, 0)$, by analogy with (3.41), we conclude that Eq. (3.39) is equivalent to the equation

$$y = Jy + d, \quad \text{where } d \in D, \ y = Mx_0.$$

In view of the "cellular" structure of the matrix J, this equation decomposes into s independent equations of the form

$$y^i = J_i y^i + d^i, \quad i = \overline{1, s}. \tag{3.57}$$

For the numbers i such that $|\lambda_i| > 1$, the vector d^i is chosen from a one-point set. Moreover, by virtue of systems (3.45), (3.46), and (3.47), the sets $Y_0 = MX_0$ and $Z^* = MR^*$ coincide in the variables corresponding to a given eigenvalue. In addition, they can be regarded as sets of lower dimensions.

Let i be such that $|\lambda_i| = 1$. Then the vector d^i is also chosen from a one-point set, the component of the set Z_0 in the corresponding variables is not unique, and the union of the sets Z_0 covers the entire space in a part of variables (for which the property of uniqueness is absent).

At the same time, Y_0 is a one-point set in the case where $\lambda_i \neq 1$ and, hence, Y_0 is a strict subset of Z^* in the variables corresponding to a given eigenvalue. However, if $\lambda_i = 1$, then a part of variables in system (3.57) turns out to be free (these are the variables for which Z_0 is not unique). Therefore, Y_0 coincides with Z^* in the variables corresponding to the given eigenvalue.

Consider the set I of all i such that $|\lambda_i| < 1$. By analogy with (3.42), Eqs. (3.57) with $i \in I$ are equivalent to the equation

$$y^I(\varepsilon) = J_I(\varepsilon) y^I(\varepsilon) + d^I(\varepsilon), \tag{3.58}$$

where $d^I(\varepsilon) \in D^I(\varepsilon) = LD^I$, $y^I(\varepsilon) = Ly^I$, y^I is a vector formed by the vectors y^i, $i \in I$, and D^I and J_I are defined similarly.

In these variables, for any $d^I(\varepsilon) \in D^I(\varepsilon)$, the solution of Eq. (3.58) is unique because the matrix $E - J_I(\varepsilon)$ is nondegenerate. We now show that $Y(\varepsilon) = LY_0$ is a subset of the set $Z^*(\varepsilon) = LZ^*$ in the corresponding variables.

The set $Z^I(\varepsilon)$ is a nonempty convex compact set. Hence, this set is a complete metric space. A mapping $\phi(y) = J_I(\varepsilon)y + d^I(\varepsilon)$, $d^I(\varepsilon) \in D^I(\varepsilon)$ is contracting because

$$\begin{aligned} \rho(\phi(y_1), \phi(y_2)) &= \rho(J_I(\varepsilon)y_1 + d^I(\varepsilon), J_I(\varepsilon)y_2 + d^I(\varepsilon)) \\ &\leq \|J_I(\varepsilon)\| \rho(y_1, y_2), \end{aligned}$$

and $\|J_I(\varepsilon)\| < 1$. Therefore, ϕ has a unique fixed point in $Z^I(\varepsilon)$. Thus, in the corresponding variables, $Y(\varepsilon) \subset Z^*(\varepsilon)$ and, hence, $X_0 \subset R^*$. As follows from Example 7, the sets $Y(\varepsilon)$ and $Z^*(\varepsilon)$ do not coincide in the general case.

Remark 9. The accumulated results can readily be extended to the case of measurable $A(t)$ and $F(t)$.

Remark 10. A linear differential inclusion ($B_i = 0$, $P_i = 0$ for all i) and a linear discrete inclusion ($A(t) \equiv 0$, $F(t) \equiv 0$) are special cases of the impulsive differential inclusion (3.37).

Remark 11. If $F(t)$ and P_i are one-point sets [i.e., (3.37) is a linear impulsive differential equation], then Eqs. (3.38) and (3.39) specify different objects: Eq. (3.39) specifies a periodic solution, whereas Eq. (3.38) specifies a periodic bundle of solutions.

Example 12. Consider an impulsive differential equation

$$\dot{x} = 0, \quad t \neq 2k - 1,$$
$$\Delta x|_{t=2k-1} = -2x.$$

This equation possesses a unique 2-periodic ordinary solution $x(t,0) \equiv 0$ and a one-parameter family of 2-periodic bundles of solutions $R(t,[-a,a]) = [-a,a]$, $a \geq 0$.

Remark 12. The accumulated results can be extended to the case where the segment $[0,T]$ contains infinite many times of pulses. In this case, for the set R_0, we get an equation of the form (3.38) with the sole difference that the sum in the definition of the set G contains infinitely many terms [95].

We now consider sufficient conditions for the existence of periodic R-solutions of linear inhomogeneous periodic differential inclusions with pulse action of the form

$$\dot{x} \in \mathcal{A}(t)x + F(t), \quad t \neq \tau_i, \tag{3.59}$$
$$\Delta x|_{t=\tau_i} \in \mathcal{B}_i x + P_i, \tag{3.60}$$

where $t \in \mathbb{R}$ is time, $x \in \mathbb{R}^n$ is the phase vector, $\mathcal{A}: \mathbb{R} \to \text{comp}(\mathbb{R}^{n\times n})$ and $F: \mathbb{R} \to \text{comp}(\mathbb{R}^n)$ are measurable T-periodic set-valued mappings, $|\mathcal{A}(t)| \leq \alpha(t)$, $|F(t)| \leq \mu(t)$, where $\alpha(t)$ and $\mu(t)$ are summable on $[0,T]$, and the compact sets $\mathcal{B}_i$ of $(n\times n)$-matrices, sets $P_i \in \text{conv}(\mathbb{R}^n)$, and times τ_i are such that

$$\mathcal{B}_{i+r} = \mathcal{B}_i, \quad P_{i+r} = P_i, \quad \tau_{i+r} = \tau_i + T$$

for all $i \in \mathbb{Z}$ and some natural r.

We also assume that $0 \leq \tau_1 < \cdots < \tau_r < T$ and $\det(E + B_i) \neq 0$ for all $B_i \in \mathcal{B}_i$, $i = \overline{1,r}$.

We now prove several auxiliary assertions.

Let $\Phi_A(t,s)$ be the matrizant of system (3.59) corresponding to a measurable branch $A(t)$ of the set-valued mapping $\mathcal{A}(t)$, i.e., the solution of the matrix Cauchy problem

$$\frac{dX}{dt} = A(t)X, \quad X(s,s) = E. \tag{3.61}$$

Lemma 2. *A set of matrizants* $\Phi_{\mathcal{A}}(t,s) = \{\Phi_A(t,s)\colon A(t) \in \mathcal{A}(t)\}$ *is a nonempty compact set in the space* $\mathbb{R}^{n\times n}$ *for any fixed* $t, s \in \mathbb{R}$, $t \geq s$.

Proof. We choose arbitrary real t and s such that $t \geq s$. The set $\Phi_{\mathcal{A}}(t,s)$ is nonempty because, by virtue of the Filippov theorem, there exists a summable branch of the set-valued mapping $\mathcal{A}(t)$ and, by virtue of the Carathéodory theorem for linear systems, there exists a solution of the matrix problem (3.61) .

We now show that the set $\Phi_{\mathcal{A}}(t,s)$ is bounded. The matrix $\Phi_A(t,s)$ is represented in the form

$$\Phi_A(t,s) = E + \int_s^t A(t_1)dt + \int_s^t A(t_1)\int_s^{t_1} A(t_2)dt_2dt_1 + \cdots .$$

As a result, we get the following sequence of estimates:

$$\begin{aligned}\|\Phi_A(t,s)\| &\leq 1 + \int_s^t \|A(t_1)\|dt + \int_s^t \|A(t_1)\| \int_s^{t_1} \|A(t_2)\|dt_2dt_1 + \cdots \\ &\leq 1 + \int_s^t \alpha(t_1)dt + \int_s^t \alpha(t_1)\int_s^{t_1} \alpha(t_2)dt_2dt_1 + \cdots .\end{aligned}$$

The validity of the inequality

$$\int_s^{\tilde{t}} \alpha(t_1)\int_s^{t_1} \alpha(t_2)\cdots\int_s^{t_{k-1}} \alpha(t_k)dt_k \ldots dt_2dt_1 \leq \frac{\gamma^k(\tilde{t},s)}{k!} \tag{3.62}$$

for all $\tilde{t} \in (s,t]$ is proved by induction.

For $k=1$, this inequality is true. Assume that (3.62) holds for $k=m$. Then, for $k=m+1$, we get

$$\begin{aligned}&\int_s^{\tilde{t}} \alpha(t_1)\int_s^{t_1} \alpha(t_2)\cdots\int_s^{t_m} \alpha(t_{m+1})dt_{m+1} \ldots dt_2dt_1 \\ &\leq \int_s^{\tilde{t}} \alpha(t_1)\frac{\gamma^m(t_1,s)}{m!}dt_1 = \frac{\gamma^{m+1}(\tilde{t},s)}{(m+1)!},\end{aligned}$$ □

Thus,

$$\|\Phi_A(t,s)\| \leq \sum_{k=0}^{\infty} \frac{\gamma^k(t,s)}{k!} = e^{\gamma(t,s)}, \tag{3.63}$$

and the boundedness of the set $\Phi_{\mathcal{A}}(t,s)$ is proved.

We now show that the set $\Phi_{\mathcal{A}}(t,s)$ is closed, i.e., the limit of any convergent sequence of matrices $\Phi_{A_k}(t,s) \in \Phi_{\mathcal{A}}(t,s)$ also belongs to the set $\Phi_{\mathcal{A}}(t,s)$. In view of the equivalence of the differential equation to a Volterra integral equation, we get the following representation:

$$\Phi_{A_k}(t,s) = E + \int_s^t A_k(\sigma)\Phi_{A_k}(\sigma,s)d\sigma \tag{3.64}$$

By analogy with (3.63), we get

$$\|\Phi_{A_k}(\sigma,s)\| \leq e^{\gamma(t,s)}$$

for all $\sigma \in [s,t]$. Since

$$\begin{aligned}\|\Phi_{A_k}(t_2,s) - \Phi_{A_k}(t_1,s)\| &= \left\| \int_{t_1}^{t_2} A_k(\sigma)\Phi_{A_k}(\sigma,s)d\sigma \right\| \\ &\leq \left| \int_{t_1}^{t_2} \alpha(\sigma)e^{\gamma(t,s)}d\sigma \right| = e^{\gamma(t,s)}|\gamma(t_2,s) - \gamma(t_1,s)|,\end{aligned}$$

where the function $\gamma(\sigma,s)$ is absolutely continuous on $[s,t]$, for any $\varepsilon > 0$, one can find $\delta(\varepsilon) > 0$ such that the inequality

$$|\gamma(t_2,s) - \gamma(t_1,s)| \leq \varepsilon e^{-\gamma(t,s)}$$

is true for any $t_1, t_2 \in [s,t]$: $|t_2 - t_1| < \delta$ and, hence,

$$\|\Phi_{A_k}(t_2,s) - \Phi_{A_k}(t_1,s)\| < \varepsilon.$$

Thus, the sequence of functions $\Phi_{A_k}(\sigma,s)$ is uniformly bounded and equicontinuous on $[s,t]$. Hence, by the Arzelà theorem, this sequence contains a subsequence uniformly convergent to a continuous matrix function $\Phi_*(\sigma,s)$.

This means that, for any $\varepsilon > 0$, one can find k_0 such that the inequality

$$\|\Phi_{A_k}(\sigma,s) - \Phi * (\sigma,s)\| < \frac{\varepsilon}{\gamma(t,s)}$$

holds for all $k > k_0$ and $\sigma \in [s,t]$.

Since

$$\begin{aligned}\int_s^t A_k(\sigma)\Phi_{A_k}(\sigma,s)d\sigma &= \int_s^t A_k(\sigma)[\Phi_{A_k}(\sigma,s) - \Phi_*(\sigma,s)]d\sigma \\ &\quad + \int_s^t A_k(\sigma)\Phi_*(\sigma,s)d\sigma,\end{aligned}$$

where

$$\left\| \int_s^t A_k(\sigma)[\Phi_{A_k}(\sigma,s) - \Phi_*(\sigma,s)]d\sigma \right\|$$
$$\leq \int_s^t \alpha(\sigma)\|\Phi_{A_k}(\sigma,s) - \Phi_*(\sigma,s)\|d\sigma < \varepsilon,$$

and, by virtue of the Lyapunov theorem, there exists a subsequence $A_{k_1}(\sigma)$ of the sequence $A_k(\sigma)$ weakly convergent to the matrix $A_*(\sigma) \in \mathcal{A}(\sigma)$ on $[s,t]$, we conclude that

$$\int_s^t A_{k_1}(\sigma)\Phi_{A_k}(\sigma,s)d\sigma \to \int_s^t A_*(\sigma)\Phi_*(\sigma,s)d\sigma \quad \text{as } k_1 \to \infty.$$

Passing to the limit in (3.64), we find

$$\Phi_*(t,s) = E + \int_s^t A_*(\sigma)\Phi_*(\sigma,s)d\sigma,$$

i.e., $\Phi_*(t,s) = \Phi_{A_*}(t,s) \in \Phi_{\mathcal{A}}(t,s)$, as required.

Thus, the set $\Phi_{\mathcal{A}}(t,s) \in \text{comp}(\mathbb{R}^{n\times n})$.

Let $\Phi_{AB_i}(t,s)$ be the matrizant of system (3.59), (3.60) corresponding to the matrices $A(t) \in \mathcal{A}(t)$, $B_i \in \mathcal{B}_i$, i.e., the solution of the matrix Cauchy problem for the system with pulse action

$$\dot{X} = A(t)X, \quad t \neq \tau_i,$$
$$\Delta X|_{t=\tau_i} = B_i X, \quad X(s,s) = E.$$

By virtue of [143], for the matrizant $\Phi_{AB_i}(t,s)$, we get

$$\Phi_{AB_i}(t,s) = \Phi_A(t,\tau_k)(E+B_k)\Phi_A(\tau_k,\tau_{k-1})\cdots(E+B_p)\Phi_A(\tau_p,s), \tag{3.65}$$
$$\tau_p < s \leq \tau_{p+1}, \quad \tau_k < t \leq \tau_{k+1}.$$

Lemma 3. *The set of matrizants $\Phi_{\mathcal{A}\mathcal{B}_i}(t,s) = \{\Phi_{AB_i}(t,s)\colon A(t) \in \mathcal{A}(t),\ B_i \in \mathcal{B}_i\}$ is a nonempty compact set in the space $\mathbb{R}^{n\times n}$ for any fixed $t,s \in \mathbb{R}$, $t \geq s$.*

Proof. We choose arbitrary real t and s such that $t \geq s$. For the sets $F, G \in \text{comp}(\mathbb{R}^{n\times n})$, we define the operation of multiplication as follows:

$$F \cdot G = \{f \cdot g \colon f \in F, g \in G\}.$$

It is clear that $F \cdot G \in \text{comp}(\mathbb{R}^{n\times n})$. Indeed, $F \cdot G$ is nonempty (in view of the fact that the sets F and G are nonempty) and bounded because, for any matrix $M \in F \cdot G$, there exist $f_0 \in F$ and $g_0 \in G$ such that $M = f_0 \cdot g_0$ and, hence,

$$\|M\| \leq \|f_0\| \cdot \|g_0\| \leq |F| \cdot |G| < \infty,$$

where $|F|$ is the modulus of the set F.

We now show that the set $F \cdot G$ is closed. We choose an arbitrary sequence of matrices $M_k \in F \cdot G$ convergent to a matrix $M_* \in \mathbb{R}^{n\times n}$. It is necessary to show that $M_* \in F \cdot G$. By the definition of the operation of multiplication of sets, for any k, there exist $f_k \in F$ and $g_k \in G$ such that the following representation is true: $M_k = f_k \cdot g_k$. Since the sets F and G are compact, there exist subsequences of the sequences $\{f_k\}$ and $\{g_k\}$ convergent to $f_* \in F$ and $g_* \in G$ respectively. Then $M_* = f_* g_* \in F \cdot G$. □

By using representation (3.65) for the matrizant, we represent the set $\Phi_{\mathcal{A}\mathcal{B}_i}(t,s)$ in the form

$$\Phi_{\mathcal{A}\mathcal{B}_i}(t,s) = \Phi_{\mathcal{A}}(t,\tau_k)(E+\mathcal{B}_k)\Phi_{\mathcal{A}}(\tau_k,\tau_{k-1})\cdots(E+\mathcal{B}_p)\Phi_{\mathcal{A}}(\tau_p,s),$$
$$\tau_p < s \le \tau_{p+1}, \quad \tau_k < t \le \tau_{k+1}.$$

By virtue of Lemma 2 and the fact that the sets $\mathcal{B}_i$ are compact, the set $\Phi_{\mathcal{A}\mathcal{B}_i}(t,s) \in \mathrm{comp}(\mathbb{R}^{n\times n})$.

In view of the T-periodicity of the right-hand sides of inclusion (3.59), (3.60), the existence of T-periodic R-solutions is directly connected with the existence of solutions to the equation

$$R_0 = \bigcup_{\substack{A(t)\in\mathcal{A}(t)\\ B_i\in\mathcal{B}_i}} \{\Phi_{AB_i}(T,0)R_0 + G_{AB_i}\} \tag{3.66}$$

in the space $\mathrm{comp}(\mathbb{R}^n)$, where

$$G_{AB_i} = \int_0^T \Phi_{AB_i}(T,\tau)F(\tau)d\tau + \sum_{0\le\tau_i<T} \Phi_{AB_i}(T,\tau_i)P_i \in \mathrm{comp}(\mathbb{R}^n).$$

Theorem 10. *Assume that the inequality*

$$\|\Phi_{AB_i}(T,0)\| < 1$$

is true for any $A(t) \in \mathcal{A}(t)$ and $B_i \in \mathcal{B}_i$.

Then inclusions (3.59), (3.60) *possess a unique T-periodic R-solution.*

Proof. Consider an operator ψ mapping the space $\mathrm{comp}(\mathbb{R}^n)$ into itself as follows:

$$\psi(R) = \bigcup_{\substack{A(t)\in\mathcal{A}(t)\\ B_i\in\mathcal{B}_i}} \{\Phi_{AB_i}(T,0)R + G_{AB_i}\}.$$

It is necessary to show that this operator is contracting. To do this, we choose arbitrary sets $R_1, R_2 \in \mathrm{comp}(\mathbb{R}^n)$. Then $h(\psi(R_1), \psi(R_2)) = \sup(d_1, d_2)$, where

$$d_1 = \rho\Bigg(\Phi_{AB_i}(T,0)r_1 + g_{AB_i}, \bigcup_{\substack{A(t)\in\mathcal{A}(t)\\ B_i\in\mathcal{B}_i}} \{\Phi_{AB_i}(T,0)R_2 + G_{AB_i}\}\Bigg),$$

$$d_2 = \rho\Bigg(\Phi_{AB_i}(T,0)r_2 + g_{AB_i}, \bigcup_{\substack{A(t)\in\mathcal{A}(t)\\ B_i\in\mathcal{B}_i}} \{\Phi_{AB_i}(T,0)R_1 + G_{AB_i}\}\Bigg),$$

$r_1 \in R_1,\ r_2 \in R_2$.

Further, we estimate d_1. We choose $r_2 \in R_2$ such that $\|r_1 - r_2\| = \rho(r_1, R_2)$. Then

$$\begin{aligned} d_1 &\le \|\Phi_{AB_i}(T,0)r_1 + g_{AB_i} - [\Phi_{AB_i}(T,0)r_2 + g_{AB_i}]\| \\ &\le \|\Phi_{AB_i}(T,0)\| \cdot \rho(r_1, R_2) \le \|\Phi_{AB_i}(T,0)\| \cdot h(R_1, R_2). \end{aligned}$$

Similarly, $d_2 \le \|\Phi_{AB_i}(T,0)\| \cdot h(R_1, R_2)$. Hence,

$$h(\psi(R_1), \psi(R_2)) \le h(R_1, R_2) \sup_{\substack{A(t)\in\mathcal{A}(t)\\ B_i\in\mathcal{B}_i}} \|\Phi_{AB_i}(T,0)\|.$$

We now show that

$$\sup_{\substack{A(t)\in\mathcal{A}(t)\\ B_i\in\mathcal{B}_i}} \|\Phi_{AB_i}(T,0)\| < 1.$$

Since, by virtue of Lemma 2, the set of matrizants $\Phi_{\mathcal{A}\mathcal{B}_i}(T,0)$ is compact in the space $\mathrm{comp}(\mathbb{R}^{n\times n})$ and $\|\cdot\|$ is a continuous function, by the Weierstrass theorem, we conclude that

$$\sup_{\substack{A(t)\in\mathcal{A}(t)\\ B_i\in\mathcal{B}_i}} \|\Phi_{AB_i}(T,0)\| = \max_{\substack{A(t)\in\mathcal{A}(t)\\ B_i\in\mathcal{B}_i}} \|\Phi_{AB_i}(T,0)\| < 1.$$

Thus, $\psi(R)$ is a contracting operator and, by the Banach theorem, has a single fixed point $R_0 \in \mathrm{comp}(\mathbb{R}^n)$ [14], i.e., Eq. (3.66) is uniquely solvable.

Hence, inclusion (3.59), (3.60) possesses a unique T-periodic R-solution. □

Example 13. Consider a controlled system

$$\dot{x} = ux + 1, \quad u \in [-2, -1]$$

which can be represented in the form of a differential inclusion

$$\dot{x} \in \mathcal{A}x + 1, \quad \mathcal{A} = \{u : u \in [-2, -1]\}. \tag{3.67}$$

The right-hand side is T-periodic, where $T > 0$ is an arbitrary constant. The initial sets of T-periodic R-solutions are determined by Eq. (3.66). In this case, Eq. (3.66) takes the form

$$R_0 = \bigcup_{u\in[-2,-1]} \left\{ e^{uT} R_0 + \frac{e^{uT}-1}{u} \right\}. \tag{3.68}$$

Since $\|\Phi_A(T,0)\| = e^{uT} < 1$ for all $u \in [-2,-1]$, by virtue of Theorem 10, Eq. (3.68) possesses a unique solution $R_0 \in \mathrm{comp}(\mathbb{R})$. In view of the fact that R_0 is a connected set in $\mathbb{R}$, we conclude that $R_0 = [a,b]$. Then Eq. (3.68) is reduced to

$$[a,b] = \left[e^{-2T}a - \frac{e^{-2T}-1}{2}, e^{-T}b - (e^{-T}-1) \right]$$
$$\Leftrightarrow \begin{cases} a = e^{-2T}a - \frac{e^{-2T}-1}{2}, \\ b = e^{-T}b - (e^{-T}-1) \end{cases} \Leftrightarrow \begin{cases} a = \frac{1}{2}, \\ b = 1. \end{cases}$$

Hence, $R_0 = [\frac{1}{2}, 1]$. Thus, for any $T > 0$, inclusion (3.67) possesses a unique T-periodic R-solution $R(t, [\frac{1}{2}, 1])$.

We now establish sufficient conditions for the existence of periodic ordinary solutions of nonlinear impulsive differential inclusions of the form

$$\dot{x} \in A(t)x + F(t,x), \quad t \neq \tau_i, \qquad \Delta x|_{t=\tau_i} \in B_i x + I_i(x), \tag{3.69}$$

where $x \in \mathbb{R}^n$ is the phase vector, $t \in \mathbb{R}$ is time, $A(t)$ is a continuous T-periodic matrix, $F\colon \mathbb{R}\times\mathbb{R}^n \to \mathrm{conv}(\mathbb{R}^n)$ is a set-valued mapping continuous in its variables, T-periodic in t, and bounded (i.e., there exists a set-valued mapping $Q\colon \mathbb{R} \to \mathrm{conv}(\mathbb{R}^n)$ such that the inclusion $F(t,x) \subseteq Q(t)$ is true for any fixed $t \in \mathbb{R}$ and all $x \in \mathbb{R}^n$), the matrices B_i, the set-valued mappings $I_i\colon \mathbb{R}^n \to \mathrm{conv}(\mathbb{R}^n)$, and the times of pulses τ_i are such that

$$B_{i+p} = B_i, \quad I_{i+p}(x) \equiv I_i(x), \quad \tau_{i+p} = \tau_i + T \tag{3.70}$$

for all $i \in \mathbb{Z}$ and some natural p, and the sets $I_i(x)$ are bounded, i.e., there exist sets $P_i \in \mathrm{conv}(\mathbb{R}^n)$ such that the inclusion $I_i(x) \subseteq P_i$ is true for all $i = \overline{1,p}$. We also assume that $0 \le \tau_1 < \cdots < \tau_p < T$ and $\det(E + B_i) \neq 0$ for all $i = \overline{1,p}$.

Let $R(T,x)$ be the set of attainability (R-solution) of inclusion (3.69) from the initial point $(0,x)$ at time $t = T$.

Together with inclusion (3.69), we consider the inclusion

$$\dot{x} \in A(t)x + Q(t), \quad t \neq \tau_i, \qquad \Delta x|_{t=\tau_i} \in B_i x + P_i. \tag{3.71}$$

Assume that inclusion (3.71) possesses a T-periodic R-solution $\overline{R}(t, R_0)$. Then the mapping $R(x) = R(T, x)$ maps the set R_0 into itself because

$$R(x) \subseteq \overline{R}(T, R_0) = R_0$$

for all $x \in R_0$.

The set-valued mapping $R(x)$ is upper semicontinuous [23]. Hence, by the Kakutani theorem [14], there exists a fixed point $x_0 \in R_0$ of the given mapping. Thus, there exists a periodic solution of the initial inclusion (3.69).

3.4 Linear Differential Equations with Pulse Action at Indefinite Times

Consider a linear differential equation with pulses at indefinite times:

$$\begin{aligned} \dot{x} &= A(t)x + f(t), \quad t \neq \tau_i, \\ \Delta x|_{t=\tau_i} &= B_i x + p_i, \end{aligned} \tag{3.72}$$

where $x \in \mathbb{R}^n$ is the phase vector, $t \in I = [t_0, T]$ is time, $A(t)$ is a matrix function continuous on I, $f(t)$ is a vector function continuous on I, $\tau_i \in [\tau_i^-, \tau_i^+] \subset I$, $i = \overline{1, m}$, are the times of pulses, $[\tau_i^-, \tau_i^+]$ are disjoint segments, B_i are $(n \times n)$-matrices, and $p_i \in \mathbb{R}^n$.

An equation of the form (3.72) describes, e.g., the physical processes subjected to pulse actions at times known with certain errors.

Let $\tau(t_0, T) = (\tau_1, \ldots, \tau_m)$ and let $\Phi(t, t_0, \tau(t_0, T))$ be the matrizant of the homogeneous impulsive differential equation corresponding to (3.72):

$$\begin{aligned} \dot{x} &= A(t)x, \quad t \neq \tau_i, \\ \Delta x|_{t=\tau_i} &= B_i x. \end{aligned}$$

By virtue of relation (1.50) in Chapter 1, the following representation is true for $\tau_k \leq t < \tau_{k+1}$:

$$\Phi(t, t_0, \tau(t_0, T)) = e^{\int_{\tau_k}^{t} A(s)ds}(E + B_k) \prod_{j=1}^{k-1} e^{\int_{\tau_j}^{\tau_{j+1}} A(s)ds}(E + B_j)e^{\int_{t_0}^{\tau_1} A(s)ds}.$$

Thus, every solution $x(t, x_0, \tau(t_0, T))$, $x(t_0, x_0, \tau(t_0, T)) = x_0$ of Eq. (3.72) for $t \in I$ can be represented in the form

$$\begin{aligned} x(t, x_0, \tau(t_0, T)) &= \Phi(t, t_0, \tau(t_0, T))x_0 \\ &+ \int_{t_0}^{t} \Phi(t, s, \tau(t_0, T)) f(s)ds + \sum_{t_0 \leq \tau_i < t} \Phi(t, \tau_i, \tau(t_0, T)) p_i. \end{aligned} \tag{3.73}$$

By $X(t, t_0, x_0)$ we denote the bundle of solutions originating from the initial point (t_0, x_0) and corresponding to different times of pulses $\tau_1, \dots, \tau_m$.

Theorem 11 ([119]). *$X(t, t_0, x_0) \in \mathrm{comp}(\mathbb{R}^n)$ for any fixed $t \in I$.*

Proof. By virtue of [142], the set $X(t, t_0, x_0)$ is nonempty. Denote

$$M_1 = \max_{t \in I} \|A(t)\|, \quad M_2 = \max_{t \in I} \|f(t)\|.$$

Then

$$\begin{aligned}
\|x(t, x_0, \tau(t_0, T))\| &\le \|\Phi(t, t_0, \tau(t_0, T))\| \cdot \|x_0\| \\
&\quad + \int_{t_0}^{t} \|\Phi(t, s, \tau(t_0, T))\| \cdot \|f(s)\| ds \\
&\quad + \sum_{t_0 \le \tau_i < t} \|\Phi(t, \tau_i, \tau(t_0, T))\| \cdot \|p_i\| \\
&\le \prod_{i=1}^{m} (1 + \|B_i\|) e^{M_1 (T - t_0)} \left\{ \|x_0\| + M_2 (T - t_0) + \sum_{i=1}^{m} \|p_i\| \right\} \\
&= K.
\end{aligned}$$

Hence, the set $X(t, t_0, x_0)$ is bounded.

We now show that the set $X(t, t_0, x_0)$ is closed. Consider a convergent sequence

$$x_k(t, x_0, \tau^k(t_0, T)) \in X(t, t_0, x_0).$$

It is necessary to show that its limit $x(t) \in X(t, t_0, x_0)$. Since $\tau_1^k \in [\tau_1^-, \tau_1^+]$, there exists a subsequence $\{k_1\}$ of the sequence $\{k\}$ such that $\tau_1^{k_1} \to \tau_1^0 \in [\tau_1^-, \tau_1^+]$.

We choose a subsequence $\{k_2\}$ of the sequence $\{k_1\}$ such that $\tau_2^{k_2} \to \tau_2^0 \in [\tau_2^-, \tau_2^+]$. Further, we continue this process and, finally, construct a sequence $\{k_m\}$ such that $\tau_m^{k_m} \to \tau_m^0 \in [\tau_m^-, \tau_m^+]$.

Since $\Phi(t, t_0, \tau(t_0, T))$ is a continuous function of the last argument, we get

$$\Phi(t, t_0, \tau^{k_m}(t_0, T)) \to \Phi(t, t_0, \tau^0(t_0, T))$$

as $k_m \to \infty$. Thus, by virtue of relation (3.73), we find

$$\begin{aligned}
x(t) &= \lim_{k \to \infty} x_k(t, x_0, \tau^k(t_0, T)) = \lim_{k_m \to \infty} x_{k_m}(t, x_0, \tau^{k_m}(t_0, T)) \\
&= x(t, x_0, \tau^0(t_0, T)) \in X(t, t_0, x_0).
\end{aligned}$$

Hence, the compactness of the set $X(t, t_0, x_0)$ is proved. □

Remark 13. The assertion of the theorem remains true if $I = [t_0, +\infty)$ and $\tau_i^{\pm} \to +\infty$ as $i \to +\infty$. To prove this, we successively consider Eq. (3.72) on the segments $[t_0, T_p]$, where $T_p \to +\infty$, and conclude that the set $X(t, t_0, x_0)$ is compact for any fixed $t \in [t_0, T_p]$ and all $p \in N$.

We now consider the problem of existence of an ω-periodic solution $x(t, x_0)$ and a periodic bundle of solutions $X(t, 0, X_0)$, $X_0 \in \mathrm{comp}(\mathbb{R}^n)$, for equations of the form (3.72) in the case where $I = [0, +\infty)$ and the right-hand side is ω-periodic, i.e., the functions $A(t)$ and $f(t)$ are assumed to be ω-periodic and the matrices B_i, vectors p_i, and times $\tau_i^{\pm}$ are such that

$$B_{i+r} = B_i, \quad p_{i+r} = p_i, \quad \tau_{i+r}^{\pm} = \tau_i^{\pm} + \omega$$

for all $i \in N$ and a natural number r.

The problem of existence of periodic solutions and periodic bundles of solutions is reduced to the problem of existence of solutions of the equations

$$x_0 = \Phi(\omega, 0, \tau(0, \omega))x_0 + d(\tau(0, \omega)), \tag{3.74}$$

where

$$d(\tau(0, \omega)) = \int_0^{\omega} \Phi(\omega, s, \tau(0, \omega)) f(s) ds + \sum_{t_0 \le \tau_i < \omega} \Phi(\omega, \tau_i, \tau(0, \omega)) p_i$$

and

$$X_0 = \bigcup_{\tau_i \in [\tau_i^-, \tau_i^+]} \{\Phi(\omega, 0, \tau(0, \omega)) X_0 + d(\tau(0, \omega))\}. \tag{3.75}$$

Equation (3.74) is solvable if and only if either the matrix $E - \Phi(\omega, 0, \tau(0, \omega))$ is nondegenerate for at least one collection $\tau_i \in [\tau_i^-, \tau_i^+]$ or the rank of the extended matrix $(E - \Phi(\omega, 0, \tau(0, \omega)), d(\tau(0, \omega)))$ is equal to the rank of the initial matrix. In this case, different $\tau_i \in [\tau_i^-, \tau_i^+]$ correspond to different matrices $\Phi(\omega, 0, \tau(0, \omega))$ and vectors $d(\tau(0, \omega))$ and, hence, a single initial point may correspond to several periodic solutions.

Consider the problem of solvability of Eq. (3.75).

Theorem 12 ([119]). *If the inequality*

$$\|\Phi(\omega, 0, \tau(0, \omega))\| \le M < 1$$

holds for all $\tau_i \in [\tau_i^-, \tau_i^+]$, $i = \overline{1, r}$, *then Eq.* (3.72) *possesses a single* ω*-periodic bundle of solutions.*

Proof. Consider a mapping

$$\xi(X_0) = \bigcup_{\tau_i \in [\tau_i^-, \tau_i^+]} \{\Phi(\omega, 0, \tau(0, \omega)) X_0 + d(\tau(0, \omega))\}.$$

By using the properties of the Hausdorff distance, we obtain

$$h(\xi(X_0), \xi(Y_0)) = \Bigg(\bigcup_{\tau_i \in [\tau_i^-, \tau_i^+]} \{\Phi(\omega, 0, \tau(0, \omega))X_0 + d(\tau(0, \omega))\},$$
$$\bigcup_{\tau_i \in [\tau_i^-, \tau_i^+]} \{\Phi(\omega, 0, \tau(0, \omega))Y_0 + d(\tau(0, \omega))\} \Bigg) = \sup\{d_1, d_2\},$$

where

$$\begin{aligned}
d_1 &= \rho\Bigg(\Phi(\omega, 0, \tau(0, \omega))x_0 + d(\tau(0, \omega)), \\
&\qquad \bigcup_{\tau_i \in [\tau_i^-, \tau_i^+]} \{\Phi(\omega, 0, \tau(0, \omega))Y_0 + d(\tau(0, \omega))\} \Bigg) \\
&\le \rho(\Phi(\omega, 0, \tau(0, \omega))x_0 + d(\tau(0, \omega)), \\
&\qquad \Phi(\omega, 0, \tau(0, \omega))Y_0 + d(\tau(0, \omega))) \\
&\le \|\Phi(\omega, 0, \tau(0, \omega))\| \rho(x_0, Y_0) \le M h(X_0, Y_0) < h(X_0, Y_0), \qquad (3.76) \\
d_2 &= \rho\Bigg(\Phi(\omega, 0, \tau(0, \omega))y_0 + d(\tau(0, \omega)), \\
&\qquad \bigcup_{\tau_i \in [\tau_i^-, \tau_i^+]} \{\Phi(\omega, 0, \tau(0, \omega))X_0 + d(\tau(0, \omega))\} \Bigg) \\
&\le \rho(\Phi(\omega, 0, \tau(0, \omega))y_0 + d(\tau(0, \omega)), \\
&\qquad \Phi(\omega, 0, \tau(0, \omega))X_0 + d(\tau(0, \omega))) \\
&\le \|\Phi(\omega, 0, \tau(0, \omega))\| \rho(y_0, X_0) \le M h(X_0, Y_0) < h(X_0, Y_0). \qquad (3.77)
\end{aligned}$$

Thus, the mapping $\xi(X_0)$ is contracting. Hence, all conditions of the Banach theorem [14] are satisfied. This means that the mapping $\xi(X_0)$ has a single fixed point in $\text{comp}(\mathbb{R}^n)$.

We now consider the problem of stability of a bundle of solutions of the impulsive differential Eq. (3.72) [119]. □

Theorem 13. *Assume that the matrizants $\Phi(t, t_0, \tau(0, +\infty))$ of Eq.* (3.72) *are uniformly bounded. Then the bundle of solutions $X(t, t_0, x_0)$ is Lyapunov stable in the positive direction.*

Proof. We now show that, for any $\varepsilon > 0$, there exists $\delta(\varepsilon) > 0$ such that the estimate

$$h(X(t, t_0, x_0), X(t, t_0, y_0)) < \varepsilon$$

is true for $\|x_0 - y_0\| < \delta$.

Further, we estimate

$$h(X(t, t_0, x_0), X(t, t_0, y_0)) = \sup\{d_1(t), d_2(t)\},$$

where, by analogy with estimates (3.76) and (3.77),

$$d_1(t) = \rho(x(t, x_0), X(t, t_0, y_0)) \leq M \|x_0 - y_0\|,$$
$$d_1(t) = \rho(x(t, y_0), X(t, t_0, x_0)) \leq M \|x_0 - y_0\|.$$

Finally, we choose $\delta = \frac{\varepsilon}{M}$ and arrive at the required assertion. □

Chapter 4

Linear Systems with Multivalued Trajectories

4.1 Differential Equations with Hukuhara Derivative

The spaces comp($\mathbb{R}^n$) and conv($\mathbb{R}^n$) are not linear because they do not contain the opposite element and, hence, the operation of subtraction. However, the operation of subtraction is urgently required in numerous cases and, hence, one can use several approaches for its definition. In what follows, we consider the notion of Hukuhara difference [58] in more detail.

Definition 1 ([15]). Let $X,\ Y \in \text{conv}(\mathbb{R}^n)$. A set $Z \in \text{conv}(\mathbb{R}^n)$ such that $X = Y + Z$ is called the difference of sets X and Y and denoted as follows: $X \overset{h}{-} Y$.

Remark 1. If the difference exists, then it is uniquely defined.

The operation of subtraction has following properties [25]:

(1) $(X+Y) \overset{h}{-} (U+V) = (X \overset{h}{-} U) + (Y \overset{h}{-} V)$ if the differences $X \overset{h}{-} U$ and $Y \overset{h}{-} V$ exist.

(2) $(X \overset{h}{-} U) \overset{h}{-} (Y \overset{h}{-} V) = (X \overset{h}{-} Y) + (V \overset{h}{-} U)$ if the differences $X \overset{h}{-} Y$, $V \overset{h}{-} U$ and $Y \overset{h}{-} V$ exist.

(3) $(X \overset{h}{-} U) \overset{h}{-} (Y \overset{h}{-} V) = (X \overset{h}{-} Y) \overset{h}{-} (U \overset{h}{-} V)$ if the differences $(X \overset{h}{-} Y) \overset{h}{-} (U \overset{h}{-} V)$ and $Y \overset{h}{-} V$ exist.

(4) $(\lambda X + \mu Y) \overset{h}{-} (\varkappa U + \sigma V) = (\lambda - \varkappa)X + \varkappa(X \overset{h}{-} U) + (\mu - \sigma)Y + {}+\sigma(Y \overset{h}{-} V)$ if $\lambda - \varkappa \geq 0,\ \mu - \sigma \geq 0$, and the differences $X \overset{h}{-} U$ and $Y \overset{h}{-} V$ exist.

(5) $X \overset{h}{-} U = (X \overset{h}{-} V) + (V \overset{h}{-} U)$ if the differences $X \overset{h}{-} V$ and $V \overset{h}{-} U$ exist.

(6) $h(X \overset{h}{-} U, Y \overset{h}{-} V) \leq h(X,Y) + h(U,V)$ if the differences $X \overset{h}{-} U$ and $Y \overset{h}{-} V$ exist.

(7) $h(\lambda X, \lambda Y) = \lambda h(X \overset{h}{-} Y, 0)$ if the difference $X \overset{h}{-} Y$ exists.

(8) $h(\lambda X, \mu Y) \leq \beta h(X,Y) + |\lambda - \mu|(h(X,0) + h(Y,0))$ provided that $\beta = \max\{\lambda, \mu\}$.

(9) The operation of subtraction is continuous relative to the Hausdorff metric: If sequences $\{X_n\}_{n=1}^{\infty}$ and $\{Y_n\}_{n=1}^{\infty}$ converge to X and Y and the differences $X_n \overset{h}{-} Y_n$ exist for all $n \in N$, then the difference $X \overset{h}{-} Y$ exists and the sequence $\{X_n \overset{h}{-} Y_n\}_{n=1}^{\infty}$ converges to $X \overset{h}{-} Y$.

Definition 2 ([24]). A set-valued mapping $X : [0, T] \to \text{conv}(\mathbb{R}^n)$ is differentiable in Hukuhara's sense at a point $t_0 \in (0, T)$ if there exists $D_h X(t_0) \in \text{conv}(\mathbb{R}^n)$ such that the limits

$$\lim_{\Delta t \downarrow 0} \frac{X(t_0 + \Delta t) \overset{h}{-} X(t_0)}{\Delta t} \quad \text{and} \quad \lim_{\Delta t \downarrow 0} \frac{X(t_0) \overset{h}{-} X(t_0 - \Delta t)}{\Delta t}$$

exist and are equal to $D_h X(t_0)$.

Note that, in this definition, it is assumed that, for all sufficiently small $\Delta t > 0$, the differences $X(t_0) \overset{h}{-} X(t_0 - \Delta t)$, $X(t_0 + \Delta t) \overset{h}{-} X(t_0)$ exist.

It makes sense to speak about unilateral derivatives at the points $t = 0$ and $t = T$.

A differential equation with Hukuhara derivative was considered for the first time in [25]:

$$D_h X = F(t, X), \quad X(0) = X_0, \tag{4.1}$$

where $F : [0, T] \times \text{conv}(\mathbb{R}^n) \to \text{conv}(\mathbb{R}^n)$ is a set-valued mapping, $X_0 \in \text{conv}(\mathbb{R}^n)$ is an initial state, and $D_h X$ is the Hukuhara derivative of a set-valued mapping $X : [0, T] \to \text{conv}(\mathbb{R}^n)$.

Definition 3 ([29]). A set-valued mapping $X(\cdot)$ is called a solution of Eq. (4.1) if it is continuously differentiable in Hukuhara's sense and satisfies system (4.1) everywhere on $[0, T]$.

The differential Eq. (4.1) is equivalent to the integral equation [25]

$$X(t) = X_0 + \int_0^t F(s, X(s))\, ds,$$

the integral in which is understood in Hukuhara's sense [58].

The following theorem on existence and uniqueness is true:

Theorem 1 ([29]). *Assume that $F(\cdot, \cdot)$ satisfies the conditions:*

(1) *$F(\cdot, \cdot)$ is continuous in (t, X) on $[0, T] \times \text{conv}(\mathbb{R}^n)$;*

(2) *$F(t, \cdot)$ has the Lipschitz property with respect to X on $\text{conv}(\mathbb{R}^n)$, i.e., there exists a constant $L > 0$ such that*

$$h(F(t, X), F(t, Y)) \leq L h(X, Y).$$

Then system (4.1) *is uniquely solvable.*

Example 1. Consider a linear differential equation of the form

$$D_h X = \lambda(t) X + F(t), \quad X(0) = X_0, \tag{4.2}$$

where $\lambda : \mathbb{R} \to \mathbb{R}_+$ is summable, $F : \mathbb{R} \to \operatorname{conv}(\mathbb{R}^n)$ is measurable, $h(F(t), 0) \le k(t)$, $k : \mathbb{R} \to \mathbb{R}_+$ is summable, and $X_0 \in \operatorname{conv}(\mathbb{R}^n)$.

By using the properties of the Hukuhara derivative, one can easily show that the set-valued mapping $X(\cdot)$ defined, for any $t \ge 0$, by the formula

$$X(t) = e^{\int_0^t \lambda(s)\,ds}\left(X_0 + \int_0^t F(s)e^{-\int_0^s \lambda(\tau)\,d\tau} ds\right)$$

is a solution of Eq. (4.2).

The other interesting result established for differential equations with Hukuhara derivative is the construction of an Euler broken line and estimation of the error.

We split the segment $[0, T]$ into N parts as follows:

$$0 = t_0 < t_2 < \cdots < t_N = T, \quad t_{k+1} - t_k = \delta,$$
$$I_k = [t_k, t_{k+1}], \quad k = \overline{0, N-1},$$

and construct the Euler broken line

$$X_k(t) = X_{k-1}(t_{k-1}) + (t - t_{k-1})F(t_{k-1}, X_{k-1}(t_{k-1})),$$
$$t \in I_{k-1}, \quad X_0(t_0) = X_0, \quad k = \overline{1, N}.$$

Denote

$$R = \sup_k D(X, X_k), \quad D(X, X_k) = \max_{I_{k-1}} h(X(t), X_k(t)), \quad k = \overline{1, N}.$$

Theorem 2 ([26]). *Assume that $F(\cdot, \cdot)$ satisfies the conditions:*

(1) *$F(\cdot, \cdot)$ is continuous in (t, X) on $[0, T] \times \operatorname{conv}(\mathbb{R}^n)$;*

(2) *$F(t, \cdot)$ satisfies the Lipschitz condition in X with constant L;*

(3) *the solution $X(\cdot)$ of system (4.1) has the second continuous derivative on $[0, T]$ such that*

$$h(D_h(D_h X(t)), 0) < K, \quad t \in [0, T].$$

Then the error R satisfies the inequality

$$R < \frac{\delta K}{2}[(1/L + \delta)(e^{T \cdot L} - 1) + \delta].$$

In [145], the differential equations with Hukuhara derivative were essentially used for the analysis of some properties of the integral funnel of a differential inclusion. Thus, it was shown that the integral funnel is a subset of the solution of the corresponding equation with Hukuhara derivative.

In recent years, the interest in differential equations with Hukuhara derivative increases in connection with their applications to differential equations under the conditions of uncertainty [38, 75, 76].

Consider a system of linear differential equations with Hukuhara derivative

$$D_h X_i(t) = \sum_{j=1}^{n} a_{ij}(t) X_j(t) + F_i(t), \tag{4.3}$$
$$X_i(0) = X_i^0, \quad i = \overline{1,n},$$

where $t \in [0,T]$ is time and $a_{ij} : [0,T] \to \mathbb{R}$ and $F_i : [0,T] \to \operatorname{conv}(\mathbb{R}^m)$ are continuous functions.

Definition 4 ([125]). Set-valued mappings $X_i : [0,T] \to \operatorname{conv}(\mathbb{R}^m)$, $i = \overline{1,n}$, are called the solution of problem (4.3) if they are continuously differentiable in Hukuhara's sense and satisfy system (4.3) everywhere on $[0,T]$.

We now show that the solution of system (4.3) can be reduced to a solution of a system of ordinary differential equations.

Let $c(A,\psi)$ be the support function of a set A. By using the properties of the support functions, we conclude that

$$c\left(\sum_{j=1}^{n} a_{ij}(t)X_j(t) + F_i(t), \psi\right) = \sum_{j=1}^{n} c(X_j(t), a_{ij}(t)\psi) + c(F_i(t),\psi)$$
$$= \sum_{j=1}^{n} |a_{ij}(t)| c(X_j(t), \operatorname{sign}(a_{ij}(t))\psi) + c(F_i(t),\psi).$$

We now find the support function of the left-hand side of Eq. (4.3):

$$c(D_h X_i(t),\psi) = c\left(\lim_{\Delta t \downarrow 0} \frac{X_i(t+\Delta t) \overset{h}{-} X_i(t)}{\Delta t}, \psi\right)$$
$$= \lim_{\Delta t \downarrow 0} c\left(\frac{X_i(t+\Delta t) \overset{h}{-} X_i(t)}{\Delta t}, \psi\right)$$
$$= \lim_{\Delta t \downarrow 0} \frac{1}{\Delta t} c\left(X_i(t+\Delta t) \overset{h}{-} X_i(t), \psi\right)$$
$$= \lim_{\Delta t \downarrow 0} \frac{1}{\Delta t} [c(X_i(t+\Delta t),\psi) - c(X_i(t),\psi)] = \frac{d}{dt} c(X_i(t+0),\psi).$$

Similarly, we get

$$c(D_h X_i(t),\psi) = \frac{d}{dt} c(X_i(t-0),\psi),$$

i.e.,

$$c(D_h X_i(t),\psi) = \frac{d}{dt} c(X_i(t),\psi).$$

In view of the convexity of the sets $X_i(t)$ and $F_i(t)$, system (4.3) is equivalent to the system

$$\frac{d}{dt}c(X_i(t),\psi) = \sum_{j=1}^{n} |a_{ij}(t)| c(X_j(t), \operatorname{sign}(a_{ij}(t))\psi) + c(F_i(t),\psi), \tag{4.4}$$
$$c(X_i(0),\psi) = c(X_i^0,\psi), \quad i=\overline{1,n},$$

for all $\psi \in S_1(0) \subset \mathbb{R}^m$.

We now introduce the following functions:

$$f_i^+(t,\psi) = c(F_i(t),\psi), \qquad f_i^-(t,\psi) = c(F_i(t),-\psi),$$
$$x_i^+(t,\psi) = c(X_i(t),\psi), \qquad x_i^-(t,\psi) = c(X_i(t),-\psi).$$

This enables us to rewrite system (4.4) in the form

$$\frac{d}{dt}x_i^+(t,\psi) = \sum_{j=1}^{n} |a_{ij}(t)| \Big[\frac{1+\operatorname{sign}(a_{ij}(t))}{2} x_j^+(t,\psi) + \frac{1-\operatorname{sign}(a_{ij}(t))}{2} x_j^-(t,\psi)\Big] + f_i^+(t,\psi),$$
$$\frac{d}{dt}x_i^-(t,\psi) = \sum_{j=1}^{n} |a_{ij}(t)| \Big[\frac{1-\operatorname{sign}(a_{ij}(t))}{2} x_j^+(t,\psi) + \frac{1+\operatorname{sign}(a_{ij}(t))}{2} x_j^-(t,\psi)\Big] + f_i^-(t,\psi), \tag{4.5}$$
$$x_i^+(t,\psi) = c(X_i^0,\psi), \quad x_i^-(t,\psi) = c(X_i^0,-\psi), \quad i=\overline{1,n}.$$

Thus, we arrive at a system of $2n$ ordinary differential equations with a parameter $\psi \in S_1^+(0) = \{\psi \in S_1(0) : \psi_1 \geq 0\}$.

Since the sets $X_i(t)$, $i=\overline{1,n}$, are convex, they are uniquely determined by their support function:

$$X_i(t) = \bigcap_{\psi \in S_1^+(0)} \{x \in \mathbb{R}^m : (x,\psi) \leq x_i^+(t,\psi), (x,-\psi) \leq x_i^-(t,\psi)\}. \tag{4.6}$$

However, the efficient reconstruction of a set according to its support function can be realized only for small values of m.

Thus, we have proved the following assertion:

Theorem 3 ([125]). *A solution of system* (4.3) *is given by n functions specified by equalities* (4.6)*, where* $x_i^{\pm}(t,\psi), i=\overline{1,n}$*, is the solution of the system of ordinary differential Eqs.* (4.5).

Example 2. Let $m = 2$. Consider a system of differential equations with Hukuhara derivative of the form

$$\begin{aligned} D_h X_1(t) &= aX_1(t) + bX_2(t) + S_{r_1}(0), \\ D_h X_2(t) &= cX_1(t) + dX_2(t) + S_{r_2}(0), \\ X_1(0) &= S_{\delta_1}(0), \quad X_2(0) = S_{\delta_2}(0). \end{aligned} \tag{4.7}$$

For this system, we write a system of equations of the form (4.5):

$$\begin{aligned} \frac{d}{dt}x_1^+(t,\psi) &= |a|\frac{1+\operatorname{sign} a}{2}x_1^+(t,\psi) + |a|\frac{1-\operatorname{sign} a}{2}x_1^-(t,\psi) \\ &\quad + |b|\frac{1+\operatorname{sign} b}{2}x_2^+(t,\psi) + |b|\frac{1-\operatorname{sign} b}{2}x_2^-(t,\psi) + r_1, \\ \frac{d}{dt}x_1^-(t,\psi) &= |a|\frac{1-\operatorname{sign} a}{2}x_1^+(t,\psi) + |a|\frac{1+\operatorname{sign} a}{2}x_1^-(t,\psi) \\ &\quad + |b|\frac{1-\operatorname{sign} b}{2}x_2^+(t,\psi) + |b|\frac{1+\operatorname{sign} b}{2}x_2^-(t,\psi) + r_1, \\ \frac{d}{dt}x_2^+(t,\psi) &= |c|\frac{1+\operatorname{sign} c}{2}x_1^+(t,\psi) + |c|\frac{1-\operatorname{sign} c}{2}x_1^-(t,\psi) \\ &\quad + |d|\frac{1+\operatorname{sign} d}{2}x_2^+(t,\psi) + |d|\frac{1-\operatorname{sign} d}{2}x_2^-(t,\psi) + r_2, \\ \frac{d}{dt}x_2^-(t,\psi) &= |c|\frac{1-\operatorname{sign} c}{2}x_1^+(t,\psi) + |c|\frac{1+\operatorname{sign} c}{2}x_1^-(t,\psi) \\ &\quad + |d|\frac{1-\operatorname{sign} d}{2}x_2^+(t,\psi) + |d|\frac{1+\operatorname{sign} d}{2}x_2^-(t,\psi) + r_2, \\ x_1^+(0,\psi) &= x_1^-(0,\psi) = \delta_1, \quad x_2^+(0,\psi) = x_2^-(0,\psi) = \delta_2. \end{aligned}$$

As a result of the term-by-term summation and subtraction of equations of this system, we find

$$\begin{aligned} \frac{d}{dt}(x_1^+(t,\psi) + x_1^-(t,\psi)) &= |a|(x_1^+(t,\psi) + x_1^-(t,\psi)) \\ &\quad + |b|(x_2^+(t,\psi) + x_2^-(t,\psi)) + 2r_1, \\ \frac{d}{dt}(x_1^+(t,\psi) - x_1^-(t,\psi)) &= a(x_1^+(t,\psi) - x_1^-(t,\psi)) \\ &\quad + b(x_2^+(t,\psi) - x_2^-(t,\psi)), \\ \frac{d}{dt}(x_2^+(t,\psi) + x_2^-(t,\psi)) &= |c|(x_1^+(t,\psi) + x_1^-(t,\psi)) \\ &\quad + |d|(x_2^+(t,\psi) + x_2^-(t,\psi)) + 2r_2, \end{aligned}$$

$$\frac{d}{dt}(x_2^+(t,\psi) - x_2^-(t,\psi)) = c(x_1^+(t,\psi) - x_1^-(t,\psi))$$
$$+ d(x_2^+(t,\psi) - x_2^-(t,\psi)),$$
$$x_1^+(0,\psi) + x_1^-(0,\psi) = 2\delta_1, \quad x_1^+(0,\psi) - x_1^-(0,\psi) = 0,$$
$$x_2^+(0,\psi) + x_2^-(0,\psi) = 2\delta_2, \quad x_2^+(0,\psi) - x_2^-(0,\psi) = 0.$$

Further, we solve the second and fourth equations separately and obtain

$$x_1^+(t,\psi) \equiv x_1^-(t,\psi) \quad \text{and} \quad x_2^+(t,\psi) \equiv x_2^-(t,\psi).$$

Substituting these relations in the first and third equations, we get

$$\frac{d}{dt}x_1^+(t,\psi) = |a|x_1^+(t,\psi) + |b|x_2^+(t,\psi) + r_1,$$
$$\frac{d}{dt}x_2^+(t,\psi) = |c|x_1^+(t,\psi) + |d|x_2^+(t,\psi) + r_2,$$
$$x_1^+(0,\psi) = \delta_1, \quad x_2^+(0,\psi) = \delta_2.$$

Since the right-hand sides of the system are independent of ψ, we find

$$x_1^+(t,\psi) = x_1(t) \quad \text{and} \quad x_2^+(t,\psi) = x_2(t),$$

and the required sets are

$$X_1(t) = S_{x_1(t)}(0) \quad \text{and} \quad X_2(t) = S_{x_2(t)}(0).$$

By the direct substitution in Eq. (4.7), we show that the required solution is indeed given by the obtained set-valued mappings.

4.2 Approximation of the Integral Funnel of a Linear Differential Inclusion with the Help of Systems of Differential Equations with Hukuhara Derivative

Consider a linear differential inclusion

$$\dot{x} \in A(t)x + F(t), \quad x(0) \in X_0 \in \text{conv}(\mathbb{R}^n), \tag{4.8}$$

where $t \in [0,T]$, $x \in \mathbb{R}^n$ is the phase vector, $A(t)$ is a continuous $n \times n$ matrix, and $F : [0,T] \to \text{conv}(\mathbb{R}^n)$ is a continuous set-valued mapping.

Inclusion (4.8) is equivalent, e.g., to a linear control system

$$\dot{x} = A(t)x + D(t)u, \quad x(0) \in X_0, \tag{4.9}$$

where $u(t) \in U(t) \in \mathrm{conv}(\mathbb{R}^p)$ is a control vector, $U(t)$ is a continuous set-valued mapping, and $D(t)$ is a continuous $n \times p$ matrix. In this case, $F(t) = \{y \in \mathbb{R}^n : y = D(t)u(t), u(t) \in U(t)\}$.

The investigation of the properties of the integral funnel of inclusion (4.8) [the set of attainability of system (4.9)] is of high significance for the qualitative theory and the problems of control. In this connection, numerous authors studied the properties of the set of attainability [6, 91, 101, 145] and various approximate methods used for its construction: the method of ellipsoids for linear systems [30, 73, 98], asymptotic methods [41, 115], and numerical methods [91, 96, 149].

We now write the differential equation with Hukuhara derivative corresponding to the differential inclusion (4.8):

$$D_h X(t) = A(t)X(t) + F(t), \quad X(0) = X_0. \tag{4.10}$$

Here, the solution $X : [0, T] \to \mathrm{conv}(\mathbb{R}^n)$ is a set-valued mapping continuously differentiable in Hukuhara's sense.

In [145], it is proved that

$$R(t) \subset X(t), \tag{4.11}$$

where $R(t)$ is the set of attainability of system (4.8).

The practical construction of a given approximation for the bundle of solutions of inclusion (4.8) is connected with significant computational difficulties encountered in finding the solutions of Eq. (4.10) for $n > 2$.

The matrix $A(t)$ is represented in the form

$$A(t) = \begin{pmatrix} A_{11}(t) & A_{12}(t) & \cdots & A_{1m}(t) \\ A_{21}(t) & A_{22}(t) & \cdots & A_{2m}(t) \\ \vdots & \vdots & \ddots & \vdots \\ A_{m1}(t) & A_{m2}(t) & \cdots & A_{mm}(t) \end{pmatrix}, \tag{4.12}$$

where $A_{ij}(t) \in \mathbb{R}^{n_i \times n_j}$, $\sum_{i=1}^m n_i = n$.

Equation (4.10) is associated with the following system of linear equations with Hukuhara derivative:

$$D_h X_i(t) = \sum_{j=1}^m A_{ij}(t)X_j(t) + F_i(t), \quad X_i(0) = X_i^0 \in \mathrm{conv}(\mathbb{R}^{n_i}),\ i = \overline{1, m}, \tag{4.13}$$

where $F(t) \subset \overline{F}(t) = F_1(t) \times \cdots \times F_m(t)$, $F_i : [0, T] \to \mathrm{conv}(\mathbb{R}^{n_i})$ are continuous set-valued mappings, $X_0 \subset X_1^0 \times \cdots \times X_m^0$, and $X_i : [0, T] \to \mathrm{conv}(\mathbb{R}^{n_i})$ are functions continuously differentiable in Hukuhara's sense.

Consider a set $\overline{X}(t) = X_1(t) \times \cdots \times X_m(t)$.

Theorem 4 ([124]). *The following inclusion is true for Eqs.* (4.10) *and* (4.13) *and any* $t \in [0, T]$*:*

$$X(t) \subset \overline{X}(t).$$

Proof. The sets $X(t)$ and $\overline{X}(t)$ are convex compact sets in $\mathbb{R}^n$. Hence, it suffices to show that the inequality

$$c(X(t), \psi) \leq c(\overline{X}(t), \psi)$$

holds for all $\psi \in \mathbb{R}^n$,

We split the segment $[0, T]$ by the points $t_k = \frac{kT}{N}$, $k = \overline{0, N}$. Consider a family of Euler broken lines for Eqs. (4.10) and (4.13):

$$X^N(t_{k+1}) = X^N(t_k) + h[A(t_k)X^N(t_k) + F(t_k)], \quad X^N(0) = X_0, \tag{4.14}$$

$$X_i^N(t_{k+1}) = X_i^N(t_k) + h\left[\sum_{j=1}^{m} A_{ij}(t_k)X_j^N(t_k) + F_i(t_k)\right], \tag{4.15}$$

$$X_i^N(0) = X_i^0, \quad i = \overline{1, m}, \quad k = \overline{0, N-1}.$$

Let

$$\overline{X}^N(t) = X_1^N(t) \times \cdots \times X_m^N(t).$$

Since $X^N(0) \subset \overline{X}^N(0)$, we have

$$c(X^N(0), \psi) \leq c(\overline{X}^N(0), \psi)$$

for all $\psi \in \mathbb{R}^n$. Assume that the inequality

$$c(X^N(t_k), \psi) \leq c(\overline{X}^N(t_k), \psi)$$

holds for all $\psi \in \mathbb{R}^n$ and let

$$\psi = \begin{pmatrix} \psi_1 \\ \vdots \\ \psi_m \end{pmatrix}, \quad \psi_i \in \mathbb{R}^{n_i}.$$

Then

$$\begin{aligned}
&c(X^N(t_{k+1}), \psi) \\
&= c(X^N(t_k) + hA(t_k)X^N(t_k) + hF(t_k), \psi) \\
&= c(X^N(t_k), \psi) + hc(X^N(t_k), A^T(t_k)\psi) + hc(F(t_k), \psi) \\
&\leq c(\overline{X}^N(t_k), \psi) + hc(\overline{X}^N(t_k), A^T(t_k)\psi) + hc(\overline{F}(t_k), \psi)
\end{aligned}$$

$$= \sum_{i=1}^{m} [c(X_i^N(t_k), \psi_i) + hc(X_i^N(t_k), (A^T(t_k)\psi)_i) + hc(F_i(t_k), \psi_i)]$$

$$= \left\{ (A^T(t_k)\psi)_i = \left(\begin{pmatrix} A_{11}^T(t_k) & A_{21}^T(t_k) & \cdots & A_{m1}^T(t_k) \\ A_{12}^T(t_k) & A_{22}^T(t_k) & \cdots & A_{m2}^T(t_k) \\ \vdots & \vdots & \ddots & \vdots \\ A_{1m}^T(t_k) & A_{2m}^T(t_k) & \cdots & A_{mm}^T(t_k) \end{pmatrix} \begin{pmatrix} \psi_1 \\ \psi_2 \\ \vdots \\ \psi_m \end{pmatrix} \right)_i \right.$$

$$\left. = \begin{pmatrix} \sum_{j=1}^m A_{j1}^T(t_k)\psi_j \\ \sum_{j=1}^m A_{j2}^T(t_k)\psi_j \\ \vdots \\ \sum_{j=1}^m A_{jm}^T(t_k)\psi_j \end{pmatrix}_i = \sum_{j=1}^{m} A_{ji}^T(t_k)\psi_j \right\}$$

$$= \sum_{i=1}^{m} \left[c(X_i^N(t_k), \psi_i) + hc\left(X_i^N(t_k), \sum_{j=1}^{m} A_{ji}^T(t_k)\psi_j \right) + hc(F_i(t_k), \psi_i) \right]$$

$$\leq \sum_{i=1}^{m} \left[c(X_i^N(t_k), \psi_i) + h \sum_{j=1}^{m} c(X_i^N(t_k), A_{ji}^T(t_k)\psi_j) + hc(F_i(t_k), \psi_i) \right]$$

$$= \sum_{i=1}^{m} c(X_i^N(t_k), \psi_i) + h \sum_{i=1}^{m} c(F_i(t_k), \psi_i) + h \sum_{i=1}^{m} \sum_{j=1}^{m} c(X_i^N(t_k), A_{ji}^T(t_k)\psi_j). \tag{4.16}$$

We now find the support function of the set $\overline{X}^N(t_{k+1})$:

$$c(\overline{X}^N(t_{k+1}), \psi)$$

$$= \sum_{i=1}^{m} c(X_i^N(t_{k+1}), \psi_i)$$

$$= \sum_{i=1}^{m} c\left(X_i^N(t_k) + h \sum_{j=1}^{m} A_{ij}(t_k) X_j^N(t_k) + hF_i(t_k), \psi_i \right)$$

$$= \sum_{i=1}^{m} \left[c(X_i^N(t_k), \psi_i) + h \sum_{j=1}^{m} c(X_j^N(t_k), A_{ij}^T(t_k)\psi_i) + hc(F_i(t_k), \psi_i) \right]$$

$$= \sum_{i=1}^{m} c(X_i^N(t_k), \psi_i) + h \sum_{i=1}^{m} c(F_i(t_k), \psi_i) + h \sum_{i=1}^{m} \sum_{j=1}^{m} c(X_j^N(t_k), A_{ij}^T(t_k)\psi_i). \tag{4.17}$$

In view of relations (4.16) and (4.17), we obtain

$$c(X^N(t_{k+1}), \psi) \le c(\overline{X}^N(t_{k+1}), \psi)$$

for all $\psi \in \mathbb{R}^n$. As $N \to \infty$, the Euler broken lines (4.14) and (4.15) converge to solutions of (4.10) and (4.13), respectively [26]. Hence, passing to the limit and using the property of continuity of the support functions, we conclude that $c(X(t), \psi) \le c(\overline{X}(t), \psi)$ for all $\psi \in \mathbb{R}^n$. Thus, $X(t) \subset \overline{X}(t)$ for all $t \in [0, T]$. The theorem is proved. □

We now study the problem of variation of the set $\overline{X}(t)$ in the case of subsequent decomposition of the matrix. Assume that the νth row and the νth column of the matrix $A(t)$ $\big(\nu \in \overline{1, m}\big)$ are split into the following matrices:

$$A_{\nu\nu}(t) = \begin{pmatrix} A_{\nu\nu}^{11}(t) & A_{\nu\nu}^{12}(t) & \cdots & A_{\nu\nu}^{1s}(t) \\ A_{\nu\nu}^{21}(t) & A_{\nu\nu}^{22}(t) & \cdots & A_{\nu\nu}^{2s}(t) \\ \vdots & \vdots & \ddots & \vdots \\ A_{\nu\nu}^{s1}(t) & A_{\nu\nu}^{s2}(t) & \cdots & A_{\nu\nu}^{ss}(t) \end{pmatrix},$$

where $A_{\nu\nu}^{pq}(t) \in \mathbb{R}^{l_p \times l_q}$ and $\sum_{p=1}^{s} l_p = n_\nu$;

$$A_{i\nu}(t) = (\, A_{i\nu}^1(t) \;\; A_{i\nu}^2(t) \;\; \cdots \;\; A_{i\nu}^s(t)\,), \tag{4.18}$$

where $A_{i\nu}^p(t) \in \mathbb{R}^{n_i \times l_p}$ and $i = \overline{1, m}$, $i \ne \nu$;

$$A_{\nu i}(t) = \begin{pmatrix} A_{\nu i}^1(t) \\ A_{\nu i}^2(t) \\ \vdots \\ A_{\nu i}^s(t) \end{pmatrix}, \quad \text{where } A_{\nu i}^p(t) \in \mathbb{R}^{l_p \times n_i} \text{ and } i = \overline{1, m},\ i \ne \nu.$$

Together with system (4.13), we consider a system

$$D_h \tilde{X}_i(t) = \sum_{\substack{j=1 \\ j \ne \nu}}^{m} A_{ij}(t)\tilde{X}_j(t) + \sum_{p=1}^{s} A_{i\nu}^p(t) X_{\nu p}(t) + \tilde{F}_i(t),$$

$$\tilde{X}_i(0) = X_i^0,\ i = \overline{1, m},\ i \ne \nu,$$

$$D_h X_{\nu q}(t) = \sum_{\substack{j=1 \\ j \ne \nu}}^{m} A_{\nu j}^q(t)\tilde{X}_j(t) + \sum_{p=1}^{s} A_{\nu\nu}^{qp}(t) X_{\nu p}(t) + F_{\nu q}(t), \tag{4.19}$$

$$X_{\nu q}(0) = X_{\nu q}^0,\ q = \overline{1, s},$$

where $F_i(t) = \tilde{F}_i(t), i \ne \nu$, $F_\nu(t) \subset \tilde{F}_\nu(t) = F_{\nu 1}(t) \times \cdots \times F_{\nu s}(t)$, $F_{\nu q} : [0, T] \to \operatorname{conv}(\mathbb{R}^{l_q})$ are continuous set-valued mappings, $X_\nu^0 \in \tilde{X}_\nu^0 = X_{\nu 1}^0 \times \cdots \times X_{\nu s}^0$, $\tilde{X}_i$:

$[0,T] \to \operatorname{conv}(\mathbb{R}^{n_i})$ $(i \neq \nu)$, and $X_{\nu q} : [0,T] \to \operatorname{conv}(\mathbb{R}^{l_q})$ $(q = \overline{1,s})$ are set-valued mappings continuously differentiable in Hukuhara's sense.

Consider a set

$$\tilde{X}(t) = \tilde{X}_1(t) \times \cdots \times \tilde{X}_{\nu-1}(t) \times X_{\nu 1}(t) \times \cdots \times X_{\nu s}(t) \\ \times \tilde{X}_{\nu+1}(t) \times \cdots \times \tilde{X}_m(t).$$

Theorem 5 ([124]). *The following inclusion holds for systems* (4.13) *and* (4.19) *and any* $t \in [0,T]$: $\overline{X}(t) \subset \tilde{X}(t)$.

Proof. The sets $\overline{X}(t)$ and $\tilde{X}(t)$ are convex compact sets in $\mathbb{R}^n$. Hence, it is sufficient to show that the inequality $c(\overline{X}(t), \psi) \le c(\tilde{X}(t), \psi)$ is true for all vectors $\psi \in \mathbb{R}^n$.

We split the segment $[0,T]$ by the points $t_k = \frac{kT}{N}$, $k = \overline{0,N}$ and consider the family of Euler broken lines for Eqs. (4.13) and (4.19), i.e., equalities (4.15) and

$$\tilde{X}_i^N(t_{k+1}) = \tilde{X}_i^N(t_k) + h\Bigg[\sum_{\substack{j=1\\ j\neq\nu}}^{m} A_{ij}(t_k)\tilde{X}_j^N(t_k) + \sum_{p=1}^{s} A_{i\nu}^{p}(t_k)X_{\nu p}^N(t_k) + \tilde{F}_i(t_k)\Bigg],$$
$$\tilde{X}_i^N(0) = X_i^0,\ i = \overline{1,m},\ i \neq \nu,$$
$$X_{\nu q}^N(t_{k+1}) = X_{\nu q}^N(t_k) + h\Bigg[\sum_{\substack{j=1\\ j\neq\nu}}^{m} A_{\nu j}^{q}(t_k)\tilde{X}_j^N(t) + \sum_{p=1}^{s} A_{\nu\nu}^{qp}(t_k)X_{\nu p}^N(t_k) + F_{\nu q}(t_k)\Bigg],$$
$$X_{\nu q}^N(0) = X_{\nu q}^0,\ q = \overline{1,s},\ k = \overline{0,N-1}. \tag{4.20}$$

Let

$$\tilde{X}^N(t) = \tilde{X}_1^N(t) \times \cdots \times \tilde{X}_{\nu-1}^N(t) \times X_{\nu 1}^N(t) \times \cdots \times X_{\nu s}^N(t) \times \tilde{X}_{\nu+1}^N(t) \times \cdots \times \tilde{X}_m^N(t).$$

Since $\overline{X}^N(0) \subset \tilde{X}^N(0)$, we have $c(\overline{X}^N(0), \psi) \le c(\tilde{X}^N(0), \psi)$ for all vectors $\psi \in \mathbb{R}^n$. Assume that the inequality

$$c(\overline{X}^N(t_k), \psi) \le c(\tilde{X}^N(t_k), \psi)$$

holds for all $\psi \in \mathbb{R}^n$. Then, by virtue of (4.17), we get

$$c(\overline{X}^N(t_{k+1}), \psi) \\ = \sum_{i=1}^{m} c(X_i^N(t_k), \psi_i) + h\sum_{i=1}^{m} c(F_i(t_k), \psi_i) + h\sum_{i=1}^{m}\sum_{j=1}^{m} c(X_j^N(t_k), A_{ij}^T(t_k)\psi_i) \\ \le \sum_{i=1}^{m} c(\tilde{X}_i^N(t_k), \psi_i) + h\sum_{i=1}^{m} c(\tilde{F}_i(t_k), \psi_i) + h\sum_{i=1}^{m}\sum_{j=1}^{m} c(\tilde{X}_j^N(t_k), A_{ij}^T(t_k)\psi_i)$$

$$= \sum_{\substack{i=1\\ i\neq\nu}}^{m} c(\tilde{X}_i^N(t_k), \psi_i) + c(\tilde{X}_\nu^N(t_k), \psi_\nu) + h \sum_{\substack{i=1\\ i\neq\nu}}^{m} c(F_i(t_k), \psi_i) + hc(\tilde{F}_\nu(t_k), \psi_\nu)$$

$$+ h \sum_{i=1}^{m} \Bigg(\sum_{\substack{j=1\\ j\neq\nu}}^{m} c(\tilde{X}_j^N(t_k), A_{ij}^T(t_k)\psi_i) + c(\tilde{X}_\nu^N(t_k), A_{i\nu}^T(t_k)\psi_i) \Bigg)$$

$$= \left\{ \psi_\nu = \begin{pmatrix} \psi_{\nu 1} \\ \vdots \\ \psi_{\nu s} \end{pmatrix}, \ \psi_{\nu q} \in \mathbb{R}^{l_q} \right\}$$

$$= \sum_{\substack{i=1\\ i\neq\nu}}^{m} c(\tilde{X}_i^N(t_k), \psi_i) + \sum_{p=1}^{s} c(X_{\nu p}^N(t_k), \psi_{\nu p}) + h \sum_{\substack{i=1\\ i\neq\nu}}^{m} c(F_i(t_k), \psi_i)$$

$$+ h \sum_{p=1}^{s} c(F_{\nu p}(t_k), \psi_{\nu p}) + h \sum_{\substack{i=1\\ i\neq\nu}}^{m} \sum_{\substack{j=1\\ j\neq\nu}}^{m} c(\tilde{X}_j^N(t_k), A_{ij}^T(t_k)\psi_i)$$

$$+ h \sum_{i=1}^{m} c(\tilde{X}_\nu^N(t_k), A_{i\nu}^T(t_k)\psi_i) + h \sum_{\substack{j=1\\ j\neq\nu}}^{m} c(\tilde{X}_j^N(t_k), A_{\nu j}^T(t_k)\psi_\nu)$$

$$= \sum_{\substack{i=1\\ i\neq\nu}}^{m} c(\tilde{X}_i^N(t_k), \psi_i) + \sum_{p=1}^{s} c(X_{\nu p}^N(t_k), \psi_{\nu p}) + h \sum_{\substack{i=1\\ i\neq\nu}}^{m} c(F_i(t_k), \psi_i)$$

$$+ h \sum_{p=1}^{s} c(F_{\nu p}(t_k), \psi_{\nu p}) + h \sum_{\substack{i=1\\ i\neq\nu}}^{m} \sum_{\substack{j=1\\ j\neq\nu}}^{m} c(\tilde{X}_j^N(t_k), A_{ij}^T(t_k)\psi_i)$$

$$+ h \sum_{\substack{i=1\\ i\neq\nu}}^{m} \sum_{p=1}^{s} c(X_{\nu p}^N(t_k), (A_{i\nu}^T(t_k)\psi_i)_p) + h \sum_{p=1}^{s} c(X_{\nu p}^N(t_k), (A_{\nu\nu}^T(t_k)\psi_\nu)_p)$$

$$+ h \sum_{\substack{j=1\\ j\neq\nu}}^{m} c\left(\tilde{X}_j^N(t_k), \sum_{p=1}^{s} (A_{\nu j}^p(t_k))^T \psi_{\nu p} \right)$$

$$= \left\{ (A_{i\nu}^T(t_k)\psi_i)_p = \left(\begin{pmatrix} (A_{i\nu}^1(t_k))^T \\ (A_{i\nu}^2(t_k))^T \\ \vdots \\ (A_{i\nu}^s(t_k))^T \end{pmatrix} \psi_i \right)_p = \left(\begin{pmatrix} (A_{i\nu}^1(t_k))^T \psi_i \\ (A_{i\nu}^2(t_k))^T \psi_i \\ \vdots \\ (A_{i\nu}^s(t_k))^T \psi_i \end{pmatrix} \right)_p \right.$$

$$= (A_{i\nu}^p(t_k))^T \psi_i; \quad (A_{\nu\nu}^T(t_k)\psi_\nu)_p$$

$$= \left(\left(\begin{array}{ccc} (A_{\nu\nu}^{11}(t_k))^T & \cdots & (A_{\nu\nu}^{s1}(t_k))^T \\ \vdots & \ddots & \vdots \\ (A_{\nu\nu}^{1s}(t_k))^T & \cdots & (A_{\nu\nu}^{ss}(t_k))^T \end{array} \right) \left(\begin{array}{c} \psi_{\nu 1} \\ \vdots \\ \psi_{\nu s} \end{array} \right) \right)_p$$

$$\left. = \left(\begin{array}{c} \sum_{q=1}^{s} (A_{\nu\nu}^{q1}(t_k))^T \psi_{\nu q} \\ \vdots \\ \sum_{q=1}^{s} (A_{\nu\nu}^{qs}(t_k))^T \psi_{\nu q} \end{array} \right)_p = \sum_{q=1}^{s} (A_{\nu\nu}^{qp}(t_k))^T \psi_{\nu q} \right\}$$

$$= \sum_{\substack{i=1 \\ i \neq \nu}}^{m} c(\tilde{X}_i^N(t_k), \psi_i) + \sum_{p=1}^{s} c(X_{\nu p}^N(t_k), \psi_{\nu p}) + h \sum_{\substack{i=1 \\ i \neq \nu}}^{m} c(F_i(t_k), \psi_i)$$

$$+ h \sum_{p=1}^{s} c(F_{\nu p}(t_k), \psi_{\nu p}) + h \sum_{\substack{i=1 \\ i \neq \nu}}^{m} \sum_{\substack{j=1 \\ j \neq \nu}}^{m} c(\tilde{X}_j^N(t_k), A_{ij}^T(t_k)\psi_i)$$

$$+ h \sum_{\substack{i=1 \\ i \neq \nu}}^{m} \sum_{p=1}^{s} c(X_{\nu p}^N(t_k), (A_{i\nu}^p(t_k))^T \psi_i) + h \sum_{p=1}^{s} c\left(X_{\nu p}^N(t_k), \sum_{q=1}^{s} (A_{\nu\nu}^{qp}(t_k))^T \psi_{\nu q} \right)$$

$$+ h \sum_{\substack{j=1 \\ j \neq \nu}}^{m} c\left(\tilde{X}_j^N(t_k), \sum_{p=1}^{s} (A_{\nu j}^p(t_k))^T \psi_{\nu p} \right)$$

$$\leq \sum_{\substack{i=1 \\ i \neq \nu}}^{m} c(\tilde{X}_i^N(t_k), \psi_i) + \sum_{p=1}^{s} c(X_{\nu p}^N(t_k), \psi_{\nu p}) + h \sum_{\substack{i=1 \\ i \neq \nu}}^{m} c(F_i(t_k), \psi_i)$$

$$+ h \sum_{p=1}^{s} c(F_{\nu p}(t_k), \psi_{\nu p}) + h \sum_{\substack{i=1 \\ i \neq \nu}}^{m} \sum_{\substack{j=1 \\ j \neq \nu}}^{m} c(\tilde{X}_j^N(t_k), A_{ij}^T(t_k)\psi_i)$$

$$+ h \sum_{\substack{i=1 \\ i \neq \nu}}^{m} \sum_{p=1}^{s} c(X_{\nu p}^N(t_k), (A_{i\nu}^p(t_k))^T \psi_i) + h \sum_{p=1}^{s} \sum_{q=1}^{s} c(X_{\nu p}^N(t_k), (A_{\nu\nu}^{qp}(t_k))^T \psi_{\nu q})$$

$$+ h \sum_{\substack{j=1 \\ j \neq \nu}}^{m} \sum_{p=1}^{s} c(\tilde{X}_j^N(t_k), (A_{\nu j}^p(t_k))^T \psi_{\nu p}). \tag{4.21}$$

We now find the support function of the set $\tilde{X}^N(t_{k+1})$:

$$c(\tilde{X}^N(t_{k+1}), \psi)$$

$$= \sum_{\substack{i=1 \\ i \neq \nu}}^{m} c(\tilde{X}_i^N(t_{k+1}), \psi_i) + \sum_{q=1}^{s} c(X_{\nu q}^N(t_{k+1}), \psi_{\nu q})$$

$$= \sum_{\substack{i=1\\ i\neq\nu}}^{m}\Big[c(\tilde{X}_i^N(t_k),\psi_i) + h\sum_{\substack{j=1\\ j\neq\nu}}^{m} c(\tilde{X}_j^N(t_k), A_{ij}^T(t_k)\psi_i)$$

$$+ h\sum_{p=1}^{s} c(X_{\nu p}^N(t_k),(A_{i\nu}^p(t_k))^T\psi_i) + hc(F_i(t_k),\psi_i)\Big]$$

$$+ \sum_{q=1}^{s}\Big[c(X_{\nu q}^N(t_k),\psi_{\nu q}) + h\sum_{\substack{j=1\\ j\neq\nu}}^{m} c(\tilde{X}_j^N(t),(A_{\nu j}^q(t_k))^T\psi_{\nu q})$$

$$+ h\sum_{p=1}^{s} c(X_{\nu p}^N(t_k),(A_{\nu\nu}^{qp}(t_k))^T\psi_{\nu q}) + hc(F_{\nu q}(t_k),\psi_{\nu q})\Big]$$

$$= \sum_{\substack{i=1\\ i\neq\nu}}^{m} c(\tilde{X}_i^N(t_k),\psi_i) + h\sum_{\substack{i=1\\ i\neq\nu}}^{m}\sum_{\substack{j=1\\ j\neq\nu}}^{m} c(\tilde{X}_j^N(t_k), A_{ij}^T(t_k)\psi_i)$$

$$+ h\sum_{\substack{i=1\\ i\neq\nu}}^{m}\sum_{p=1}^{s} c(X_{\nu p}^N(t_k),(A_{i\nu}^p(t_k))^T\psi_i) + h\sum_{\substack{i=1\\ i\neq\nu}}^{m} c(F_i(t_k),\psi_i)$$

$$+ \sum_{p=1}^{s} c(X_{\nu p}^N(t_k),\psi_{\nu p}) + h\sum_{\substack{j=1\\ j\neq\nu}}^{m}\sum_{p=1}^{s} c(\tilde{X}_j^N(t_k),(A_{\nu j}^p(t_k))^T\psi_{\nu p})$$

$$+ h\sum_{p=1}^{s}\sum_{q=1}^{s} c(X_{\nu p}^N(t_k),(A_{\nu\nu}^{qp}(t_k))^T\psi_{\nu q}) + h\sum_{p=1}^{s} c(F_{\nu p}(t_k),\psi_{\nu p}). \tag{4.22}$$

In view of relations (4.21) and (4.22), we obtain

$$c(\overline{X}^N(t_{k+1}),\psi) \le c(\tilde{X}^N(t_{k+1}),\psi)$$

for all $\psi \in \mathbb{R}^n$. As $N \to \infty$, the Euler broken lines (4.15) and (4.20) converge to the solutions of (4.13) and (4.19), respectively [26]. Hence, passing to the limit and using the property of continuity of the support functions, we conclude that $c(\overline{X}(t),\psi) \le c(\tilde{X}(t),\psi)$ for all $\psi \in \mathbb{R}^n$. Thus, $\overline{X}(t) \subset \tilde{X}(t)$ for all $t \in [0,T]$. The theorem is proved. □

Corollary 1 ([124]). *Let Δ_1 and Δ_2 be decompositions of the matrix $A(t)$ for Eq.* (4.10). *Assume that Δ_2 can be obtained from Δ_1 by additional decomposition. Let $\overline{X}_{\Delta_1}(t)$ and $\overline{X}_{\Delta_2}(t)$ be solutions of systems of the form* (4.13) *corresponding to the indicated decompositions. Then $\overline{X}_{\Delta_1}(t) \subset \overline{X}_{\Delta_2}(t)$ for all $t \in [0,T]$.*

Corollary 2 ([124]). *Denote by* $Y_m(t)$ *the intersection of all sets* $\overline{X}(t)$ *of solutions of systems of the form* (4.13) *for all possible decompositions of the matrix* $A(t)$ *into the matrices* $A_{ij}(t)$, $i, j = \overline{1, m}$. *Then, by Theorems* 4 *and* 5,

$$X(t) = Y_1(t) \subset Y_2(t) \subset \cdots \subset Y_n(t) \quad \textit{for all } t \in [0, T].$$

The inclusion $X(t) \subset Y_n(t)$ is proved and illustrated by model examples in [116, 120].

Remark 2 ([116]). For $m = n$, system (4.13) can be simplified.

Let $X_i(t) = x_i(t) + y_i(t)[-1, 1]$ and $F_i(t) = f_i(t) + g_i(t)[-1, 1]$. Then system (4.13) can be represented in the form

$$D_h\{x_i(t) + y_i(t)[-1, 1]\} = \sum_{j=1}^{n} a_{ij}(t)\{x_j(t) + y_j(t)[-1, 1]\} + f_i(t) + g_i(t)[-1, 1],$$

$$x_i(0) + y_i(0)[-1, 1] = x_i^0 + y_i^0[-1, 1], \quad i = \overline{1, n}.$$

By definition, the Hukuhara derivative

$$\begin{aligned}
&D_h\{x_i(t) + y_i(t)[-1, 1]\} \\
&= \lim_{\Delta \to 0} \frac{1}{\Delta}\{[x_i(t + \Delta) - y_i(t + \Delta), x_i(t + \Delta) + y_i(t + \Delta)] \\
&\qquad - [x_i(t) - y_i(t), x_i(t) + y_i(t)]\} \\
&= \lim_{\Delta \to 0} \frac{1}{\Delta}[x_i(t + \Delta) - y_i(t + \Delta) - (x_i(t) - y_i(t)), \\
&\qquad + x_i(t + \Delta) + y_i(t + \Delta) - (x_i(t) + y_i(t))] \\
&= \Bigg[\lim_{\Delta \to 0} \frac{x_i(t + \Delta) - x_i(t) - (y_i(t + \Delta) - y_i(t))}{\Delta}, \\
&\qquad \lim_{\Delta \to 0} \frac{x_i(t + \Delta) - x_i(t) + (y_i(t + \Delta) - y_i(t))}{\Delta} \Bigg] \\
&= [(x_i(t) - y_i(t))', (x_i(t) + y_i(t))'] = \dot{x}_i(t) + \dot{y}_i(t)[-1, 1].
\end{aligned}$$

Thus, system (4.13) is decomposed into two linear inhomogeneous systems of ordinary differential equations

$$\begin{cases} \dot{x}_i(t) = \sum_{j=1}^{n} a_{ij}(t) x_j(t) + f_i(t), \\ x_i(0) = x_i^0, \quad i = \overline{1, n}, \end{cases}$$

$$\begin{cases} \dot{y}_i(t) = \sum_{j=1}^{n} |a_{ij}(t)| y_j(t) + g_i(t), \\ y_i(0) = y_i^0, \quad i = \overline{1, n}, \end{cases}$$

whose solutions are obtained in the explicit form.

Remark 3. As mentioned above, the construction of solutions of the Hukuhara equation in spaces with dimensionality $n > 2$ encounters serious computational difficulties. Therefore, it is reasonable to decompose the matrix $A(t)$ into blocks such that $n_i \leq 2,\ i = \overline{1,m}$.

Example 3. Consider a controlled system

$$\dot{x}_1 = x_2, \quad \dot{x}_2 = u, \quad u \in [-1,1], \quad x_1(0) = x_2(0) = 0. \tag{4.23}$$

The set of attainability for this system takes the form

$$R(t) = \left\{(x_1, x_2) : \frac{x_2^2}{4} + \frac{x_2 t}{2} - \frac{t^2}{4} \leq x_1 \leq -\frac{x_2^2}{4} + \frac{x_2 t}{2} + \frac{t^2}{4}\right\}.$$

System (4.23) corresponds to the following equation with Hukuhara derivative:

$$D_h X(t) = \begin{pmatrix} 0 & 1 \\ 0 & 0 \end{pmatrix} X(t) + F(t), \quad X(0) = 0,$$

where $F(t) = 0 \times [-1,1]$.

We decompose the matrix A into blocks of dimensionality 1×1 and consider a system of differential equations with Hukuhara derivative of the form (4.13). For any $i = 1,2$, the sum $\sum_{j=1}^{2} A_{ji}^T \psi_j$ contains at most one nonzero element and, moreover, $X_0 = \overline{X}_0$ and $F(t) \equiv \overline{F}(t)$. Hence, by virtue of (4.16) and (4.17), the identity $X(t) \equiv \overline{X}(t)$ is true for all $t \in [0,1]$. According to Remark 2, system (4.13) is reduced to the following two systems of ordinary differential equations:

$$\begin{cases} \dot{x}_1 = x_2,\ \dot{x}_2 = 0, \\ x_1(0) = x_2(0) = 0, \end{cases} \qquad \begin{cases} \dot{y}_1 = y_2,\ \dot{y}_2 = 1, \\ y_1(0) = y_2(0) = 0. \end{cases}$$

As a result of the solution of these systems, we find (Figure 1):

$$x_1(t) = x_2(t) = 0, \quad y_1(t) = \frac{t^2}{2}, \quad \text{and} \quad y_2(t) = t.$$

Thus, we get an approximation of the set of attainability for problem (4.23) in the following form:

$$R(t) \subset X(t) = \overline{X}(t) = \left[-\frac{t^2}{2}, \frac{t^2}{2}\right] \times [-t, t].$$

Example 4. Consider a linear control problem on the segment $[0,1]$

$$\dot{x} = Ax + u, \quad x(0) \in X_0, \quad u(t) \in U, \tag{4.24}$$

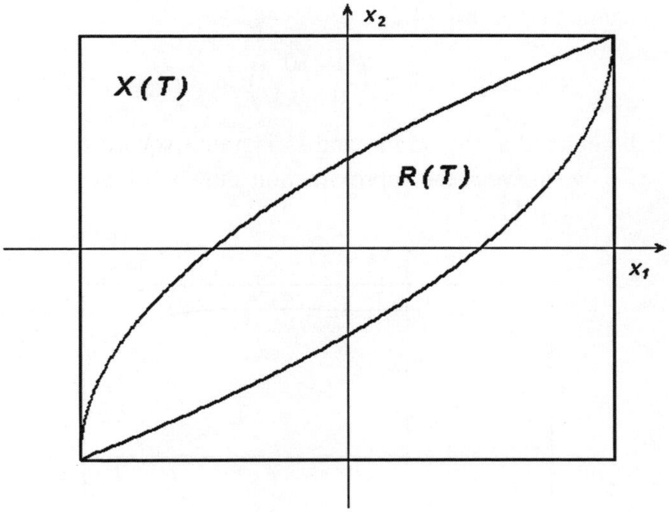

Figure 1. Approximation of the set of attainability for system (4.23).

where

$$A = \begin{pmatrix} -1 & 0 \\ 0 & -1 \end{pmatrix}, \tag{4.25}$$

X_0 is a unit ball centered at the point $(1, 2)$, and U is a unit ball centered at the origin. For $t = 1$, we get the approximation depicted in Figure 2.

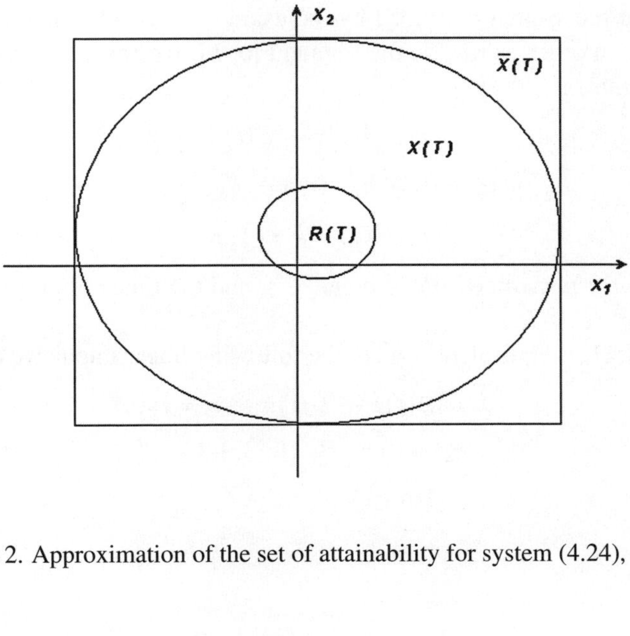

Figure 2. Approximation of the set of attainability for system (4.24), (4.25).

Example 5. Assume that, in Eq. (4.24),

$$A = \begin{pmatrix} 1 & 0 \\ 0 & -0.1 \end{pmatrix}, \tag{4.26}$$

X_0 is a unit ball centered at the origin, and U is a unit square centered at the point $(1, 0)$. In this case, we arrive at the approximation shown in Figure 3.

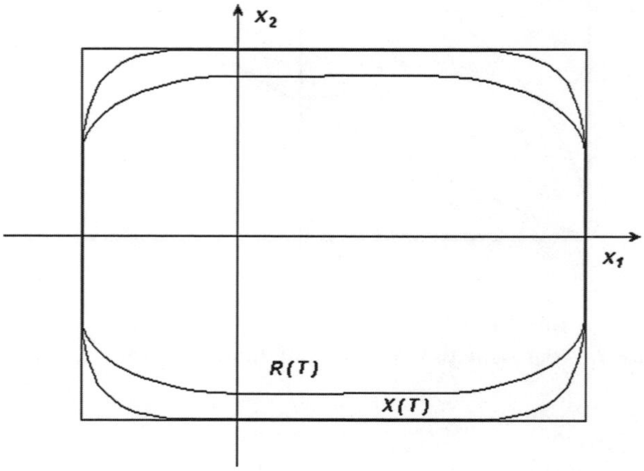

Figure 3. Approximation of the set of attainability for system (4.24), (4.26).

It is natural to study the problem of construction of approximations to the bundles of solutions for linear impulsive differential inclusions.

Assume that the system described by inclusion (4.8) is subjected to pulse actions at fixed times. In other words, on the segment $[0, T]$, we consider a linear impulsive differential inclusion

$$\begin{gathered} \dot{x} \in A(t)x + F(t), \quad t \neq \tau_k, \\ x(\tau_k + 0) \in B_k x(\tau_k) + P_k, \\ x(0) \in X_0, \quad k = \overline{1, K}, \end{gathered} \tag{4.27}$$

where B_k are $n \times n$ matrices, $P_k \in \operatorname{conv}(\mathbb{R}^n)$, and the times of pulses are such that $0 \le \tau_1 < \cdots < \tau_K < T$.

Inclusion (4.27) is equivalent, e.g., to the following linear impulsive control system

$$\begin{gathered} \dot{x} = A(t)x + D(t)u, \quad t \neq \tau_k, \\ x(\tau_k + 0) = B_k x(\tau_k) + C_k v_k, \\ x(0) \in X_0, \quad k = \overline{1, K}, \end{gathered}$$

where $u(t) \in U(t)$ and $v_k \in V_k \in \operatorname{conv}(\mathbb{R}^r)$ are control vectors, C_k are $(n \times r)$-matrices, and in addition, $P_k = \{y \in \mathbb{R}^n : y = C_k v_k, v_k \in V_k\}$.

The equation with Hukuhara derivative corresponding to the impulsive differential inclusion (4.27) takes the form

$$\begin{aligned} D_h X(t) &= A(t)X(t) + F(t), \quad t \neq \tau_k, \\ X(\tau_k + 0) &= B_k X(\tau_k) + P_k, \\ X(0) &= X_0, \quad k = \overline{1, K}, \end{aligned} \tag{4.28}$$

where the solution $X : [0, T] \to \operatorname{conv}(\mathbb{R}^n)$ is a set-valued mapping piecewise continuously differentiable in Hukuhara's sense.

Assume that the matrix $A(t)$ can be represented in the form (4.12) and that the matrices B_k $(k = \overline{1, K})$ admit the following representations:

$$B_k = \begin{pmatrix} B_{11}^k & B_{12}^k & \cdots & B_{1m}^k \\ B_{21}^k & B_{22}^k & \cdots & B_{2m}^k \\ \vdots & \vdots & \ddots & \vdots \\ B_{m1}^k & B_{m2}^k & \cdots & B_{mm}^k \end{pmatrix}, \quad \text{where } B_{ij}^k \in \mathbb{R}^{n_i \times n_j}.$$

Equation (4.28) is associated the following system of linear impulsive differential equations with Hukuhara derivative:

$$\begin{aligned} D_h X_i(t) &= \sum_{j=1}^{m} A_{ij}(t) X_j(t) + F_i(t), \quad t \neq \tau_k, \\ X_i(\tau_k + 0) &= \sum_{j=1}^{m} B_{ij}^k X_j(\tau_k) + P_i^k, \\ X_i(0) &= X_i^0 \in \operatorname{conv}(\mathbb{R}^{n_i}), \quad i = \overline{1, m}, \ k = \overline{1, K}, \end{aligned} \tag{4.29}$$

where $P_k \subset \overline{P}_k = P_1^k \times \cdots \times P_m^k$, $P_i^k \in \operatorname{conv}(\mathbb{R}^{n_i})$, and $X_i : [0, T] \to \operatorname{conv}(\mathbb{R}^{n_i})$ are functions piecewise continuously differentiable in Hukuhara's sense.

Consider a set $\overline{X}(t) = X_1(t) \times \cdots \times X_m(t)$.

Theorem 6 ([124]). *The inclusion $R(t) \subset X(t) \subset \overline{X}(t)$, where $R(t)$ is the set of attainability of* (4.27) *holds for Eqs.* (4.27), (4.28), *and* (4.29) *for any $t \in [0, T]$.*

Proof. Denote $\tau_0 = 0$ and $\tau_{K+1} = T$. Suppose that the inclusion

$$R(\tau_{k-1} + 0) \subset X(\tau_{k-1} + 0) \subset \overline{X}(\tau_{k-1} + 0)$$

holds for some $k \in \overline{1, K + 1}$. By virtue of Theorem 4 and inclusion (4.11), we get $R(t) \subset X(t) \subset \overline{X}(t)$ for all $\tau_{k-1} < t \leq \tau_k$.

We now show that $R(\tau_k+0)\subset X(\tau_k+0)\subset \overline{X}(\tau_k+0)$, $k\in\overline{1,K}$. The first part of the inclusion directly follows from (4.27) and (4.28). We now prove the second inclusion. In view of the convexity of the sets $X(\tau_k+0)$ and $\overline{X}(\tau_k+0)$, it suffices to show that the inequality $c(X(\tau_k+0),\psi)\le c(\overline{X}(\tau_k+0),\psi)$ holds for all $\psi\in\mathbb{R}^n$.

Equations (4.28) and (4.29) now imply that

$$
\begin{aligned}
c(X(\tau_k+0),\psi) &= c(B_kX(\tau_k)+P_k,\psi)\\
&= c(X(\tau_k),B_k^T\psi)+c(P_k,\psi)\le c(\overline{X}(\tau_k),B_k^T\psi)+c(\overline{P}_k,\psi)\\
&= \sum_{i=1}^m c(X_i(\tau_k),(B_k^T\psi)_i)+\sum_{i=1}^m c(P_i^k,\psi_i)\\
&= \left\{ (B_k^T\psi)_i = \left(\begin{pmatrix} (B_{11}^k)^T & (B_{21}^k)^T & \cdots & (B_{m1}^k)^T\\ (B_{12}^k)^T & (B_{22}^k)^T & \cdots & (B_{m2}^k)^T\\ \vdots & \vdots & \ddots & \vdots\\ (B_{1m}^k)^T & (B_{2m}^k)^T & \cdots & (B_{mm}^k)^T \end{pmatrix} \begin{pmatrix}\psi_1\\ \psi_2\\ \vdots\\ \psi_m\end{pmatrix}\right)_i \right.\\
&\qquad \left. = \begin{pmatrix} \sum_{j=1}^m (B_{j1}^k)^T\psi_j\\ \sum_{j=1}^m (B_{j2}^k)^T\psi_j\\ \vdots\\ \sum_{j=1}^m (B_{jm}^k)^T\psi_j \end{pmatrix}_i = \sum_{j=1}^m (B_{ji}^k)^T\psi_j \right\}\\
&= \sum_{i=1}^m c\Bigl(X_i(\tau_k),\sum_{j=1}^m (B_{ji}^k)^T\psi_j\Bigr)+\sum_{i=1}^m c(P_i^k,\psi_i)\\
&\le \sum_{i=1}^m\sum_{j=1}^m c(X_i(\tau_k),(B_{ji}^k)^T\psi_j)+\sum_{i=1}^m c(P_i^k,\psi_i);
\end{aligned}
\tag{4.30}
$$

$$
\begin{aligned}
c(\overline{X}(\tau_k+0),\psi) &= \sum_{i=1}^m c(X_i(\tau_k+0),\psi_i)\\
&= \sum_{i=1}^m c\Bigl(\sum_{j=1}^m B_{ij}^kX_j(\tau_k)+P_i^k,\psi_i\Bigr)\\
&= \sum_{i=1}^m\sum_{j=1}^m c(X_j(\tau_k),(B_{ij}^k)^T\psi_i)+\sum_{i=1}^n c(P_i^k,\psi_i).
\end{aligned}
\tag{4.31}
$$

In view of (4.30) and (4.31), we conclude that the inequality

$$c(X(\tau_k+0),\psi)\le c(\overline{X}(\tau_k+0),\psi)$$

is true for all $\psi \in \mathbb{R}^n$ and, hence, $X(\tau_k + 0) \subset \overline{X}(\tau_k + 0)$. The theorem is thus proved.

We now study the problem of variation of the set $\overline{X}(t)$ in the case of subsequent decomposition of the matrix. Assume that the νth row and νth column of the matrix $A(t)$ $\left(\nu \in \overline{1,m}\right)$ are decomposed as in (4.18) and the νth rows and νth columns of the matrices B_k are decomposed into the following matrices:

$$B^k_{\nu\nu} = \begin{pmatrix} B^{k_{11}}_{\nu\nu} & B^{k_{12}}_{\nu\nu} & \cdots & B^{k_{1s}}_{\nu\nu} \\ B^{k_{21}}_{\nu\nu} & B^{k_{22}}_{\nu\nu} & \cdots & B^{k_{2s}}_{\nu\nu} \\ \vdots & \vdots & \ddots & \vdots \\ B^{k_{s1}}_{\nu\nu} & B^{k_{s2}}_{\nu\nu} & \cdots & B^{k_{ss}}_{\nu\nu} \end{pmatrix}, \quad \text{where } B^{k_{pq}}_{\nu\nu} \in \mathbb{R}^{l_p \times l_q},\ \sum_{p=1}^{s} l_p = n_\nu;$$

$$B^k_{i\nu} = \begin{pmatrix} B^{k1}_{i\nu} & B^{k2}_{i\nu} & \cdots & B^{ks}_{i\nu} \end{pmatrix}, \quad \text{where } B^{kp}_{i\nu} \in \mathbb{R}^{n_i \times l_p},\ i = \overline{1,m},\ i \neq \nu; \tag{4.32}$$

$$B^k_{\nu i} = \begin{pmatrix} B^{k1}_{\nu i} \\ B^{k2}_{\nu i} \\ \vdots \\ B^{ks}_{\nu i} \end{pmatrix}, \quad \text{where } B^{kp}_{\nu i} \in \mathbb{R}^{l_p \times n_i},\ i = \overline{1,m},\ i \neq \nu.$$

Parallel with system (4.13), we consider a system

$$\begin{aligned} D_h \tilde{X}_i(t) &= \sum_{\substack{j=1 \\ j\neq\nu}}^{m} A_{ij}(t)\tilde{X}_j(t) + \sum_{p=1}^{s} A^p_{i\nu}(t) X_{\nu p}(t) + \tilde{F}_i(t), \quad t \neq \tau_k, \\ \tilde{X}_i(\tau_k + 0) &= \sum_{\substack{j=1 \\ j\neq\nu}}^{m} B^k_{ij}\tilde{X}_j(\tau_k) + \sum_{p=1}^{s} B^{kp}_{i\nu} X_{\nu p}(\tau_k) + \tilde{P}^k_i, \\ \tilde{X}_i(0) &= X^0_i, \quad i = \overline{1,m},\ i \neq \nu, \\ D_h X_{\nu q}(t) &= \sum_{\substack{j=1 \\ j\neq\nu}}^{m} A^q_{\nu j}(t)\tilde{X}_j(t) + \sum_{p=1}^{s} A^{qp}_{\nu\nu}(t) X_{\nu p}(t) + F_{\nu q}(t), \\ X_{\nu q}(\tau_k + 0) &= \sum_{\substack{j=1 \\ j\neq\nu}}^{m} B^{kq}_{\nu j}\tilde{X}_j(\tau_k) + \sum_{p=1}^{s} B^{k_{qp}}_{\nu\nu} X_{\nu p}(\tau_k) + P^k_{\nu q}, \\ X_{\nu q}(0) &= X^0_{\nu q}, \quad q = \overline{1,s},\ k = \overline{1,K}, \end{aligned} \tag{4.33}$$

where $P^k_i = \tilde{P}^k_i$, $i \neq \nu$, $P^k_\nu \subset \tilde{P}^k_\nu = P^k_{\nu 1} \times \cdots \times P^k_{\nu s}$, $P^k_{\nu q} \in \operatorname{conv}(\mathbb{R}^{l_q})$, $X^0_\nu \in \tilde{X}^0_\nu = X^0_{\nu 1} \times \cdots \times X^0_{\nu s}$, $\tilde{X}_i : [0,T] \to \operatorname{conv}(\mathbb{R}^{n_i})$ $(i \neq \nu)$, and $X_{\nu q} : [0,T] \to \operatorname{conv}(\mathbb{R}^{l_q})$ $(q = \overline{1,s})$ are set-valued mappings piecewise continuously differentiable in Hukuhara's sense.

Consider a set

$$\tilde{X}(t) = \tilde{X}_1(t) \times \cdots \times \tilde{X}_{\nu-1}(t) \times X_{\nu 1}(t) \times \cdots \times X_{\nu s}(t)$$
$$\times \tilde{X}_{\nu+1}(t) \times \cdots \times \tilde{X}_m(t). \qquad \square$$

Theorem 7 ([124]). *The following inclusion is true for systems* (4.29) *and* (4.33) *for any* $t \in [0, T]$*:*

$$\overline{X}(t) \subset \tilde{X}(t).$$

Proof. Denote $\tau_0 = 0$ and $\tau_{K+1} = T$. Assume that the inclusion

$$\overline{X}(\tau_{k-1} + 0) \subset \tilde{X}(\tau_{k-1} + 0)$$

holds for some $k \in \overline{1, K+1}$. By Theorem 5, for all $\tau_{k-1} < t \le \tau_k$, we have $\overline{X}(t) \subset \tilde{X}(t)$.

We now show that $\overline{X}(\tau_k + 0) \subset \tilde{X}(\tau_k + 0)$, $k \in \overline{1, K}$. In view of the convexity of the sets $\overline{X}(\tau_k + 0)$ and $\tilde{X}(\tau_k + 0)$, it suffices to show that the inequality

$$c(\overline{X}(\tau_k + 0), \psi) \le c(\tilde{X}(\tau_k + 0), \psi)$$

is true for all $\psi \in \mathbb{R}^n$. By virtue of (4.31), Eqs. (4.29) and (4.33) imply that

$$\begin{aligned}
c(\overline{X}(\tau_k + 0), \psi) &= \sum_{i=1}^{m} \sum_{j=1}^{m} c(X_j(\tau_k), (B_{ij}^k)^T \psi_i) + \sum_{i=1}^{m} c(P_i^k, \psi_i) \\
&\le \sum_{i=1}^{m} \sum_{j=1}^{m} c(\tilde{X}_j(\tau_k), (B_{ij}^k)^T \psi_i) + \sum_{i=1}^{m} c(\tilde{P}_i^k, \psi_i) \\
&= \sum_{i=1}^{m} \Bigg[\sum_{\substack{j=1 \\ j \ne \nu}}^{m} c(\tilde{X}_j(\tau_k), (B_{ij}^k)^T \psi_i) + c(\tilde{X}_\nu(\tau_k), (B_{i\nu}^k)^T \psi_i) \Bigg] \\
&\quad + \sum_{\substack{i=1 \\ i \ne \nu}}^{m} c(P_i^k, \psi_i) + c(\tilde{P}_\nu^k, \psi_\nu) \\
&= \sum_{\substack{i=1 \\ i \ne \nu}}^{m} \sum_{\substack{j=1 \\ j \ne \nu}}^{m} c(\tilde{X}_j(\tau_k), (B_{ij}^k)^T \psi_i) + \sum_{i=1}^{m} c(\tilde{X}_\nu(\tau_k), (B_{i\nu}^k)^T \psi_i) \\
&\quad + \sum_{\substack{j=1 \\ j \ne \nu}}^{m} c(\tilde{X}_j(\tau_k), (B_{\nu j}^k)^T \psi_\nu) + \sum_{\substack{i=1 \\ i \ne \nu}}^{m} c(P_i^k, \psi_i) + \sum_{p=1}^{s} c(P_{\nu p}^k, \psi_{\nu p})
\end{aligned}$$

$$= \sum_{\substack{i=1 \\ i \neq \nu}}^{m} \sum_{\substack{j=1 \\ j \neq \nu}}^{m} c(\tilde{X}_j(\tau_k), (B_{ij}^k)^T \psi_i) + \sum_{\substack{i=1 \\ i \neq \nu}}^{m} \sum_{p=1}^{s} c(X_{\nu p}(\tau_k), ((B_{i\nu}^k)^T \psi_i)_p)$$

$$+ \sum_{p=1}^{s} c(X_{\nu p}(\tau_k), ((B_{\nu\nu}^k)^T \psi_\nu)_p) + \sum_{\substack{j=1 \\ j \neq \nu}}^{m} c(\tilde{X}_j(\tau_k), (B_{\nu j}^k)^T \psi_\nu)$$

$$+ \sum_{\substack{i=1 \\ i \neq \nu}}^{m} c(P_i^k, \psi_i) + \sum_{p=1}^{s} c(P_{\nu p}^k, \psi_{\nu p})$$

$$= \left\{ ((B_{i\nu}^k)^T \psi_i)_p = \left(\begin{pmatrix} (B_{i\nu}^{k1})^T \\ \vdots \\ (B_{i\nu}^{ks})^T \end{pmatrix} \psi_i \right)_p = (B_{i\nu}^{kp})^T \psi_i; \right.$$

$$(B_{\nu j}^k)^T \psi_\nu = \sum_{p=1}^{s} (B_{\nu j}^{kp})^T \psi_{\nu p};$$

$$((B_{\nu\nu}^k)^T \psi_\nu)_p = \left(\begin{pmatrix} (B_{\nu\nu}^{k_{11}})^T & \cdots & (B_{\nu\nu}^{k_{s1}})^T \\ \vdots & \ddots & \vdots \\ (B_{\nu\nu}^{k_{1s}})^T & \cdots & (B_{\nu\nu}^{k_{ss}})^T \end{pmatrix} \begin{pmatrix} \psi_{\nu 1} \\ \vdots \\ \psi_{\nu s} \end{pmatrix} \right)_p$$

$$\left. = \begin{pmatrix} \sum_{q=1}^{s} (B_{\nu\nu}^{k_{q1}})^T \psi_{\nu q} \\ \vdots \\ \sum_{q=1}^{s} (B_{\nu\nu}^{k_{qs}})^T \psi_{\nu q} \end{pmatrix}_p = \sum_{q=1}^{s} (B_{\nu\nu}^{k_{qp}})^T \psi_{\nu q} \right\}$$

$$= \sum_{\substack{i=1 \\ i \neq \nu}}^{m} \sum_{\substack{j=1 \\ j \neq \nu}}^{m} c(\tilde{X}_j(\tau_k), (B_{ij}^k)^T \psi_i) + \sum_{\substack{i=1 \\ i \neq \nu}}^{m} \sum_{p=1}^{s} c(X_{\nu p}(\tau_k), (B_{i\nu}^{kp})^T \psi_i)$$

$$+ \sum_{p=1}^{s} c\left(X_{\nu p}(\tau_k), \sum_{q=1}^{s} (B_{\nu\nu}^{k_{qp}})^T \psi_{\nu q} \right)$$

$$+ \sum_{\substack{j=1 \\ j \neq \nu}}^{m} c\left(\tilde{X}_j(\tau_k), \sum_{p=1}^{s} (B_{\nu j}^{kp})^T \psi_{\nu p} \right)$$

$$+ \sum_{\substack{i=1 \\ i \neq \nu}}^{m} c(P_i^k, \psi_i) + \sum_{p=1}^{s} c(P_{\nu p}^k, \psi_{\nu p})$$

$$
\begin{aligned}
&\leq \sum_{\substack{i=1\\ i\neq\nu}}^{m}\sum_{\substack{j=1\\ j\neq\nu}}^{m} c(\tilde{X}_j(\tau_k),(B_{ij}^{k})^T\psi_i)\\
&\quad+\sum_{\substack{i=1\\ i\neq\nu}}^{m}\sum_{p=1}^{s} c(X_{\nu p}(\tau_k),(B_{i\nu}^{kp})^T\psi_i)\\
&\quad+\sum_{p=1}^{s}\sum_{q=1}^{s} c(X_{\nu p}(\tau_k),(B_{\nu\nu}^{k_{qp}})^T\psi_{\nu q})\\
&\quad+\sum_{\substack{j=1\\ j\neq\nu}}^{m}\sum_{p=1}^{s} c(\tilde{X}_j(\tau_k),(B_{\nu j}^{kp})^T\psi_{\nu p})\\
&\quad+\sum_{\substack{i=1\\ i\neq\nu}}^{m} c(P_i^k,\psi_i)+\sum_{p=1}^{s} c(P_{\nu p}^k,\psi_{\nu p}).
\end{aligned}
\tag{4.34}
$$

We now find the support function of the set $\tilde{X}(\tau_k+0)$:

$$
\begin{aligned}
c(\tilde{X}(\tau_k+0),\psi) &= \sum_{\substack{i=1\\ i\neq\nu}}^{m} c(\tilde{X}_i(\tau_k+0),\psi_i)+\sum_{q=1}^{s} c(X_{\nu q}(\tau_k+0),\psi_{\nu q})\\
&= \sum_{\substack{i=1\\ i\neq\nu}}^{m} c\bigg(\sum_{\substack{j=1\\ j\neq\nu}}^{m} B_{ij}^{k}\tilde{X}_j(\tau_k)+\sum_{p=1}^{s} B_{i\nu}^{kp}X_{\nu p}(\tau_k)+\tilde{P}_i^k,\psi_i\bigg)\\
&\quad+\sum_{q=1}^{s} c\bigg(\sum_{\substack{j=1\\ j\neq\nu}}^{m} B_{\nu j}^{kq}\tilde{X}_j(\tau_k)+\sum_{p=1}^{s} B_{\nu\nu}^{k_{qp}}X_{\nu p}(\tau_k)+P_{\nu q}^k,\psi_{\nu q}\bigg)\\
&= \sum_{\substack{i=1\\ i\neq\nu}}^{m}\sum_{\substack{j=1\\ j\neq\nu}}^{m} c(\tilde{X}_j(\tau_k),(B_{ij}^{k})^T\psi_i)+\sum_{\substack{i=1\\ i\neq\nu}}^{m}\sum_{p=1}^{s} c(X_{\nu p}(\tau_k),(B_{i\nu}^{kp})^T\psi_i)\\
&\quad+\sum_{\substack{i=1\\ i\neq\nu}}^{m} c(P_i^k,\psi_i)+\sum_{q=1}^{s}\sum_{\substack{j=1\\ j\neq\nu}}^{m} c(\tilde{X}_j(\tau_k),(B_{\nu j}^{kq})^T\psi_{\nu q})\\
&\quad+\sum_{q=1}^{s}\sum_{p=1}^{s} c(X_{\nu p}(\tau_k),(B_{\nu\nu}^{k_{qp}})^T\psi_{\nu q})+\sum_{q=1}^{s} c(P_{\nu q}^k,\psi_{\nu q}).
\end{aligned}
\tag{4.35}
$$

By using (4.34) and (4.35), we conclude that the inequality $c(\overline{X}(\tau_k+0),\psi)\le c(\tilde{X}(\tau_k+0),\psi)$ holds for all $\psi\in\mathbb{R}^n$. Hence, $\overline{X}(\tau_k+0)\subset\tilde{X}(\tau_k+0)$. The theorem is thus proved. □

Corollary 3 ([124]). *Let Δ_1 and Δ_2 be decompositions of the matrices $A(t)$ and B_k for Eq.* (4.28). *Assume that Δ_2 can be obtained from Δ_1 by additional decomposition. Let $\overline{X}_{\Delta_1}(t)$ and $\overline{X}_{\Delta_2}(t)$ be solutions of systems of the form* (4.29) *corresponding to the given partitions. Then $\overline{X}_{\Delta_1}(t)\subset\overline{X}_{\Delta_2}(t)$ for all $t\in[0,T]$.*

Corollary 4 ([124]). *Let $Y_m(t)$, be the intersection of all sets $\overline{X}(t)$ of solutions of systems of the form* (4.29) *for all possible decompositions of the matrices $A(t)$ and B_k into the matrices $A_{ij}(t)$ and B^k_{ij} $i,j=\overline{1,m}$. Hence, by virtue of Theorems* 6 *and* 7*,*

$$X(t)=Y_1(t)\subset Y_2(t)\subset\cdots\subset Y_n(t)\quad \textit{for all}\ t\in[0,T].$$

The estimate $X(t)\subset Y_n(t)$ is proved in [116].

Remark 4 ([116]). For $m=n$, system (4.29) decomposes into two systems of linear impulsive differential equations. Let $X_i(t)=x_i(t)+y_i(t)[-1,1]$, $F_i(t)=f_i(t)+g_i(t)[-1,1]$, and $P_i^k=p_i^k+q_i^k[-1,1]$.

Thus, by analogy with Remark 2, we find

$$\begin{cases}\dot{x}_i=\sum_{j=1}^n a_{ij}(t)x_j(t)+f_i(t), & t\neq\tau_k,\ k=\overline{1,N},\\ x_i(\tau_k+0)=\sum_{j=1}^n b_{ij}^k x_j(\tau_k)+p_i^k, & \\ x_i(0)=x_i^0. & \end{cases}\tag{4.36}$$

$$\begin{cases}\dot{y}_i=\sum_{j=1}^n |a_{ij}(t)|y_j(t)+g_i(t), & t\neq\tau_k,\ k=\overline{1,N},\\ y_i(\tau_k+0)=\sum_{j=1}^n |b_{ij}^k| y_j(\tau_k)+q_i^k, & \\ y_i(0)=y_i^0. & \end{cases}\tag{4.37}$$

Example 6. Consider the following linear impulsive equation with Hukuhara derivative:

$$D_hX(t)=\begin{pmatrix}2 & 1\\ -1 & -2\end{pmatrix}X(t),\quad t\neq i,\ i\in N,$$

$$X(i+0)=\begin{pmatrix}\frac{2}{3e^4} & -\frac{1}{3e^4}\\ \frac{1}{3e^4} & -\frac{2}{3e^4}\end{pmatrix}X(i),\tag{4.38}$$

$$X(0)=X_0\in\operatorname{conv}(\mathbb{R}^2).$$

Let $X_0 \subset \tilde{X}_0 = X_1^0 \times X_2^0$, where $X_k^0 = x_k^0 + y_k^0[-1, 1]$, $k = 1, 2$. In this case, systems (4.36) and (4.37) take the form

$$\begin{cases} \dot{x}_1 = 2x_1 + x_2, \\ \dot{x}_2 = -x_1 - 2x_2, \quad t \neq i, \\ x_1(i+0) = \frac{2}{3e^4} x_1(i) - \frac{1}{3e^4} x_2(i), \\ x_2(i+0) = \frac{1}{3e^4} x_1(i) - \frac{2}{3e^4} x_2(i), \quad i \in N, \\ x_1(0) = x_1^0, \\ x_2(0) = x_2^0, \end{cases} \tag{4.39}$$

$$\begin{cases} \dot{y}_1 = 2y_1 + y_2, \\ \dot{y}_2 = y_1 + 2y_2, \quad t \neq i, \\ y_1(i+0) = \frac{2}{3e^4} y_1(i) + \frac{1}{3e^4} y_2(i), \\ y_2(i+0) = \frac{1}{3e^4} y_1(i) + \frac{2}{3e^4} y_2(i), \quad i \in N, \\ y_1(0) = y_1^0, \\ y_2(0) = y_2^0. \end{cases} \tag{4.40}$$

We now apply Theorem 9.4 from [143] to systems (4.39) and (4.40). For system (4.39), we have $\gamma = 2$, $\alpha = e^{-4}$, and $p = 1$. At the same time, for system (4.40), we get $\gamma = 3$, $\alpha = e^{-4}$, and $p = 1$. In both cases, $\gamma + p \ln \alpha < 0$ and, hence, all solutions of systems (4.39) and (4.40) are asymptotically stable. Therefore, $|\overline{X}(t)| \to 0$ as $t \to \infty$. By virtue of Theorem 6, this means that $|X(t)| \to 0$ as $t \to \infty$. Thus, we have constructed the asymptotics of the solution of the impulsive equation with Hukuhara derivative (4.38) as $t \to \infty$.

Remark 5. In systems (4.8) and (4.27), the matrix $A(t)$ and the set-valued mapping $F(t)$ can be regarded as measurable.

Remark 6. According to the definition of the Hukuhara derivative, the set-valued mapping $X(t)$ always extends in the course of time. However, the set of attainability of inclusion (4.8) [or (4.27)] may shrink or extend for different periods of time. In this connection, we consider the application of systems of differential equations with π-derivative for the construction of approximations to the set of attainability of these systems.

4.3 Linear Differential Equations with π-Derivative

As already indicated, the spaces $\operatorname{comp}(\mathbb{R}^n)$ and $\operatorname{conv}(\mathbb{R}^n)$ are not linear. In [146] and [15], the operation of subtraction in the space $\operatorname{conv}(\mathbb{R}^n)$ is introduced with the help of imbedding of this space in a linear space. By the theorem of imbedding [129], there exist a normalized linear space $\mathcal{B}$ and an isometric mapping $\pi : \operatorname{conv}(\mathbb{R}^n) \to \mathcal{B}$ such that $\pi(\operatorname{conv}(\mathbb{R}^n))$ is a convex cone in $\mathcal{B}$. In addition, the operations of summation and multiplication by a nonnegative scalar in $\mathcal{B}$ induce the corresponding operations in $\operatorname{conv}(\mathbb{R}^n)$.

In order to define the space $\mathcal{B}$, we introduce the relation of equivalence in the space $\operatorname{conv}(\mathbb{R}^n) \times \operatorname{conv}(\mathbb{R}^n)$ as follows:

$$(A, B) \sim (C, D) \quad \text{for } A + D = B + C.$$

By $\langle A, B\rangle$ we denote the class of equivalence containing (A, B). The space $\mathcal{B}$ is defined as the quotient space

$$\operatorname{conv}(\mathbb{R}^n) \times \operatorname{conv}(\mathbb{R}^n)/\sim.$$

In the space $\mathcal{B}$, we introduce the operations of summation and multiplication by a scalar as follows:

$$\langle A, B\rangle + \langle C, D\rangle = \langle A + C, B + D\rangle,$$

$$\alpha\langle A, B\rangle = \begin{cases} \langle \alpha A, \alpha B\rangle, & \alpha \geq 0, \\ \langle |\alpha| B, |\alpha| A\rangle, & \alpha < 0. \end{cases}$$

The space $\mathcal{B}$ thus defined is linear.

The imbedding $\pi : \operatorname{conv}(\mathbb{R}^n) \to \mathcal{B}$ is defined as follows:

$$\pi(A) = \langle A, 0\rangle, \quad A \in \operatorname{conv}(\mathbb{R}^n),$$

i.e., $\langle A, 0\rangle$ is the class of equivalence $\{(A + D, D) : D \in \operatorname{conv}(\mathbb{R}^n)\}$.

The metric and norm in the space $\mathcal{B}$ are given by the formulas

$$\rho(\langle A, B\rangle, \langle C, D\rangle) = h(A + D, C + B),$$

$$\|\langle A, B\rangle\| = \rho(\langle A, B\rangle, \langle 0, 0\rangle).$$

The space $\mathcal{B}$ is not a complete metric space [127]. At the same time, the sharp convex closed cone corresponding to the complete metric space $\operatorname{conv}(\mathbb{R}^n)$ [35] is complete [127].

Let $A, B \in \operatorname{conv}(\mathbb{R}^n)$. The difference of the spaces A and B is defined as the element of the space $\mathcal{B}$ equal to $\langle A, B\rangle$. Since the space $\mathcal{B}$ is linear, the indicated difference possesses all properties of difference in linear spaces.

In the general case, the difference of two sets from $\operatorname{conv}(\mathbb{R}^n)$ is not an element of the space $\operatorname{conv}(\mathbb{R}^n)$. At the same time, if the Hukuhara difference of two sets from the space $\operatorname{conv}(\mathbb{R}^n)$ exists, then

$$\langle A, B\rangle = \langle A \overset{h}{-} B, 0\rangle.$$

Definition 5 ([15, 146]). A set-valued mapping $F : \mathbb{R} \to \operatorname{conv}(\mathbb{R}^n)$ is called π-differentiable at a point $t_0 \in \mathbb{R}$ if there exists a continuous linear mapping $G(t_0) : \mathbb{R} \to \mathcal{B}$ such that

$$\langle F(t), F(t_0)\rangle = G(t_0)(t - t_0) + o(\|t - t_0\|). \tag{4.41}$$

If $G(t_0)(t - t_0) = \langle A(t_0)(t - t_0), B(t_0)(t - t_0)\rangle, A(t_0), B(t_0) \in \operatorname{conv}(\mathbb{R}^n)$, then, in terms of the Hausdorff metric, relation (4.41) means that

$$h(F(t) + B(t_0)(t - t_0), F(t_0) + A(t_0)(t - t_0)) = o(\|t - t_0\|).$$

Moreover, if $G(t_0)(t - t_0) = \langle A(t_0)(t - t_0), 0\rangle$, then we say that $F(\cdot)$ is conically differentiable for $t_0 \in \mathbb{R}$.

We now consider the relationship between the differentiability in Hukuhara's sense and π-conic differentiability.

Theorem 8 ([15]). *If a set-valued mapping $F : \mathbb{R} \to \operatorname{conv}(\mathbb{R}^n)$ is differentiable in Hukuhara's sense for $t_0 \in \mathbb{R}$ with the derivative $D_h F(t_0)$, then $F(\cdot)$ is conically differentiable and*

$$G(t_0)(t - t_0) = (t - t_0)\langle D_h F(t_0), 0\rangle.$$

It is also possible to show that if $F : \mathbb{R} \to \operatorname{conv}(\mathbb{R}^n)$ is conically differentiable for $t_0 \in \mathbb{R}$ and the Hukuhara differences exist for sufficiently small Δt, then $F(\cdot)$ is differentiable in Hukuhara's set at the point $t_0 \in \mathbb{R}$. Moreover, if $G(t)\Delta t = \Delta t\langle \widehat{F}(t), 0\rangle, t \in \mathbb{R}$, then $D_h F(t) = \widehat{F}(t)$.

However, generally speaking, the π-differentiability of $F : \mathbb{R} \to \operatorname{conv}(\mathbb{R}^n)$ at a point $t_0 \in \mathbb{R}$ does not imply the differentiability of $F(\cdot)$ in Hukuhara's sense.

Example 7. Consider a set-valued mapping

$$F(t) = (2 + \sin(t))S_1(0), \quad t \in [0, 2\pi].$$

This mapping is π-differentiable on $[0, 2\pi]$ and

$$G(t) = \left\langle \frac{1 + \operatorname{sign}(\cos(t))}{2} |\cos(t)| S_1(0), \frac{1 - \operatorname{sign}(\cos(t))}{2} |\cos(t)| S_1(0)\right\rangle.$$

At the same time, it is not differentiable in Hukuhara's sense because the function $\operatorname{diam}(F(\cdot))$ is not nondecreasing on $[0, 2\pi]$.

The differential equations with π-derivative can be regarded as a natural generalization of the notion of differential equations with Hukuhara derivative.

Consider a system of linear differential equations with π-derivative

$$D_\pi \langle X_i(t), Y_i(t)\rangle = \sum_{j=1}^{n} a_{ij}(t)\langle X_j(t), Y_j(t)\rangle + \langle F_i(t), G_i(t)\rangle, \tag{4.42}$$

$$\langle X_i(0), Y_i(0)\rangle = \langle X_i^0, Y_i^0\rangle, \quad i = \overline{1,n},$$

where $t \in [0, T]$ is time and $a_{ij} : [0, T] \to \mathbb{R}$ and $F_i, G_i : [0, T] \to \mathrm{conv}(\mathbb{R}^m)$ are continuous functions.

Definition 6 ([121]). The couples $\langle X_i(t), Y_i(t)\rangle$, $X_i, Y_i : [0, T] \to \mathrm{conv}(\mathbb{R}^m)$, $i = \overline{1,n}$, are called the solution of problem (4.42) if they are continuously π-differentiable and satisfy (4.42) everywhere on $[0, T]$.

We now show that the solution of system (4.42) can be reduced to the solution of a system of ordinary differential equations.

Definition 7 ([121]). A function

$$c(\langle A, B\rangle, \psi) = c(A, \psi) - c(B, \psi)$$

is called the support function of the couple $\langle A, B\rangle$ in the direction of a vector $\psi \in \mathbb{R}^m$.

This definition is correct, i.e., independent of the choice of a representative (A, B) of the class of equivalence $\langle A, B\rangle$:

Let $(A, B), (C, D) \in \langle A, B\rangle$. Then

$$c(A, \psi) + c(D, \psi) = c(B, \psi) + c(C, \psi)$$
$$\Rightarrow c(A, \psi) - c(B, \psi) = c(C, \psi) - c(D, \psi).$$

Moreover, the class $\langle A, B\rangle$ is uniquely defined by its support function. Assume the contrary, i.e., that there exist two classes $\langle A, B\rangle$ and $\langle C, D\rangle$ with the same support function. According to the definition of the support function of the couple, we have

$$c(A, \psi) - c(B, \psi) = c(C, \psi) - c(D, \psi)$$

for all $\psi \in \mathbb{R}^m$. Hence,

$$c(A, \psi) + c(D, \psi) = c(B, \psi) + c(C, \psi) \quad \text{for all } \psi \in \mathbb{R}^m$$
$$\Rightarrow c(A + D, \psi) = c(B + C, \psi) \quad \text{for all } \psi \in \mathbb{R}^m.$$

In view of the convexity of the sets A, B, C, and D, we find $A + D = = B + C$. Hence, the classes $\langle A, B\rangle$ and $\langle C, D\rangle$ coincide.

Lemma 1 ([121]). *The following assertions are true:*

(a) $c(\langle A, B\rangle + \langle C, D\rangle, \psi) = c(\langle A, B\rangle, \psi) + c(\langle C, D\rangle, \psi)$;

(b) $c(\lambda\langle A, B\rangle, \psi) = \lambda c(\langle A, B\rangle, \psi)$;

(c) $c(\langle A, B\rangle, \psi)$ *is a continuous function of its arguments;*

(d) $c(D_\pi\langle X(t), Y(t)\rangle, \psi) = \frac{d}{dt}c(\langle X(t), Y(t)\rangle, \psi)$.

Proof. (a)

$$
\begin{aligned}
&c(\langle A, B\rangle + \langle C, D\rangle, \psi) = c(\langle A + C, B + D\rangle, \psi)\\
&= c(A + C, \psi) - c(B + D, \psi)\\
&= c(A, \psi) + c(C, \psi) - (c(B, \psi) + c(D, \psi))\\
&= c(A, \psi) - c(B, \psi) + c(C, \psi) - c(D, \psi) = c(\langle A, B\rangle, \psi) + c(\langle C, D\rangle, \psi).
\end{aligned}
$$

(b) Let $\lambda \geq 0$. Then

$$
\begin{aligned}
c(\lambda\langle A, B\rangle, \psi) &= c(\langle \lambda A, \lambda B\rangle, \psi) = c(\lambda A, \psi) - c(\lambda B, \psi)\\
&= \lambda c(A, \psi) - \lambda c(B, \psi) = \lambda c(\langle A, B\rangle, \psi).
\end{aligned}
$$

At the same time, if $\lambda < 0$, then

$$
\begin{aligned}
c(\lambda\langle A, B\rangle, \psi) &= c(\langle |\lambda| B, |\lambda| A\rangle, \psi) = c(|\lambda| B, \psi) - c(|\lambda| A, \psi)\\
&= |\lambda|(c(B, \psi) - c(A, \psi)) = \lambda c(\langle A, B\rangle, \psi).
\end{aligned}
$$

(c) This assertion follows from the definition of the support function of a couple and from the continuity of the support function $c(\cdot, \cdot)$ in its arguments.

(d)

$$
\begin{aligned}
&c(D_\pi\langle X(t), Y(t)\rangle, \psi)\\
&= c\Big(\lim_{\Delta t \to 0} \frac{\langle X(t + \Delta t), Y(t + \Delta t)\rangle - \langle X(t), Y(t)\rangle}{\Delta t}, \psi\Big)\\
&= c\Big(\lim_{\Delta t \to 0} \frac{1}{\Delta t}\langle X(t + \Delta t) + Y(t), Y(t + \Delta t) + X(t)\rangle, \psi\Big)\\
&= \lim_{\Delta t \to 0} \frac{1}{\Delta t} c(\langle X(t + \Delta t) + Y(t), Y(t + \Delta t) + X(t)\rangle, \psi)\\
&= \lim_{\Delta t \to 0} \frac{1}{\Delta t}[c(X(t + \Delta t), \psi) + c(Y(t), \psi)\\
&\qquad - (c(Y(t + \Delta t), \psi) + c(X(t), \psi))]
\end{aligned}
$$

$$= \lim_{\Delta t \to 0} \frac{1}{\Delta t}[c(X(t+\Delta t), \psi) - c(Y(t+\Delta t), \psi) - (c(X(t), \psi) - c(Y(t), \psi))]$$

$$= \frac{d}{dt}(c(X(t), \psi) - c(Y(t), \psi)) = \frac{d}{dt} c(\langle X(t), Y(t) \rangle, \psi).$$

The lemma is proved. □

By virtue of the lemma, system (4.42) is equivalent to the following system:

$$\frac{d}{dt} c(\langle X_i(t), Y_i(t) \rangle, \psi) = \sum_{j=1}^{n} a_{ij}(t) c(\langle X_j(t), Y_j(t) \rangle, \psi) + c(\langle F_i(t), G_i(t) \rangle, \psi),$$

$$c(\langle X_i(0), Y_i(0) \rangle, \psi) = c(\langle X_i^0, Y_i^0 \rangle, \psi), \quad i = \overline{1, n},$$

for all $\psi \in \mathbb{R}^m$.

Denote

$$z_i(t, \psi) = c(\langle X_i(t), Y_i(t) \rangle, \psi), \quad f_i(t, \psi) = c(\langle F_i(t), G_i(t) \rangle, \psi).$$

As a result, the previous system can be rewritten in the form

$$\frac{d}{dt} z_i(t, \psi) = \sum_{j=1}^{n} a_{ij}(t) z_j(t, \psi) + f_i(t, \psi), \tag{4.43}$$

$$z_i(0, \psi) = c(\langle X_i^0, Y_i^0 \rangle, \psi), \quad i = \overline{1, n}.$$

Thus, we get a system of n ordinary differential equations with parameter ψ.

Hence, we have proved the following assertion:

Theorem 9 ([121]). *The solution of system* (4.42) *is given by n couples $\langle X_i(t), Y_i(t) \rangle$, $i = \overline{1, n}$, uniquely determined by their support functions $z_i(t, \psi)$ obtained as solutions of the system of ordinary differential Eqs.* (4.43).

4.4 Extension of the Space conv($\mathbb{R}^n$) for $n = 1$

The elements of the space conv($\mathbb{R}$) are segments. Hence, we present each class of equivalence in the form $\langle [x, x+\delta], 0 \rangle, \delta \geq 0$ or $\langle x, [0, \delta] \rangle$, $\delta > 0$. This presentation of the class of equivalence is called canonical.

Indeed, any element $([a, b], [c, d])$ belongs to a class of the first kind for $b - a \geq d - c$ [in this case, $x = a - c$ and $\delta = (b - a) - (d - c)$] or to a class of the second kind for $b - a < d - a$ [in this case, $x = a - c$ and $\delta = (d - c) - (b - a)$].

We represent a class of equivalence in the canonical form as (x, δ). In this case, $\langle [x, x+\delta], 0 \rangle$ for $\delta \geq 0$ and $\langle x, [0, -\delta] \rangle = \langle 0, [-x, -x-\delta] \rangle$ for $\delta < 0$.

Lemma 2 ([121]). *The following assertions are true:*

(a) $(x,\delta) = (y,\Delta) \Leftrightarrow x = y,\ \delta = \Delta$;

(b) $(x,\delta) + (y,\Delta) = (x + y, \delta + \Delta)$;

(c) $a(x,\delta) = (ax, a\delta)$ *for any* $a \in R$.

Proof. (a) The sufficiency of this assertion is obvious. Thus, we prove its necessity.

Let $\delta \geq 0$ and $\Delta \geq 0$. Then

$$(x,\delta) = \langle [x, x+\delta], 0\rangle = \langle [y, y+\Delta], 0\rangle = (y,\Delta).$$

This yields $x = y$ and $\delta = \Delta$.

Let $\delta < 0$ and $\Delta < 0$. Then

$$\begin{aligned}(x,\delta) &= \langle x, [0, -\delta]\rangle = \langle 0, [-x, -x-\delta]\rangle,\\ (y,\Delta) &= \langle y, [0, -\Delta]\rangle = \langle 0, [-y, -y-\Delta]\rangle,\end{aligned}$$

whence it follows that $x = y$ and $\delta = \Delta$.

Let $\delta \geq 0$ and $\Delta < 0$. Then

$$(x,\delta) = \langle [x, x+\delta], 0\rangle, \quad (y,\Delta) = \langle 0, [-y, -y-\Delta]\rangle,$$

which means that the equality $(x,\delta) = (y,\Delta)$ is impossible. The case where $\delta < 0$ and $\Delta \geq 0$ is analyzed similarly.

(b) Let $\delta \geq 0$ and $\Delta \geq 0$. Then

$$\begin{aligned}(x,\delta) + (y,\Delta) &= \langle [x, x+\delta], 0\rangle + \langle [y, y+\Delta], 0\rangle\\ &= \langle [x+y, x+y+\delta+\Delta], 0\rangle = (x+y, \delta+\Delta).\end{aligned}$$

Let $\delta < 0$ and $\Delta < 0$. Then

$$\begin{aligned}(x,\delta) + (y,\Delta) &= \langle x, [0, -\delta]\rangle + \langle y, [0, -\Delta]\rangle\\ &= \langle x+y, [0, -\delta-\Delta]\rangle = (x+y, \delta+\Delta).\end{aligned}$$

Let $\delta \geq 0$ and $\Delta < 0$. Then

$$\begin{aligned}(x,\delta) + (y,\Delta) &= \langle [x, x+\delta], 0\rangle + \langle y, [0, -\Delta]\rangle\\ &= \langle [x+y, x+y+\delta], [0, -\Delta]\rangle\\ &= \begin{cases}\langle [x+y, x+y+\delta-(-\Delta)], 0\rangle, & \delta \geq -\Delta\\ \langle x+y, [0, -\Delta-\delta]\rangle, & \delta < -\Delta\end{cases} = (x+y, \delta+\Delta).\end{aligned}$$

The case where $\delta < 0$ and $\Delta \geq 0$ is analyzed similarly.

(c) For $a = 0$, the assertion is obvious. Consider the case $a > 0$. If $\delta \geq 0$, then

$$a(x,\delta) = a\langle[x, x+\delta], 0\rangle = \langle[ax, ax + a\delta], 0\rangle = (ax, a\delta).$$

Further, if $\delta < 0$, then

$$a(x,\delta) = a\langle x, [0, -\delta]\rangle = \langle ax, [0, -a\delta]\rangle = (ax, a\delta).$$

Now let $a < 0$. Thus, if $\delta \geq 0$, then

$$\begin{aligned} a(x,\delta) &= a\langle[x, x+\delta], 0\rangle = \langle 0, |a|[x, x+\delta]\rangle = \langle 0, [|a|x, |a|x + |a|\delta]\rangle \\ &= (-|a|x, -|a|\delta) = (ax, a\delta). \end{aligned}$$

At the same time, if $\delta < 0$, then

$$\begin{aligned} a(x,\delta) &= a\langle x, [0, -\delta]\rangle = \langle |a|[0, -\delta], |a|x\rangle \\ &= \langle[-|a|x, -|a|x - |a|\delta], 0\rangle = (-|a|x, -|a|\delta) = (ax, a\delta). \end{aligned}$$ □

Lemma 3 ([121]).

$$\lim_{k\to\infty} (x_k, \delta_k) = (x,\delta) \Leftrightarrow \lim_{k\to\infty} x_k = x, \lim_{k\to\infty} \delta_k = \delta.$$

Proof. First, we prove necessity. Assume that all $\delta_k \geq 0$ beginning with some k_0. Then

$$(x_k, \delta_k) = \langle[x_k, x_k + \delta_k], 0\rangle.$$

Further, let $\delta \geq 0$. Then

$$(x,\delta) = \langle[x, x+\delta], 0\rangle.$$

Since

$$\lim_{k\to\infty} (x_k, \delta_k) = (x,\delta),$$

we get

$$\begin{aligned} \rho((x_k, \delta_k), (x,\delta)) &= \rho(\langle[x_k, x_k + \delta_k], 0\rangle, \langle[x, x+\delta], 0\rangle) \\ &= h([x_k, x_k + \delta_k], [x, x+\delta]) \\ &= \max\{\|x_k - x\|, \|x_k + \delta_k - x - \delta\|\} \to 0 \end{aligned}$$

as $k \to \infty$. Hence,

$$\lim_{k\to\infty} x_k = x \quad \text{and} \quad \lim_{k\to\infty} \delta_k = \delta.$$

If $\delta < 0$, then $(x,\delta) = \langle 0, [-x, -x-\delta]\rangle$ and, therefore,

$$\begin{aligned} \rho((x_k, \delta_k), (x,\delta)) &= \rho(\langle[x_k, x_k + \delta_k], 0\rangle, \langle 0, [-x, -x-\delta]\rangle) \\ &= h([x_k - x, x_k - x + \delta_k - \delta], 0) \\ &= \max\{\|x_k - x\|, \|x_k + \delta_k - x - \delta\|\} \nrightarrow 0, \end{aligned}$$

which contradicts the condition.

The case where all $\delta_k < 0$ beginning with some k_0 is analyzed similarly.

Assume that there are infinitely many $\delta_k \geq 0$ and infinitely many $\delta_k < 0$. In this case, we split the sequence of numbers $\{k\}$ into two subsequences $\{k_1 : \delta_{k_1} \geq 0\}$ and $\{k_2 : \delta_{k_2} < 0\}$.

Since any subsequence of a convergent sequence converges, we have $\lim_{k_1\to\infty}(x_{k_1}, \delta_{k_1}) = (x, \delta)$ and, as shown above, $\lim_{k_1\to\infty} x_{k_1} = x$ and $\lim_{k_1\to\infty} \delta_{k_1} = \delta \geq 0$, i.e., for any $\varepsilon > 0$, there exists k_1^0 such that the following estimates are true for $k_1 > k_1^0$:

$$\|x_{k_1} - x\| < \varepsilon \quad \text{and} \quad \|\delta_{k_1} - \delta\| < \varepsilon.$$

Similarly, we obtain $\lim_{k_2\to\infty}(x_{k_2}, \delta_{k_2}) = (x, \delta)$. Hence, $\lim_{k_2\to\infty} x_{k_2} = x$, $\lim_{k_2\to\infty} \delta_{k_2} = \delta \leq 0$. In other words, for any $\varepsilon > 0$, there exists k_2^0 such that the estimates

$$\|x_{k_2} - x\| < \varepsilon \quad \text{and} \quad \|\delta_{k_2} - \delta\| < \varepsilon$$

are true for $k_2 > k_2^0$.

Thus, we get $\delta = 0$. Choosing $k_0 = \max\{k_1^0, k_2^0\}$, we conclude that $\|x_k - x\| < \varepsilon$ and $\|\delta_k\| < \varepsilon$ for $k > k_0$. Hence, necessity is proved.

Sufficiency is proved similarly. □

Lemma 4 ([121]).

$$D_\pi(x(t), \delta(t)) = (\dot{x}(t), \dot{\delta}(t)),$$

where $D_\pi(x(t), \delta(t))$ *is the* π*-derivative* [146], [15] *of the couple* $(x(t), \delta(t))$.

Proof. By using the definition of the π-derivative and Lemmas 2 and 3, we find

$$\begin{aligned}
D_\pi(x(t), \delta(t)) &= \lim_{\Delta\to 0} \frac{1}{\Delta}[(x(t+\Delta), \delta(t+\Delta)) - (x(t), \delta(t))] \\
&= \lim_{\Delta\to 0} \frac{1}{\Delta}[(x(t+\Delta), \delta(t+\Delta)) + (-x(t), -\delta(t))] \\
&= \lim_{\Delta\to 0} \frac{1}{\Delta}(x(t+\Delta) - x(t), \delta(t+\Delta) - \delta(t)) \\
&= \lim_{\Delta\to 0} \left(\frac{x(t+\Delta) - x(t)}{\Delta}, \frac{\delta(t+\Delta) - \delta(t)}{\Delta} \right) \\
&= \left(\lim_{\Delta\to 0} \frac{x(t+\Delta) - x(t)}{\Delta}, \lim_{\Delta\to 0} \frac{\delta(t+\Delta) - \delta(t)}{\Delta} \right) \\
&= (\dot{x}(t), \dot{\delta}(t)).
\end{aligned}$$

□

4.5 Approximation of the Integral Funnel of a Linear Differential Inclusion with the Help of Systems of Differential Equations with π-Derivative

Example 8. Consider a linear differential inclusion

$$\dot{x} \in ax + [-m, m], \quad x(0) = 0 \tag{4.44}$$

for $m > 0$ and various $a \neq 0$. Tolstonogov ([145], p. 232) studied the relationship between the R-solution $R(t)$ of inclusion (4.44) and a solution of the corresponding equation with Hukuhara derivative

$$D_h X(t) = aX(t) + [-m, m], \quad X(0) = 0. \tag{4.45}$$

It was shown that the following equality is true for $a > 0$ and $t \in [0, T]$:

$$R(t) = X(t) = \frac{e^{at} - 1}{a}[-m, m].$$

At the same time, if $a < 0$, then

$$R(t) = \frac{e^{at} - 1}{a}[-m, m], \quad X(t) = e^{-at} R(t)$$

and, hence, $R(t) \subset X(t)$ for $t \in (0, T]$.

Consider the equation with π-derivative corresponding to inclusion (4.44):

$$\begin{cases} D_\pi(x, \delta) = a(x, \delta) + (-m, 2m), \\ (x(0), \delta(0)) = (0, 0). \end{cases} \tag{4.46}$$

By virtue of Lemmas 2 and 4, this equation decomposes into two linear inhomogeneous equations

$$\begin{cases} \dot{x} = ax - m, \\ x(0) = 0, \end{cases} \qquad \begin{cases} \dot{\delta} = a\delta + 2m, \\ \delta(0) = 0. \end{cases}$$

As a result of the solution of these equations, we conclude that

$$x(t) = -\frac{m}{a}(e^{at} - 1) \quad \text{and} \quad \delta(t) = \frac{2m}{a}(e^{at} - 1).$$

Therefore,

$$\begin{aligned} (x(t), \delta(t)) &= \left(-\frac{m}{a}(e^{at} - 1), \frac{2m}{a}(e^{at} - 1)\right) \\ &= \left\langle \left[-\frac{m}{a}(e^{at} - 1), \frac{m}{a}(e^{at} - 1)\right], 0 \right\rangle = \left\langle \frac{e^{at} - 1}{a}[-m, m], 0 \right\rangle. \end{aligned}$$

Hence, for all a, the R-solution of inclusion (4.44) coincides with the integral funnel of inclusion (4.44) and with the solution of the equation with π-derivative (4.46).

Thus, it is reasonable to consider the problem of approximation of the R-solution of a linear differential inclusion with the help of the solution of the corresponding equation with π-derivative.

Consider a linear differential inclusion

$$\dot{x} \in A(t)x + F(t), \quad x(0) \in X_0, \tag{4.47}$$

where $t \in [0, T]$, $x \in \mathbb{R}^n$ is the phase vector, $A(t)$ is a continuous $(n \times n)$-matrix, $F(t)$ is a continuous set-valued mapping $[0, T] \to \mathrm{comp}(\mathbb{R}^n)$, and $X_0 \in \mathrm{conv}(\mathbb{R}^n)$.

The R-solution $R(t)$ of inclusion (4.47) with the initial condition $R(0) = X_0$ has the form

$$R(t) = \Phi(t, 0)X_0 + \int_0^t \Phi(t, s)F(s)ds, \tag{4.48}$$

where $\Phi(t, s)$ is the matrizant of the system $\dot{x} \in A(t)x$.

Assume that a mapping $\tilde{F}(t) = F_1(t) \times \cdots \times F_n(t)$, where

$$F_i(t) = [f_i(t), f_i(t) + r_i(t)], \quad i = \overline{1, n},$$

and a set $\tilde{X}_0 = X_1^0 \times \cdots \times X_n^0$, where $X_i^0 = [x_i^0, x_i^0 + \delta_i^0], i = \overline{1, n}$, are such that $F(t) \subseteq \tilde{F}(t)$ $(F(t) \supseteq \tilde{F}(t))$ for all $t \in [0, T]$ and $X_0 \subseteq \tilde{X}_0$ $(X_0 \supseteq \tilde{X}_0)$.

Inclusion (4.47) is associated with a system of linear differential equations with π-derivative of the form

$$D_\pi(x_i, \delta_i) = \sum_{j=1}^n a_{ij}(t)(x_j, \delta_j) + (f_i(t), r_i(t)), \tag{4.49}$$
$$(x_i(0), \delta_i(0)) = (x_i^0, \delta_i^0), \quad i = \overline{1, n}.$$

By virtue of Lemmas 2 and 4, system (4.49) decomposes into two systems of linear differential equations

$$\begin{cases} \dot{x}_i = \sum_{j=1}^n a_{ij}(t)x_j + f_i(t), \\ x_i(0) = x_i^0, \quad i = \overline{1, n}, \end{cases} \tag{4.50}$$

$$\begin{cases} \dot{\delta}_i = \sum_{j=1}^n a_{ij}(t)\delta_j + r_i(t), \\ \delta_i(0) = \delta_i^0, \quad i = \overline{1, n}, \end{cases} \tag{4.51}$$

whose solutions can be represented in the form

$$x(t) = (x_1(t), \ldots, x_n(t))^T = \Phi(t, 0)x_0 + \int_0^t \Phi(t, s)f(s)ds,$$

$$\delta(t) = (\delta_1(t), \ldots, \delta_n(t))^T = \Phi(t, 0)\delta_0 + \int_0^t \Phi(t, s)r(s)ds.$$

Now let

$$\begin{cases} [x_i(t), x_i(t) + \delta_i(t)], & \delta_i(t) \geq 0, \\ [x_i(t) + \delta_i(t), x_i(t)], & \delta_i(t) < 0, \end{cases}$$

and let $\tilde{X}(t) = X_1(t) \times \cdots \times X_n(t)$.

We now study the relationship between the sets $R(t)$ and $\tilde{X}(t)$. According to the property of the support functions, we find

$$\begin{aligned} c(R(t), \psi) &= c(\Phi(t,0)X_0, \psi) + c\bigg(\int_0^t \Phi(t,s)F(s)ds, \psi\bigg) \\ &= c(X_0, \Phi^T(t,0)\psi) + \int_0^t c(F(s), \Phi^T(t,s)\psi)ds \\ &\leq (\geq)\ c(\tilde{X}_0, \Phi^T(t,0)\psi) + \int_0^t c(\tilde{F}(s), \Phi^T(t,s)\psi)ds \\ &= \sum_{i=1}^n \bigg\{\bigg(x_i^0 + \frac{\delta_i^0}{2}\bigg)(\Phi^T(t,0)\psi)_i + \frac{\delta_i^0}{2}|(\Phi^T(t,0)\psi)_i|\bigg\} \\ &\quad + \int_0^t \sum_{i=1}^n \bigg\{\bigg(f_i(s) + \frac{r_i(s)}{2}\bigg)(\Phi^T(t,s)\psi)_i + \frac{r_i(s)}{2}|(\Phi^T(t,s)\psi)_i|\bigg\}ds \\ &= \sum_{i=1}^n \bigg\{\bigg(x_i^0 + \frac{\delta_i^0}{2}\bigg)\sum_{j=1}^n \phi_{ji}(t,0)\psi_j + \frac{\delta_i^0}{2}\bigg|\sum_{j=1}^n \phi_{ji}(t,0)\psi_j\bigg|\bigg\} \\ &\quad + \sum_{i=1}^n \int_0^t \bigg\{\bigg(f_i(s) + \frac{r_i(s)}{2}\bigg)\sum_{j=1}^n \phi_{ji}(t,s)\psi_j \\ &\qquad + \frac{r_i(s)}{2}\bigg|\sum_{j=1}^n \phi_{ji}(t,s)\psi_j\bigg|\bigg\}ds = I_1(t,\psi), \end{aligned}$$

$$\begin{aligned} c(\tilde{X}(t), \psi) &= \sum_{i=1}^n c(X_i, \psi_i) = \sum_{i=1}^n \bigg(\bigg(x_i + \frac{\delta_i}{2}\bigg)\psi_i + \frac{|\delta_i|}{2}|\psi_i|\bigg) \\ &= \sum_{i=1}^n \bigg\{\bigg(\sum_{j=1}^n \phi_{ij}(t,0)\bigg(x_j^0 + \frac{\delta_j^0}{2}\bigg) \\ &\quad + \int_0^t \sum_{j=1}^n \phi_{ij}(t,s)\bigg(f_j(s) + \frac{r_j(s)}{2}\bigg)ds\bigg)\psi_i \\ &\quad + \frac{|\psi_i|}{2}\bigg|\sum_{j=1}^n \phi_{ij}(t,0)\delta_j^0 + \sum_{j=1}^n \int_0^t \phi_{ij}(t,s)r_j(s)ds\bigg|\bigg\} = I_2(t,\psi). \end{aligned}$$

The relationship between $I_1(t,\psi)$ and $I_2(t,\psi)$ is directly connected with the relationship between

$$J_1(t,\psi) = \sum_{i=1}^{n}\left(\delta_i^0\left|\sum_{j=1}^{n}\phi_{ji}(t,0)\psi_j\right| + \int_0^t\left(r_i(s)\left|\sum_{j=1}^{n}\phi_{ji}(t,s)\psi_j\right|\right)ds\right)$$

and

$$J_2(t,\psi) = \sum_{i=1}^{n}|\psi_i|\left|\sum_{j=1}^{n}\phi_{ij}(t,0)\delta_j^0 + \sum_{j=1}^{n}\int_0^t \phi_{ij}(t,s)r_j(s)ds\right|.$$

Thus, we have proved the following assertion:

Theorem 10 ([121]). *Assume that, for the linear differential inclusion* (4.47)*, there exist a continuous function $\tilde{F}(t)$ such that $F(t) \subseteq \tilde{F}(t)$ $(F(t) \supseteq \tilde{F}(t))$ and a set $\tilde{X}_0$ such that $X_0 \subseteq \tilde{X}_0$ $(X_0 \supseteq \tilde{X}_0)$ and $J_1(t,\psi) \le J_2(t,\psi)$ $(J_1(t,\psi) \ge J_2(t,\psi))$ for all $\psi \in \mathbb{R}^n$ and $t \in [0,T]$.*

Then $R(t) \subseteq \tilde{X}(t)$ $(R(t) \supseteq \tilde{X}(t))$ for all $t \in [0,T]$.

Corollary 5. *Assume that $n = 1$. In this case, $F(t) = \tilde{F}(t)$, $X_0 = \tilde{X}_0$, and*

$$\begin{aligned}J_1(t,\psi) &= \delta_0|\phi(t,0)\psi| + \int_0^t r(s)|\phi(t,s)\psi|ds\\ &\ge |\psi|\left|\phi(t,0)\delta_0 + \int_0^t r(s)\phi(t,s)ds\right| = J_2(t,\psi)\end{aligned}$$

for all $\psi \in \mathbb{R}$ and $t \in [0,T]$. Then $R(t) \supseteq \tilde{X}(t)$.

Example 9. Consider the differential inclusion (4.47) with $n = 2$ and a diagonal matrix $\Phi(t,s)$. Assume that the mapping $\tilde{F}(t)$ and the set $\tilde{X}_0$ are such that $F(t) \supseteq \tilde{F}(t)$ and $X_0 \supseteq \tilde{X}_0$. In this case,

$$\begin{aligned}J_1(t,\psi) &= \delta_1^0|\phi_{11}(t,0)\psi_1| + \delta_2^0|\phi_{22}(t,0)\psi_2|\\ &\quad + \int_0^t r_1(s)|\phi_{11}(t,s)\psi_1|ds + \int_0^t r_2(s)|\phi_{22}(t,s)\psi_2|ds,\\ J_2(t,\psi) &= |\psi_1|\left|\int_0^t r_1(s)\phi_{11}(t,s)ds + \phi_{11}(t,0)\delta_1^0\right|\\ &\quad + |\psi_2|\left|\int_0^t r_2(s)\phi_{22}(t,s)ds + \phi_{22}(t,0)\delta_2^0\right|.\end{aligned}$$

Since $\phi_{11}(t,s)$ and $\phi_{22}(t,s)$ are sign-preserving functions (otherwise, the matrix $\Phi(t,s)$ is nondegenerate), we conclude that $J_1(t,\psi) = J_2(t,\psi)$ for all $\psi \in \mathbb{R}^n$

and $t \in [0, T]$. Thus, by virtue of the already proved theorem, $R(t) \supseteq \tilde{X}(t)$ for all $t \in [0, T]$. If $F(t) = \tilde{F}(t)$ and $X_0 = \tilde{X}_0$, then $R(t) = \tilde{X}(t)$.

In the case where the matrix $\Phi(t, s)$ is inversely diagonal, we get the same estimate.

Corollary 6. *Assume that a matrix $\Phi(t, s)$ has the following property: Each row and each column of the matrix contain a single nonzero element. Let a mapping $\tilde{F}(t)$ and a set $\tilde{X}_0$ be such that $F(t) \supseteq \tilde{F}(t)$ and $X_0 \supseteq \tilde{X}_0$. Then $R(t) \supseteq \tilde{X}(t)$ for all $t \in [0, T]$.*

It is clear that, in the general case, it is impossible to say which of the quantities $J_1(t, \psi)$ and $J_2(t, \psi)$ is larger.

Theorem 11 ([121]). *Assume that, for the linear differential inclusion* (4.47) *and any $k \notin \overline{1, n}$, all $\phi_{ki}(t, s)$ are nonnegative (nonpositive) for all $i \notin \overline{1, n}$ and $t, s \in [0, T]$. Moreover, there exist a function $\tilde{F}(t)$ such that $F(t) \subseteq \tilde{F}(t)$ and a set $\tilde{X}_0$ such that $X_0 \subseteq \tilde{X}_0$.*

Then $R(t) \subseteq \tilde{X}(t)$ for all $t \in [0, T]$.

Proof. We use Theorem 10. It suffices to check the validity of the inclusion $J_1(t, \psi) \leq J_2(t, \psi)$ for the vectors $\psi = \pm e_k$, $k = \overline{1, n}$, where e_k is a unit vector. In this case,

$$J_1(t, \pm e_k) = \sum_{i=1}^{n} \left(\delta_i^0 |\phi_{ki}(t, 0)| + \int_0^t r_i(s) |\phi_{ki}(t, s)| ds \right),$$

$$J_2(t, \pm e_k) = \left| \sum_{i=1}^{n} \phi_{ki}(t, 0) \delta_i^0 + \sum_{i=1}^{n} \int_0^t \phi_{ki}(t, s) r_i(s) ds \right|.$$

Since, for any $k = \overline{1, n}$, all $\phi_{ki}(t, s)$ are nonnegative (nonpositive) for all $i = \overline{1, n}$ and $0 \leq s \leq t \leq T$, we conclude that $J_1(t, \pm e_k) = J_2(t, \pm e_k)$ for all $k = \overline{1, n}$ and $t \in [0, T]$. □

Remark 7. If, in Theorem 10, $F(t) = \tilde{F}(t)$ and $X_0 = \tilde{X}_0$, then $J_1(t, \pm e_k) = J_2(t, \pm e_k)$ for all $k = \overline{1, n}$ and $t \in [0, T]$. Thus, the set $R(t)$ is inscribed in the set $\tilde{X}(t)$.

Example 10. In [84], the differential inclusion (4.47) was considered for $n = 1$:

$$\dot{x} \in a(t)x + F(t), \quad x(0) \in X_0. \tag{4.52}$$

In the case where $a(t) \geq 0$ and the mapping $F : [0, T] \to \text{conv}(\mathbb{R})$ is measurable and integrally bounded, it was shown in [145] that the R-solution $R(t)$ of inclusion (4.52) coincides with a solution of the equation with Hukuhara derivative

$$D_h X = a(t)X + F(t), \quad X(0) = X_0.$$

Consider the equation with π-derivative corresponding to inclusion (4.52):

$$D_\pi(x,\delta) = a(t)(x,\delta) + (f(t), r(t)), \tag{4.53}$$
$$(x(0), \delta(0)) = (x_0, \delta_0).$$

In this case, $F(t) = \tilde{F}(t)$, $X_0 = \tilde{X}_0$, and

$$\phi(t,s) = \exp \int_s^t a(\tau)d\tau > 0$$

for all $t, s \in [0, T]$.

Thus, by virtue of Theorem 11 and Corollary 5, we get

$$\begin{aligned} R(t) &= \tilde{X}(t) \\ &= \Big[x_0 e^{\int_0^t a(s)ds} + \int_0^t f(s) e^{\int_s^t a(\tau)d\tau} ds, (x_0+\delta_0)e^{\int_0^t a(s)ds} \\ &\quad + \int_0^t (f(s)+r(s)) e^{\int_s^t a(\tau)d\tau} ds \Big]. \end{aligned}$$

Example 11. Consider a linear inhomogeneous inclusion

$$\begin{pmatrix} \dot{x}_1 \\ \dot{x}_2 \end{pmatrix} \in \begin{pmatrix} -2 & 1 \\ 1 & -2 \end{pmatrix} \begin{pmatrix} x_1 \\ x_2 \end{pmatrix} + \begin{pmatrix} 0 \\ [0,2]\sin^2 t \end{pmatrix}, \tag{4.54}$$
$$x_1(0) = x_2(0) = 0.$$

The matrizant of the corresponding homogeneous system takes the form

$$\Phi(t,s) = \frac{1}{2} \begin{pmatrix} e^{-(t-s)} + e^{-3(t-s)} & e^{-(t-s)} - e^{-3(t-s)} \\ e^{-(t-s)} - e^{-3(t-s)} & e^{-(t-s)} + e^{-3(t-s)} \end{pmatrix}.$$

Since all $\phi_{ij}(t,s) > 0$ for $i, j \in 1, 2$ and all $0 \le s \le t \le T$, by virtue of Theorem 11, the R-solution $R(t)$ of inclusion (4.54) is a subset of the set $\tilde{X}(t)$ specified by the system of equations with π-derivative

$$\begin{cases} D_\pi(x_1,\delta_1) = -2(x_1,\delta_1) + (x_2,\delta_2), \\ D_\pi(x_2,\delta_2) = (x_1,\delta_1) - 2(x_2,\delta_2) + (0, 2\sin^2 t), \\ (x_1(0), \delta_1(0)) = (0,0), \\ (x_2(0), \delta_2(0)) = (0,0). \end{cases}$$

This system decomposes into two systems of linear equations

$$\begin{cases} \dot{x}_1 = -2x_1 + x_2, \\ \dot{x}_2 = x_1 - 2x_2, \\ x_1(0) = 0, \\ x_2(0) = 0, \end{cases} \qquad \begin{cases} \dot{\delta}_1 = -2\delta_1 + \delta_2, \\ \dot{\delta}_2 = \delta_1 - 2\delta_2 + 2\sin^2 t, \\ \delta_1(0) = 0, \\ \delta_2(0) = 0, \end{cases}$$

whose solutions can be represented in the form

$$x_1(t) \equiv 0, \quad x_2(t) \equiv 0,$$

$$\delta_1(t) = -\frac{2}{5}e^{-t} + \frac{10}{195}e^{-3t} - \frac{8}{65}\sin 2t + \frac{1}{65}\cos 2t + \frac{1}{3},$$

$$\delta_2(t) = -\frac{2}{5}e^{-t} - \frac{10}{195}e^{-3t} - \frac{18}{65}\sin 2t - \frac{14}{65}\cos 2t + \frac{2}{3}.$$

Thus, $\tilde{X}(t)$ approaches a periodic solution as $t \to \infty$, whereas the solution of the equation with Hukuhara derivative corresponding to inclusion (4.54) is an extending set as $t \to \infty$.

Consider a linear impulsive differential inclusion

$$\begin{aligned} \dot{x} &\in A(t)x + F(t), \quad t \neq \tau_k, \ x(0) \in X_0, \\ \Delta x|_{t=\tau_k} &\in B_k x + P_k, \end{aligned} \tag{4.55}$$

where the times of pulses τ_k are such that $0 \le \tau_1 < \cdots < \tau_N \le T$, B_k are $n \times n$-matrices, and $P_k \in \operatorname{conv}(\mathbb{R}^n)$.

Assume that the sets $\tilde{P}_k = P_1^k \times \cdots \times P_n^k$, where $P_i^k = [p_i^k, p_i^k + q_i^k]$, $i = \overline{1,n}$, are such that $P_k \subseteq \tilde{P}_k$ $\big(P_k \supseteq \tilde{P}_k\big)$. Inclusion (4.55) is associated with a system of linear differential equations with π-derivative of the form

$$\begin{aligned} D_\pi(x_i, \delta_i) &= \sum_{j=1}^{n} a_{ij}(t)(x_j, \delta_j) + (f_i(t), r_i(t)), \quad t \neq \tau_k, \\ \Delta(x_i, \delta_i)|_{t=\tau_k} &= \sum_{j=1}^{n} b_{ij}^k(x_j, \delta_j) + (p_i^k, q_i^k), \\ (x_i(0), \delta_i(0)) &= (x_i^0, \delta_i^0), \quad i = \overline{1,n}, \end{aligned} \tag{4.56}$$

where $\Delta(x_i, \delta_i)|_{t=\tau_k} = (\Delta x_i|_{t=\tau_k}, \Delta\delta_i|_{t=\tau_k})$.

This system decomposes into two systems of linear inhomogeneous impulsive differential equations

$$\begin{cases} \dot{x}_i = \sum_{j=1}^n a_{ij}(t)x_j + f_i(t), \quad t \neq \tau_k, \\ \Delta x_i|_{t=\tau_k} = \sum_{j=1}^n b_{ij}^k x_j + p_i^k, \\ x_i(0) = x_i^0, \quad i = \overline{1,n}, \end{cases} \tag{4.57}$$

$$\begin{cases} \dot{\delta}_i = \sum_{j=1}^n a_{ij}(t)\delta_j + r_i(t), \quad t \neq \tau_k, \\ \Delta\delta_i|_{t=\tau_k} = \sum_{j=1}^n b_{ij}^k \delta_j + g_i^k, \\ \delta_i(0) = \delta_i^0, \quad i = \overline{1,n}. \end{cases} \tag{4.58}$$

Let $\tilde{X}(\tau_k+0) = X_1(\tau_k+0)\times\cdots\times X_n(\tau_k+0)$, where

$$X_i(\tau_k+0) = \begin{cases} [x_i(\tau_k+0), x_i(\tau_k+0)+\delta_i(\tau_k+0)], & \delta_i(\tau_k+0)\geq 0,\\ [x_i(\tau_k+0)+\delta_i(\tau_k+0), x_i(\tau_k+0)], & \delta_i(\tau_k+0)<0,\end{cases}$$

$$x_i(\tau_k+0) = \sum_{j=1}^{n}(e_{ij}+b_{ij}^k)x_j(\tau_k)+p_i^k,$$

$$\delta_i(\tau_k+0) = \sum_{j=1}^{n}(e_{ij}+b_{ij}^k)\delta_j(\tau_k)+q_i^k,$$

and e_{ij} is the Kronecker symbol.

We now study the relationship between the sets $R(\tau_k+0)$ and $\tilde{X}(\tau_k+0)$ under the assumption that $R(\tau_k)\subseteq\tilde{X}(\tau_k)$ $\big(R(\tau_k)\supseteq\tilde{X}(\tau_k)\big)$. According to the property of the support functions, we find

$$\begin{aligned}
c(R(\tau_k+0),\psi) &= c((E+B_k)R(\tau_k)+P_k,\psi)\\
&= c(R(\tau_k),(E+B_k)^T\psi)+c(P_k,\psi)\\
&\leq (\geq)\ c(\tilde{X}(\tau_k),(E+B_k)^T\psi)+c(\tilde{P}_k,\psi)\\
&= \sum_{i=1}^{n}\{c(X_i(\tau_k),((E+B_k)^T\psi)_i)+c(P_i^k,\psi_i)\}\\
&= \sum_{i=1}^{n}\bigg\{\Big(x_i(\tau_k)+\frac{\delta_i(\tau_k)}{2}\Big)\sum_{j=1}^{n}(e_{ji}+b_{ji}^k)\psi_j+\Big(p_i^k+\frac{q_i^k}{2}\Big)\psi_i\\
&\qquad+\frac{|\delta_i(\tau_k)|}{2}\bigg|\sum_{j=1}^{n}(e_{ji}+b_{ji}^k)\psi_j\bigg|+\frac{q_i^k}{2}|\psi_i|\bigg\} = I_{k1}(\psi),
\end{aligned}$$

$$\begin{aligned}
c(\tilde{X}(\tau_k+0),\psi) &= \sum_{i=1}^{n}c(X_i(\tau_k+0),\psi_i)\\
&= \sum_{i=1}^{n}\bigg\{\Big(x_i(\tau_k+0)+\frac{\delta_i(\tau_k+0)}{2}\Big)\psi_i+\frac{|\delta_i(\tau_k+0)|}{2}|\psi_i|\bigg\}\\
&= \sum_{i=1}^{n}\bigg\{\sum_{j=1}^{n}(e_{ij}+b_{ij}^k)\Big(x_j(\tau_k)+\frac{\delta_j(\tau_k)}{2}\Big)\psi_i+\Big(p_i^k+\frac{q_i^k}{2}\Big)\psi_i\\
&\qquad+\frac{1}{2}\bigg|\sum_{j=1}^{n}(e_{ij}+b_{ij}^k)\delta_j(\tau_k)+q_i^k\bigg\|\psi_i|\bigg\} = I_{k2}(\psi).
\end{aligned}$$

The relationship between $I_{k1}(\psi)$ and $I_{k2}(\psi)$ is directly connected with the relationship between

$$J_{k1}(\psi) = \sum_{i=1}^{n}\left\{|\delta_i(\tau_k)|\left|\sum_{j=1}^{n}(e_{ji} + b_{ji}^k)\psi_j\right| + q_i^k|\psi_i|\right\}$$

and

$$J_{k2}(\psi) = \sum_{i=1}^{n}\left|\sum_{j=1}^{n}(e_{ij} + b_{ij}^k)\delta_j(\tau_k) + q_i^k\right||\psi_i|.$$

Thus, we have proved the following assertion:

Theorem 12 ([121]). *Let $P_k \subseteq \tilde{P}_k (P_k \supseteq \tilde{P}_k)$ and let $X(\tau_k) \subseteq \tilde{X}(\tau_k)$ $(X(\tau_k) \supseteq \tilde{X}(\tau_k))$. Assume that $J_{k1}(\psi) \leq (\geq) J_{k2}(\psi)$ for all $\psi \in \mathbb{R}^n$. Then the following inclusion is true: $R(\tau_k + 0) \subseteq \tilde{X}(\tau_k + 0)$ $(R(\tau_k + 0) \supseteq \tilde{X}(\tau_k + 0))$.*

Corollary 7. *For $n = 1$, we have $P_k = \tilde{P}_k$ and*

$$J_{k1}(\psi) = |\delta_i(\tau_k)(1 + b_k)\psi| + q_k|\psi| \geq |(1 + b_k)\delta_i(\tau_k) + q_k| \cdot |\psi| = J_{k2}(\psi)$$

for all $\psi \in \mathbb{R}$. Thus, $R(\tau_k + 0) \supseteq \tilde{X}(\tau_k + 0)$.

Theorem 13 ([121]). *Let $P_k \subseteq \tilde{P}_k$ and let $R(\tau_k) \subseteq \tilde{X}(\tau_k)$. If, for any $m = \overline{1,n}$, all $(e_{mi} + b_{mi}^k)\delta_i(\tau_k)$ are nonnegative for all $i = \overline{1,n}$, then the following inclusion is true: $R(\tau_k + 0) \subseteq \tilde{X}(\tau_k + 0)$.*

Proof. It is sufficient to check the validity of the inequality $J_{k1}(\psi) \leq J_{k2}(\psi)$ for the vectors $\psi = \pm e_m$, $m = \overline{1,n}$. In this case,

$$J_{k1}(\pm e_m) = \sum_{i=1}^{n}|\delta_i(\tau_k)(e_{mi} + b_{mi}^k)| + q_m^k,$$

$$J_{k2}(\pm e_m) = \left|\sum_{i=1}^{n}(e_{mi} + b_{mi}^k)\delta_i(\tau_k) + q_m^k\right|.$$

Since, for any $m = \overline{1,n}$, all $(e_{mi} + b_{mi}^k)\delta_i(\tau_k)$ are nonnegative for all $i = \overline{1,n}$, we have $J_{k1}(\pm e_m) = J_{k2}(\pm e_m)$ for all $m = \overline{1,n}$. □

Remark 8. If the equalities $P_k = \tilde{P}_k$ and $R(\tau_k) = \tilde{X}(\tau_k)$ in Theorem 13 are true, then $J_{k1}(\pm e_m) = J_{k2}(\pm e_m)$ for all $m = \overline{1,n}$. Thus, the set $R(\tau_k + 0)$ is inscribed in the set $\tilde{X}(\tau_k + 0)$.

For $n = 1$, we conclude that $R(\tau_k + 0) = \tilde{X}(\tau_k + 0)$.

Example 12. Consider an impulsive differential inclusion (4.55) for $n = 1$:

$$\begin{aligned} \dot{x} &\in a(t)x + F(t), \quad t \neq \tau_k, \\ \Delta x|_{t=\tau_k} &\in b_k x + P_k, \quad x(0) \in X_0. \end{aligned}$$

It is associated with the following linear equation with π-derivative:

$$\begin{aligned} D_\pi(x,\delta) &= a(t)(x,\delta) + (f(t), r(t)), \quad t \neq \tau_k, \\ \Delta(x,\delta)|_{t=\tau_k} &= b_k(x,\delta) + (p_k, q_k), \\ (x(0),\delta(0)) &= (x_0,\delta_0), \end{aligned}$$

which decomposes into two linear impulsive differential equations

$$\begin{cases} \dot{x} = a(t)x + f(t), \quad t \neq \tau_k, \\ \Delta x|_{t=\tau_k} = b_k x + p_k, \\ x(0) = x_0, \end{cases} \qquad \begin{cases} \dot{\delta} = a(t)\delta + r(t), \quad t \neq \tau_k, \\ \Delta \delta|_{t=\tau_k} = b_k \delta + q_k, \\ \delta(0) = \delta_0. \end{cases}$$

In this case, $F(t) = \tilde{F}(t)$ and $X_0 = \tilde{X}_0$. Assume that $b_k \geq -1$ for all $k \in \mathbb{N}$. We now check the validity of the conditions of Theorem 13,

$$(1 + b_1)\delta(\tau_1) = (1 + b_1)\left(e^{\int_0^{\tau_1} a(s)ds}\delta_0 + \int_0^{\tau_1} r(s)e^{\int_s^{\tau_1} a(s)ds}ds\right) \geq 0.$$

Assume that $(1 + b_{k-1})\delta(\tau_{k-1}) \geq 0$. Then

$$\begin{aligned} (1 + b_k)\delta(\tau_k) &= (1 + b_k)\left(e^{\int_{\tau_{k-1}}^{\tau_k} a(s)ds}\delta(\tau_{k-1} + 0) + \int_{\tau_{k-1}}^{\tau_k} r(s)e^{\int_s^{\tau_k} a(s)ds}ds\right) \\ &= (1 + b_k)\Bigg[((1 + b_{k-1})\delta(\tau_{k-1}) + q_{k-1})e^{\int_{\tau_{k-1}}^{\tau_k} a(s)ds} \\ &\qquad + \int_{\tau_{k-1}}^{\tau_k} r(s)e^{\int_s^{\tau_k} a(s)ds}ds\Bigg] \geq 0. \end{aligned}$$

Thus, by virtue of Example 10 and Remark 7, we get $R(t) \subseteq \tilde{X}(t)$. Hence, $R(t) = \tilde{X}(t)$.

Chapter 5
Method of Averaging in Systems with Pulse Action

In their monograph "Introduction to Nonlinear Mechanics" [72], Krylov and Bogolyubov used on an elegant example to illustrate the efficiency of application of the method of averaging to the investigation of oscillations of a pendulum subjected to a pulse action. Later, their ideas were developed by Mitropol'skii [88] and Samoilenko [133,136,137] for a broader class of systems subjected to the action of pulse perturbations. The cited works initiated mathematical investigations of oscillating processes in systems with pulses and stimulated the investigations in the entire theory of differential equations with pulse action. At present, these investigations are extensively carried out both in the Ukraine and abroad.

In what follows, we consider the main results obtained in the development of the method of averaging for systems of differential equations with pulse action.

5.1 Oscillating System with One Degree of Freedom

Consider an oscillating system subjected to a pulse action and described by the equation

$$\ddot{x} + \omega^2 x = \varepsilon f(t, x, \dot{x}),$$

where ε is a small parameter. Impulsive forces may act on the system at fixed times or when the image point hits certain sets of the phase (extended phase) space. It is customary to assume that the action of impulsive forces leads to changes in the velocity of motion of the phase point of the system by a certain value, generally speaking, depending on the position and velocity of the phase point at the time of pulse action.

We now consider an autonomous oscillating system subjected to the action of a periodic impulsive force

$$\ddot{x} + \omega^2 x = \varepsilon f(x, \dot{x}), \quad t \neq \tau_i, \quad \Delta \dot{x}|_{t=\tau_i} = \varepsilon I_i(x, \dot{x}). \tag{5.1}$$

It is supposed that the times of pulse action τ_i and the functions $I_i(x, \dot{x})$ are such that

$$\tau_{i+p} = \tau_i + 2\pi, \quad I_{i+p}(x, \dot{x}) = I_i(x, \dot{x}) \tag{5.2}$$

for all integers i and a natural number p.

In analyzing Eqs. (5.1), the resonance and nonresonance cases should be considered separately. In the nonresonance case, i.e., in the case where ω is an irrational number poorly approximated by rational numbers, system (5.1) was studied in [90].

Nonresonance Case. Assume that $I_i(x, \dot{x})$ are finite polynomials, the function $f(x, \dot{x})$ is continuously differentiable, and its derivatives satisfy the Lipschitz condition with respect to both variables in the domain

$$x^2 + \frac{\dot{x}^2}{\omega^2} \leq R^2. \tag{5.3}$$

By the change of variables

$$x = a \sin \psi, \quad \dot{x} = a\omega \cos \psi, \tag{5.4}$$

in the system of equations (5.1), we arrive at the equations

$$\begin{aligned} \frac{da}{dt} &= \frac{\varepsilon}{\omega} f(a \sin \psi, a\omega \cos \psi) \cos \psi, \\ \frac{d\psi}{dt} &= \omega - \frac{\varepsilon}{\omega a} f(a \sin \psi, a\omega \cos \psi) \sin \psi, \quad t \neq \tau_i, \\ \Delta a|_{t=\tau_i} &= \frac{\varepsilon}{\omega} I_i(a \sin \psi, a\omega \cos \psi) \cos \psi + \varepsilon^2 \cdots, \\ \Delta \psi|_{t=\tau_i} &= \frac{\varepsilon}{\omega a} I_i(a \sin \psi, a\omega \cos \psi) \sin \psi + \varepsilon^2 \cdots. \end{aligned} \tag{5.5}$$

These equations are analyzed by the method of averaging.

As shown in [72], the application of the method of averaging is equivalent to a certain sequence of changes aimed at the exclusion of the dependent variable (time) from the system up to quantities of arbitrarily high smallness in ε. It turns out that the process of successive changes can also be used for systems subjected to pulse actions.

We introduce the following notation:

$$\begin{aligned} f(a \sin \psi, a\omega \cos \psi) \cos \psi &= f^{(1)}(a, \psi), \\ f(a \sin \psi, a\omega \cos \psi) \sin \psi &= f^{(2)}(a, \psi), \\ I_i(a \sin \psi, a\omega \cos \psi) \cos \psi &= I_i^{(1)}(a, \psi), \\ I_i(a \sin \psi, a\omega \cos \psi) \sin \psi &= I_i^{(2)}(a, \psi) \end{aligned}$$

and assume that the functions $I_i^{(1)}(a, \psi)$ and $I_i^{(2)}(a, \psi)$ are finite trigonometric polynomials. Moreover, let

$$\frac{\partial I_i^{(j)}(a, \psi)}{\partial \psi} = \sum_{k=1}^{N} (A_{ik}^{(j)}(a) \sin k\psi + B_{ik}^{(j)}(a) \cos k\psi),$$
$$i = \overline{1, p}, \quad j = 1, 2.$$

By $z_j(a,\psi,t)$ we denote the following functions:

$$z_j(a,\psi,t)$$
$$= -\frac{1}{\pi}\sum_{i=1}^{p}\sum_{k=1}^{N}\left\{[A_{ik}^{(j)}(a)\sin k\psi + B_{ik}^{(j)}(a)\cos k\psi]\sum_{n=1}^{\infty}\frac{\cos n(t-\tau_i)}{(k\omega)^2-n^2}\right.$$
$$\left. + k\omega[B_{ik}^{(j)}(a)\sin k\psi - A_{ik}^{(j)}(a)\cos k\psi]\sum_{n=1}^{\infty}\frac{\sin n(t-\tau_i)}{n[(k\omega)^2-n^2]}\right\}.$$

By direct verification, we can show that, for $t \neq \tau_i$, these functions satisfy the relation

$$\frac{\partial z_j}{\partial\psi}\omega + \frac{\partial z_j}{\partial t} = -\sum_{i=1}^{p}\frac{\partial I_i^{(j)}(a,\psi)}{\partial\psi}\sigma\left(\frac{1}{2}-\frac{t-\tau_i}{2\pi}\right),$$

where $\sigma(\pi - s)$ is a 2π-periodic function defined, on a period, by the expression $\pi - s$, $0 < s < 2\pi$.

In the system of equations (5.5), we perform the following change of variables:

$$a = b + \frac{\varepsilon}{\omega}u^{(1)}(b,\theta,t), \quad \psi = \theta - \frac{\varepsilon}{\omega b}u^{(2)}(b,\theta,t), \tag{5.6}$$

where

$$u^{(j)}(b,\theta,t) = \frac{1}{\omega}\int\left[f^{(j)}(b,\theta) + \frac{1}{2\pi}\sum_{i=1}^{p}I_i^{(j)}(b,\theta)\right.$$
$$\left. - f_0^{(j)}(b) - \frac{1}{2\pi}\sum_{i=1}^{p}\overline{I}_i^{(j)}(b)\right] + z_j(b,\theta,t)$$
$$+\sum_{i=1}^{p}I_i^{(j)}(b,\theta)\sigma\left(\frac{1}{2}-\frac{t-\tau_i}{2\pi}\right), \quad j = 1,2.$$

Here, $f_0^{(j)}(b)$ and $\overline{I}_i^{(j)}(b)$ are the mean values of the functions $f^{(j)}(b,\theta)$ and $I_i^{(j)}(b,\theta)$, respectively, and the integrals stand for the primitives whose mean value is equal to zero.

Substituting (5.6) in (5.5), we get

$$\frac{db}{dt} = \frac{\varepsilon}{\omega}F(b) + \varepsilon^2 R^{(1)}(b,\theta,t,\varepsilon),$$
$$\frac{d\theta}{dt} = \omega - \frac{\varepsilon}{\omega b}\Phi(b) + \varepsilon^2 R^{(2)}(b,\theta,t,\varepsilon), \quad t \neq \tau_i,$$
$$\Delta b|_{t=\tau_i} = \varepsilon^2 J_i^{(1)}(b,\theta,\varepsilon), \quad \Delta\theta|_{t=\tau_i} = \varepsilon^2 J_i^{(2)}(b,\theta,\varepsilon), \tag{5.7}$$

where

$$F(b) = f_0^{(1)}(b) + \frac{1}{2\pi} \sum_{i=1}^{p} \overline{I}_i^{(1)}(b),$$

$$\Phi(b) = f_0^{(2)}(b) + \frac{1}{2\pi} \sum_{i=1}^{p} \overline{I}_i^{(2)}(b),$$

the functions $R^{(1)}(b, \theta, t, \varepsilon)$ and $R^{(2)}(b, \theta, t, \varepsilon)$ satisfy the Lipschitz condition with respect to b and θ uniformly in t and ε and are piecewise continuous in t with discontinuities of the first kind at points $t = \tau_i$. The functions $J_i^{(1)}(b, \theta, \varepsilon)$ and $J_i^{(2)}(b, \theta, \varepsilon)$ satisfy the Lipschitz condition with respect to b and θ uniformly in i and ε.

Neglecting the quantities of the order of smallness ε^2 in Eq. (5.7), we arrive at the equations of the first approximation

$$\frac{db}{dt} = \frac{\varepsilon}{\omega} F(b), \quad \frac{d\theta}{dt} = \omega - \frac{\varepsilon}{\omega b} \Phi(b). \tag{5.8}$$

Assume that the equation $F(b) = 0$ has a simple positive root $b = b_0$ such that $F'(b_0) < 0$. We now clarify what corresponds to this root in the system of equations (5.7). To this end, we perform certain transformations in this system.

Note that the equation

$$\omega \frac{db}{dt} = F(b),$$

has the general solution $b = A(ce^{\lambda t})$ in a neighborhood of the point $b = b_0$, where $\lambda = F'(b_0)$, c is an arbitrary constant and, in addition, $A(0) = b_0$. Denote

$$\overline{\Phi}(h) = -\int_0^h \frac{1}{\lambda \omega h} \left[\frac{\Phi(A(h))}{A(h)} - \frac{\Phi(b_0)}{b_0} \right] dh.$$

It is clear that

$$\lambda h \frac{\partial \overline{\Phi}}{\partial h} = -\left(\frac{\Phi(A)}{\omega A} - \frac{1}{\omega b_0} \Phi(b_0) \right).$$

In Eqs. (5.7), we perform the change of variables

$$b = A(h), \quad \theta = \phi - \overline{\Phi}(h). \tag{5.9}$$

For $t \neq \tau_i$, we get

$$\frac{dh}{dt} = \varepsilon \lambda h + \varepsilon^2 P(h, \phi, t, \varepsilon), \quad \frac{d\phi}{dt} = \omega(\varepsilon) + \varepsilon^2 Q(h, \phi, t, \varepsilon),$$

where

$$
\begin{aligned}
P(h, \phi, t, \varepsilon) &= \frac{R^{(1)}(A(h), \phi - \overline{\Phi}(h), t, \varepsilon)}{A'(h)}, \\
Q(h, \phi, t, \varepsilon) &= R^{(2)}(A(h), \phi - \overline{\Phi}(h), t, \varepsilon) \\
&\quad + \frac{\frac{\partial \overline{\Phi}}{\partial h} R^{(1)}(A(h), \phi - \overline{\Phi}(h), t, \varepsilon)}{A'(h)}, \\
\omega(\varepsilon) &= \omega - \frac{\varepsilon}{\omega b_0} \Phi(b_0).
\end{aligned}
$$

For $t = \tau_i$, we find

$$
\begin{aligned}
A(h + \Delta h) - A(h) &= \varepsilon^2 J_i^{(1)}(A(h), \phi - \overline{\Phi}(h), \varepsilon), \qquad (5.10) \\
\Delta \phi - [\overline{\Phi}(h + \Delta h) - \overline{\Phi}(h)] &= \varepsilon^2 J_i^{(2)}(A(h), \phi - \overline{\Phi}(h), \varepsilon).
\end{aligned}
$$

It is clear that these equations are solvable with respect to Δh and $\Delta \phi$ and their solutions are quantities of the same order as ε^2. Thus, by the change of variables (5.9), we reduce system (5.7) to the equations

$$
\begin{aligned}
&\frac{dh}{dt} = \varepsilon \lambda h + \varepsilon^2 P(h, \phi, t, \varepsilon), \quad \frac{d\phi}{dt} = \omega(\varepsilon) + \varepsilon^2 Q(h, \phi, t, \varepsilon), \qquad (5.11) \\
&\Delta h|_{t=\tau_i} = \varepsilon^2 H_i(h, \phi, \varepsilon), \quad \Delta \phi|_{t=\tau_i} = \varepsilon^2 G_i(h, \phi, \varepsilon).
\end{aligned}
$$

In this case, there exist positive numbers ρ_0 and ε_0 such that, in the domain

$$
|h| \le \rho_0, \quad -\infty < \phi < \infty, \quad -\infty < t < \infty, \quad 0 < \varepsilon \le \varepsilon_0, \qquad (5.12)
$$

the functions $P(h, \phi, t, \varepsilon)$, $Q(h, \phi, t, \varepsilon)$, $H_i(h, \phi, \varepsilon)$, and $G_i(h, \phi, \varepsilon)$ are defined and continuous in h, ϕ, and ε, piecewise continuous in t with discontinuities of the first kind at $t = \tau_i$, periodic in ϕ and t with period 2π, and satisfy the inequalities

$$
\begin{aligned}
&|P(h', \phi', t, \varepsilon) - P(h'', \phi'', t, \varepsilon)| + |Q(h', \phi', t, \varepsilon) - Q(h'', \phi'', t, \varepsilon)| \\
&\quad + |H_i(h', \phi', \varepsilon) - H_i(h'', \phi'', \varepsilon)| + |G_i(h', \phi', \varepsilon) - G_i(h'', \phi'', \varepsilon)| \\
&\le L(|h' - h''| + |\phi' - \phi''|), \qquad (5.13) \\
&|P(h, \phi, t, \varepsilon)| + |Q(h, \phi, t, \varepsilon)| + |H_i(h, \phi, \varepsilon)| + |G_i(h, \phi, \varepsilon)| \le M, \qquad (5.14)
\end{aligned}
$$

where L and M are positive constants.

By virtue of the assumption that the equation $F(b) = 0$ possesses a simple positive root, Eq. (5.8) has a stationary solution $b = b_0$, $\phi = \omega(\varepsilon)t + \theta_0$, where θ_0 is an arbitrary constant. The approximate stationary solution of Eq. (5.1) corresponding to this solution has the form

$$
\begin{aligned}
x &= b_0 \sin(\omega(\varepsilon)t + \theta_0) \\
&\quad + \frac{\varepsilon}{\omega}(u^{(1)}(b_0) \sin(\omega(\varepsilon)t + \theta_0) - u^{(2)}(b_0) \cos(\omega(\varepsilon)t + \theta_0)). \qquad (5.15)
\end{aligned}
$$

Theorem 1 ([90]). *All solutions of the differential equation with pulse action* (5.1) *whose initial values are sufficiently close to the initial values of the approximate stationary solutions* (5.15) *are uniformly bounded on the entire real semiaxis* ($t \geq 0$), *at least for sufficiently small values of the parameter* ε.

The proof of Theorem 1 is based on the following two lemmas:

Lemma 1. *Assume that, for* $t \geq t_0$, *a nonnegative piecewise continuous function* $y(t)$ *satisfies the inequality*

$$y(t) \leq \alpha + \int_{t_0}^{t} [\beta e^{\delta(s-t_0)} + \gamma y(s)]ds + \sum_{t_0<\tau_i<t} [\beta e^{\delta(\tau_i - t_0)} + \gamma y(\tau_i - 0)],$$

where $\alpha \geq 0$, $\beta \geq 0$, $\delta > 0$, $\gamma > 0$, *and* τ_i *are points of discontinuity of the first kind of the function* $y(t)$. *Then*

$$y(t) \leq \left[\alpha(1+\gamma)^n + \beta \sum_{i=1}^{n} e^{(\delta-\gamma)(\tau_i - \tau_{i-1})}(1+\gamma)^{i-1}\right] e^{\gamma(t-t_0)}$$
$$+ \frac{\beta(e^{\delta(t-t_0)} - e^{\gamma(t-t_0)})}{\delta - \gamma}, \qquad t_n < t < t_{n+1}.$$

The lemma is proved by induction by using Theorem 1.5 from [46] in each interval of continuity of the function $y(t)$.

Corollary 1. *If, in the conditions of the lemma,* $\delta > \gamma$ *and the sequence of times* τ_i *is such that* $\tau_{i+p} = \tau_i + T$, *then, for all* $t \geq t_0$, *the function* $y(t)$ *admits the following estimate:*

$$y(t) \leq \left[\alpha(1+\gamma)^p e^{\frac{p\gamma}{T}(t-t_0)} + \beta e^{(\delta-\gamma)T} \frac{(1+\gamma)^p e^{\frac{p\gamma}{T}(t-t_0)}}{\gamma}\right] e^{\gamma(t-t_0)}$$
$$+ \frac{\beta}{\delta-\gamma} e^{\delta(t-t_0)}.$$

Moreover, if $\delta > (1 + \frac{p}{T})\gamma$, *then*

$$e^{-\delta(t-t_0)} y(t) \leq \alpha(1+\gamma)^p + \frac{\beta p e^{(\delta-\gamma)T}}{T(\delta - (1+\frac{p}{T})\gamma)} + \frac{\beta}{\delta - \gamma}.$$

Lemma 2. *There exists a positive number* $\bar{\varepsilon} \leq \varepsilon_0$ *such that the solutions* $h_t(\phi, \varepsilon)$ *of the system of equations* (5.10) *are uniformly bounded independently of* t *if the initial values* h_0 *are sufficiently small for all* $0 < \varepsilon < \bar{\varepsilon}$ *and* $t \geq t_0$.

Indeed, we fix a sufficiently small positive number $\delta_0 < \rho_0$ and choose $\bar{\varepsilon}$ from the condition

$$(1+\bar{\varepsilon}^2 L)^p \delta_0 + \frac{\bar{\varepsilon} M p e^{-\bar{\varepsilon}\lambda T}}{T[|\lambda| - \bar{\varepsilon}(1+\frac{p}{T})L]} + \frac{\bar{\varepsilon} M}{|\lambda| - \bar{\varepsilon} L} \le \rho_0.$$

In view of inequalities (5.13) and (5.14), it follows from Eqs. (5.11) that

$$e^{-\varepsilon\lambda(t-t_0)}|h_t| \le |h_0| + \varepsilon^2 \int_{t_0}^{t} [e^{-\varepsilon\lambda(s-t_0)} M + L e^{-\varepsilon\lambda(s-t_0)}|h_s|] ds$$
$$+ \varepsilon^2 \sum_{t_0<\tau_i<t} (e^{-\varepsilon\lambda(\tau_i-t_0)} M + L e^{-\varepsilon\lambda(\tau_i-t_0)}|h_{\tau_i}|),$$

whence, by virtue of Corollary 1, we get

$$|h_t(\phi,\varepsilon)| \le (1+\varepsilon^2 L)^p |h_0| + \frac{\varepsilon M p e^{-\varepsilon\lambda T}}{T[|\lambda| - \varepsilon(1+\frac{p}{T})L]} + \frac{\varepsilon M}{|\lambda| - \varepsilon L}.$$

Therefore,

$$|h_t(\phi,\varepsilon)| \le \rho_0$$

for all $t \ge t_0$ and $0 < \varepsilon \le \bar{\varepsilon}$ provided that $|h_0| < \delta_0$, which completes the proof of Lemma 2.

In view of the changes of variables (5.4), (5.6), and (5.9), Lemma 2 implies the assertion of Theorem 1 because the initial condition h_0 in the variables b takes the form $b = b_0$. Hence, if ε is sufficiently small and h_0 is close to zero, then the initial values x_0 and $\dot{x}_0$ of the original variables are close to the initial values of the corresponding approximate stationary solutions (5.15).

We now prove the existence of an invariant toroidal set of the system of equations (5.11). We seek the required invariant set $\Gamma(\varepsilon)$ in the form

$$\Gamma(\varepsilon)\colon h = u(\phi,t,\varepsilon), \tag{5.16}$$

where $u(\phi,t,\varepsilon)$ is a function continuous in ϕ, piecewise continuous in t with discontinuities of the first kind at $t=\tau_i$, and periodic in ϕ and t with period 2π. Expression (5.16) specifies $\Gamma(\varepsilon)$ whenever

$$\frac{du(\phi_t(s,\phi),t,\varepsilon)}{dt} \equiv \varepsilon\lambda u(\phi_t(s,\phi),t,\varepsilon)$$
$$+ \varepsilon^2 P(u(\phi_t(s,\phi),t,\varepsilon),\phi_t(s,\phi),t,\varepsilon) \tag{5.17}$$

for all $t \in (-\infty,\infty)$, $t \ne \tau_i$, and

$$\Delta u(\phi_{\tau_i}(s,\phi),\tau_i,\varepsilon) = \varepsilon^2 H_i(u(\phi_{\tau_i}(s,\phi),\tau_i,\varepsilon),\phi_{\tau_i}(s,\phi),\varepsilon),$$

where $\phi_t(s,\phi)$ is a solution of the system

$$\begin{aligned}\frac{d\phi}{dt} &= \omega(\varepsilon) + \varepsilon^2 Q(u(\phi,t,\varepsilon),\phi,t,\varepsilon), \quad t \neq \tau_i, \\ \Delta\phi|_{t=\tau_i} &= \varepsilon^2 G_i(u(\phi,\tau_i,\varepsilon),\phi,\varepsilon), \quad \phi_s(s,\phi) = \phi.\end{aligned} \tag{5.18}$$

To find the invariant set $\Gamma(\varepsilon)$, we use a simple iterative method according to which $\Gamma(\varepsilon)$ is sought as the limit of a sequence of sets $\Gamma^{(1)}(\varepsilon),\dots,\Gamma^{(n)}(\varepsilon),\dots$ each of which is an invariant set

$$\Gamma^{(n+1)}(\varepsilon)\colon h = u^{(n+1)}(\phi,t,\varepsilon), \quad n = 0,1,\dots,$$

of the system of equations

$$\begin{aligned}\frac{dh}{dt} &= \varepsilon\lambda h + \varepsilon^2 P(u^{(n)}(\phi,t,\varepsilon),\phi,t,\varepsilon), \\ \frac{d\phi}{dt} &= \omega(\varepsilon) + \varepsilon^2 Q(u^{(n)}(\phi,t,\varepsilon),\phi,t,\varepsilon), \quad t \neq \tau_i, \\ \Delta h|_{t=\tau_i} &= \varepsilon^2 H_i(u^{(n)}(\phi,\tau_i,\varepsilon),\phi,\varepsilon), \\ \Delta\phi|_{t=\tau_i} &= \varepsilon^2 G_i(u^{(n)}(\phi,\tau_i,\varepsilon),\phi,\varepsilon).\end{aligned} \tag{5.19}$$

Let $\phi_t^{(n)}(s,\phi)$, $\phi_{\tau_i}^{(n)}(s,\phi) = \phi$ be the general solution of the equation obtained from (5.18) as a result of the replacement of $u(\phi,t,\varepsilon)$ with $u^{(n)}(\phi,t,\varepsilon)$. Then

$$\begin{aligned}h_t(\phi,\varepsilon) = \varepsilon^2 &\int_{-\infty}^{t} e^{\varepsilon\lambda(t-s)} P(u^{(n)}(\phi_\sigma^{(n)}(s,\phi),\sigma,\varepsilon),\phi_s^{(n)}(\sigma,\phi),\sigma,\varepsilon)d\sigma \\ &+ \varepsilon^2 \sum_{\tau_i<t} e^{\varepsilon\lambda(t-\tau_i)} H_i(u^{(n)}(\phi_{\tau_i}^{(n)}(s,\phi),\tau_i,\varepsilon),\phi_{\tau_i}^{(n)}(s,\phi),\varepsilon)\end{aligned}$$

is a family of bounded solutions of the equation

$$\begin{aligned}\frac{dh}{dt} &= \varepsilon\lambda h + \varepsilon^2 P(u^{(n)}(\phi_t^{(n)}(s,\phi),t,\varepsilon),\phi_t^{(n)}(s,\phi),t,\varepsilon), \quad t \neq \tau_i, \\ \Delta h|_{t=\tau_i} &= \varepsilon^2 H_i(u^{(n)}(\phi_{\tau_i}^{(n)}(s,\phi),\tau_i,\varepsilon),\phi_{\tau_i}^{(n)}(s,\phi),\varepsilon)\end{aligned}$$

depending on ϕ, s, and ε as on parameters. This family covers the invariant set

$$\begin{aligned}\Gamma^{(n+1)}(\varepsilon)\colon h &= u^{(n+1)}(\phi,t,\varepsilon) \\ &\equiv \varepsilon^2 \int_{-\infty}^{t} e^{-\varepsilon\lambda\sigma} P(u^{(n)}(\phi_{t+\sigma}(t,\phi),t+\sigma,\varepsilon),\phi_{t+\sigma}^{(n)}(t,\phi),t+\sigma,\varepsilon)d\sigma \\ &\quad + \varepsilon^2 \sum_{\tau_i<t} e^{\varepsilon\lambda(t-\tau_i)} H_i(u^{(n)}(\phi_{\tau_i}^{(n)}(t,\phi),\tau_i,\varepsilon),\phi_{\tau_i}(t,\phi),\varepsilon).\end{aligned} \tag{5.20}$$

It can be directly shown that the function $u^{(n+1)}(\phi, t, \varepsilon)$ is periodic in ϕ and t with period 2π.

Prior to proving the convergence of the sequence of functions $u^{(n)}(\phi, t, \varepsilon)$, we establish some properties of these functions and the functions $\phi_t^{(n)}(s, \phi)$. To this end, we need the following assertion:

Lemma 3. *There exist positive numbers ε_1 $(0 < \varepsilon_1 \le \varepsilon_0)$ and $N = N(\varepsilon)$ $(N(\varepsilon) \to 0$ as $\varepsilon \to 0)$ such that, for all $0 < \varepsilon_1 \le \varepsilon_0$, $-\infty < t < \infty$, $-\infty < s < \infty$, ϕ', ϕ'', $t \ge s$, and $n = 0, 1, \ldots,$*

$$|u^{(n+1)}(\phi', t, \varepsilon) - u^{(n+1)}(\phi'', t, \varepsilon)| \le N(\varepsilon)|\phi' - \phi''|, \tag{5.21}$$

$$|\phi_t^{(n)}(s, \phi') - \phi_t^{(n)}(s, \phi'')| \le [1 + N^0(\varepsilon)]^p e^{N^0(\varepsilon)(1+\frac{p}{T})(t-s)}|\phi' - \phi''|,$$

$$N^0(\varepsilon) = \varepsilon^2 L(1 + N(\varepsilon)).$$

The lemma is proved by induction. For $n = 0$, by using the following representation for $\phi_t^{(0)}(s, \phi)$:

$$\phi_t^{(0)}(s, \phi) = \phi + \omega(\varepsilon)(t - s) + \varepsilon^2 \int_s^t Q(0, \phi_\sigma^{(0)}(s, \phi), \sigma, \varepsilon)d\sigma$$
$$+ \varepsilon^2 \sum_{s<\tau_i<t} G_i(0, \phi_{\tau_i}^{(0)}(s, \phi), \varepsilon),$$

we find

$$|\phi_t^{(0)}(s, \phi') - \phi_t^{(0)}(s, \phi'')| \le |\phi' - \phi''| + \varepsilon^2 L \int_s^t |\phi_\sigma^{(0)}(s, \phi') - \phi_\sigma^{(0)}(s, \phi'')|d\sigma$$
$$+ \varepsilon^2 L \sum_{s<\tau_i<t} |\phi_{\tau_i}^{(0)}(s, \phi') - \phi_{\tau_i}^{(0)}(s, \phi'')|.$$

By virtue of Lemma 1 in Chapter 1, this yields

$$|\phi_t^{(0)}(s, \phi') - \phi_t^{(0)}(s, \phi'')| \le e^{\varepsilon^2 L(t-s)}(1 + \varepsilon^2 L)^{[\frac{t-s}{T}+1]p}|\phi' - \phi''|$$
$$\le (1 + \varepsilon^2 L)^p e^{\varepsilon^2 L(1+\frac{p}{T})(t-s)}|\phi' - \phi''|, \tag{5.22}$$

i.e., inequality (5.20) holds for $n = 0$. For $n = 0$, it follows from (5.20) that

$$|u^{(1)}(\phi', t, \varepsilon) - u^{(1)}(\phi'', t, \varepsilon)| \le \varepsilon^2 L \int_{-\infty}^0 e^{-\varepsilon\lambda\sigma}|\phi_{\sigma+t}^{(0)}(t, \phi') - \phi_{\sigma+t}^{(0)}(t, \phi'')|d\sigma$$
$$+ \varepsilon^2 L \sum_{\tau_i<t} e^{\varepsilon\lambda(t-\tau_i)}|\phi_{\tau_i}^{(0)}(t, \phi') - \phi_{\tau_i}^{(0)}(t, \phi'')|.$$

In view of inequality (5.22), this yields

$$|u^{(1)}(\phi',t,\varepsilon)-u^{(1)}(\phi'',t,\varepsilon)|\le \varepsilon L N_0(\varepsilon)|\phi'-\phi''|,$$

where

$$N_0(\varepsilon)=(1+\varepsilon^2L)^p\left[\frac{1}{|\lambda|-\varepsilon L(1+\frac{p}{T})}+\frac{\varepsilon p e^{\varepsilon|\lambda|T}}{1-e^{-\varepsilon[|\lambda|-\varepsilon L(1+\frac{p}{T})T]}}\right].$$

Note that

$$N_0(\varepsilon)\to\frac{1}{|\lambda|}\left(1+\frac{p}{T}\right)\quad \text{as } \varepsilon\to 0.$$

Finally, we can write

$$|u^{(1)}(\phi',t,\varepsilon)-u^{(1)}(\phi'',t,\varepsilon)|\le N_1(\varepsilon)|\phi'-\phi''|,$$

where $N_1(\varepsilon)=\varepsilon L N_0(\varepsilon)$.

Assume that the following inequalities hold for $n=\overline{1,k-1}$:

$$|u^{(n+1)}(\phi',t,\varepsilon)-u^{(n+1)}(\phi'',t,\varepsilon)|\le N_{n+1}(\varepsilon)|\phi'-\phi''|,$$

$$|\phi_t^{(n)}(s,\phi')-\phi_t^{(n)}(s,\phi'')|$$
$$\le[1+\varepsilon^2L(1+N_n(\varepsilon))]^p e^{\varepsilon^2L(1+N_n(\varepsilon))(1+\frac{p}{T})(t-s)}|\phi'-\phi''|.$$

Then, for $n=k$, the representation

$$\begin{aligned}\phi_t^{(k)}(s,\phi)&=\phi+\omega(\varepsilon)(t-s)\\&\quad+\varepsilon^2\int_s^t Q(u^{(k)}(\phi_\sigma^{(k)}(s,\phi),\sigma,\varepsilon),\phi_\sigma^{(k)}(s,\phi),\sigma,\varepsilon)d\sigma\\&\quad+\varepsilon^2\sum_{s<\tau_i<t}G_i(u^{(k)}(\phi_{\tau_i}^{(k)}(s,\phi),\tau_i,\varepsilon),\phi_{\tau_i}^{(k)}(s,\phi),\varepsilon)\end{aligned}\tag{5.23}$$

implies that

$$\begin{aligned}|\phi_t^{(k)}(s,\phi')-\phi_t^{(k)}(s,\phi'')|&\le|\phi'-\phi''|\\&\quad+\varepsilon^2L\int_s^t(1+N_k(\varepsilon))|\phi_\sigma^{(k)}(s,\phi')-\phi_\sigma^{(k)}(s,\phi'')|d\sigma\\&\quad+\varepsilon^2L(1+N_k(\varepsilon))\sum_{s<\tau_i<t}|\phi_{\tau_i}^{(k)}(s,\phi')-\phi_{\tau_i}^{(k)}(s,\phi'')|.\end{aligned}$$

By virtue of Lemma 1 in Chapter 1, this inequality yields the following estimate:

$$\begin{aligned}&|\phi_t^{(k)}(s,\phi')-\phi_t^{(k)}(s,\phi'')|\\&\le(1+\varepsilon^2L(1+N_k(\varepsilon))^p e^{\varepsilon^2L(1+N_k(\varepsilon))(1+\frac{p}{T})(t-s)}|\phi'-\phi''|.\end{aligned}\tag{5.24}$$

Further, it follows from (5.20) that

$$|u^{(k+1)}(\phi', t, \varepsilon) - u^{(k+1)}(\phi'', t, \varepsilon)|$$
$$\leq \varepsilon^2 L(1 + N_k(\varepsilon))\Big[\int_{-\infty}^{0} e^{-\varepsilon\lambda\sigma}|\phi^{(k)}_{t+\sigma}(t, \phi') - \phi^{(k)}_{t+\sigma}(t, \phi'')|d\sigma$$
$$+ \sum_{\tau_i < t} e^{\varepsilon\lambda(t-\tau_i)}|\phi^{(k)}_{\tau_i}(t, \phi') - \phi^{(k)}_{\tau_i}(t, \phi'')|\Big].$$

In view of inequality (5.24), we obtain

$$|u^{(k+1)}(\phi', t, \varepsilon) - u^{(k+1)}(\phi'', t, \varepsilon)| \leq N_{k+1}(\varepsilon)|\phi' - \phi''|,$$

where

$$N_{k+1}(\varepsilon) = \varepsilon L(1 + N_k(\varepsilon))M(N_k(\varepsilon), \varepsilon),$$
$$M(N, \varepsilon) = (1 + \varepsilon^2 L)(1 + N)\Big[\frac{1}{|\lambda| - \varepsilon L(1 + N)(1 + \frac{p}{T})}$$
$$+ \frac{\varepsilon p e^{\varepsilon|\lambda|T}}{1 - e^{-\varepsilon[|\lambda| - \varepsilon L(1+N)(1+\frac{p}{T})T]}}\Big].$$

Thus, by induction, we conclude that, for all $n = 0, 1, \ldots,$

$$|u^{(n)}(\phi', t, \varepsilon) - u^{(n)}(\phi'', t, \varepsilon)| \leq N_n(\varepsilon)|\phi' - \phi''|,$$

where $N_n(\varepsilon)$ satisfy the recurrence relation

$$N_{n+1}(\varepsilon) = \varepsilon L(1 + N_n(\varepsilon))M(N_n(\varepsilon), \varepsilon).$$

We choose $0 < \varepsilon_1 \leq \varepsilon_0$ sufficiently small to guarantee that

$$\varepsilon_1 L M(1, \varepsilon_1) < \frac{1}{2}.$$

This choice is possible because

$$M(1, \varepsilon) \to \frac{1}{|\lambda|}\left(1 + \frac{p}{T}\right) \quad \text{as } \varepsilon \to 0.$$

Thus, in view of the monotonicity of the function $M(N, \varepsilon)$, we get

$$N_{n+1}(\varepsilon) \leq 2\varepsilon L M(1, \varepsilon)$$

for all $n = 0, 1, \ldots$ and $0 < \varepsilon \leq \varepsilon_1$.

Hence, to complete the proof of Lemma 3, it suffices to set $N(\varepsilon) = 2\varepsilon L M(1, \varepsilon)$.

We now prove the uniform convergence of the functions $u^{(n)}(\phi, t, \varepsilon)$ and $\phi_t^{(n)}(s, \phi)$. To do this, we estimate the differences

$$\phi_t^{(n+1)}(s, \phi) - \phi_t^{(n)}(s, \phi) \quad \text{and} \quad u^{(n+1)}(\phi, t, \varepsilon) - u^{(n)}(\phi, t, \varepsilon).$$

By using representation (5.23) for $\phi_t^{(n)}(s, \phi)$ and inequality (5.21), we get

$$\begin{aligned}|\phi_t^{(n+1)}(s, \phi) - \phi_t^{(n)}(s, \phi)| \le \varepsilon^2 L \int_s^t [&\|u^{(n+1)}(\phi, t, \varepsilon) - u^{(n)}(\phi, t, \varepsilon)\| \\ &+ (1 + N(\varepsilon))|\phi_\sigma^{(n+1)}(s, \phi) - \phi_\sigma^{(n)}(s, \phi)|]d\sigma \\ + \varepsilon^2 L \sum_{s<\tau_i<t} [&\|u^{(n+1)}(\phi, t, \varepsilon) - u^{(n)}(\phi, t, \varepsilon)\| \\ &+ (1 + N(\varepsilon))|\phi_{\tau_i}^{(n+1)}(s, \phi) - \phi_{\tau_i}^{(n)}(s, \phi)|],\end{aligned}$$

where $\|u\| = \sup_{0\le\phi\le 2\pi, 0\le t\le 2\pi} |u(\phi, t, \varepsilon)|$. By virtue of Lemma 3 in Chapter 1, this yields

$$\begin{aligned}&|\phi_t^{(n+1)}(s, \phi) - \phi_t^{(n)}(s, \phi)| \\ &\le \frac{\|u^{(n+1)} - u^{(n)}\|}{1 + N(\varepsilon)}[(1 + \varepsilon^2 L(1 + N(\varepsilon)))^p e^{\varepsilon^2 L(1+N(\varepsilon))(1+\frac{p}{T})(t-s)} - 1], \quad t > s.\end{aligned} \tag{5.25}$$

Representation (5.20) and estimate (5.21) now imply that

$$\begin{aligned}&|u^{(n+1)}(\phi, t, \varepsilon) - u^{(n)}(\phi, t, \varepsilon)| \\ &\le \varepsilon^2 L \int_{-\infty}^0 e^{-\varepsilon\lambda\sigma}[\|u^{(n)}(\phi, t, \varepsilon) - u^{(n-1)}(\phi, t, \varepsilon)\| \\ &\qquad + (1 + N(\varepsilon))|\phi_{t+\sigma}^{(n)}(t, \phi) - \phi_{t+\sigma}^{(n-1)}(t, \phi)|]d\sigma \\ &\quad + \varepsilon^2 L \sum_{\tau_i<t} e^{\varepsilon\lambda(t-\tau_i)}[\|u^{(n)}(\phi, t, \varepsilon) - u^{(n-1)}(\phi, t, \varepsilon)\| \\ &\qquad + (1 + N(\varepsilon))|\phi_{\tau_i}^{(n)}(t, \phi) - \phi_{\tau_i}^{(n-1)}(t, \phi)|].\end{aligned}$$

Applying Lemma 3 in Chapter 1 and estimate (5.25) to the last inequality, we find

$$\|u^{(n+1)}(\phi, t, \varepsilon) - u^{(n)}(\phi, t, \varepsilon)\| \le \rho(\varepsilon)\|u^{(n)}(\phi, t, \varepsilon) - u^{(n-1)}(\phi, t, \varepsilon)\|,$$

where

$$\begin{aligned}\rho(\varepsilon) &= \varepsilon L(1 + \varepsilon^2 L(1 + N(\varepsilon)))^p\left[\frac{1}{|\lambda_1(\varepsilon)|} + \frac{\varepsilon p e^{\varepsilon|\lambda_1(\varepsilon)|T}}{1 - e^{\varepsilon\lambda_1(\varepsilon)T}}\right], \\ \lambda_1(\varepsilon) &= \lambda + \varepsilon L(1 + N(\varepsilon))\left(1 + \frac{p}{T}\right).\end{aligned}$$

Hence, for all $n = 1, 2, \ldots$, we conclude that

$$\|u^{(n+1)}(\phi, t, \varepsilon) - u^{(n)}(\phi, t, \varepsilon)\| \le \rho^n(\varepsilon)\, \|u^{(1)}(\phi, t, \varepsilon)\|, \tag{5.26}$$

$$|\phi_t^{(n)}(s, \phi) - \phi_t^{(n-1)}(s, \phi)| \tag{5.27}$$

$$\le \frac{\rho^{n-1}(\varepsilon)}{1 + N(\varepsilon)}[(1 + \varepsilon^2 L(1 + N(\varepsilon)))^p e^{\varepsilon^2 L(1+N(\varepsilon))(1+\frac{p}{T})(t-s)} - 1]\|u^{(1)}(\phi, t, \varepsilon)\|.$$

It follows from (5.20) that

$$\|u^{(1)}(\phi, t, \varepsilon)\| \le \varepsilon M_0(\varepsilon), \quad M_0(\varepsilon) = \left(\frac{1}{|\lambda|} + \frac{\varepsilon p e^{-\varepsilon\lambda T}}{1 - e^{\varepsilon\lambda T}}\right) M.$$

We now choose $0 < \varepsilon^0 \le \varepsilon_1$ sufficiently small to guarantee that the inequalities

$$\varepsilon M_0(\varepsilon) < \rho_0, \quad \rho(\varepsilon) < 1$$

are simultaneously true for all $0 < \varepsilon \le \varepsilon^0$.

Then, for all $0 < \varepsilon \le \varepsilon^0$ and $n = 1, 2, \ldots$, we get

$$\|u^{(n)}(\phi, t, \varepsilon)\| \le \rho_0.$$

It follows from inequalities (5.26) and (5.27) that, in any finite interval of the real line and, hence, for all $t \in \mathbb{R}$, the sequence of functions $\phi_t^{(n)}(s, \phi)$ is uniformly convergent and the same is true for the sequence of functions $u^{(n)}(\phi, t, \varepsilon)$. Denote

$$\lim_{n\to\infty} \phi_t^{(n)}(s, \phi) = \phi_t(s, \phi), \quad \lim_{n\to\infty} u^{(n)}(\phi, t, \varepsilon) = u(\phi, t, \varepsilon). \tag{5.28}$$

We now show that the limiting function $u(\phi, t, \varepsilon)$ specifies the invariant set $\Gamma(\varepsilon)$: $h = u(\phi, t, \varepsilon)$ of the system of equations (5.11). Indeed, since $h = u^{(n+1)}(\phi, t, \varepsilon)$ is an invariant set of the system of equations (5.19), the trajectories $\phi_t^{(n)}$ and $h_t^{(n)}$ lying in this set satisfy the relations

$$\begin{aligned}\phi_t^{(n)}(s, \phi) = \phi &+ \int_s^t [\omega(\varepsilon) + \varepsilon^2 Q(u^{(n)}(\phi_\sigma^{(n)}, \sigma, \varepsilon), \phi_\sigma^{(n)}(s, \phi), \sigma, \varepsilon)]d\sigma \\ &+ \varepsilon^2 \sum_{s<\tau_i<t} G_i(u^{(n)}(\phi_{\tau_i}^{(n)}, \tau_i, \varepsilon), \phi_{\tau_i}^{(n)}, \varepsilon),\end{aligned} \tag{5.29}$$

$$\begin{aligned}u^{(n+1)}(\phi_t^{(n)}, t, \varepsilon) = u^{(n+1)}(\phi, s, \varepsilon) &+ \int_s^t [\varepsilon\lambda u^{(n+1)}(\phi_\sigma^{(n)}, \sigma, \varepsilon) \\ &+ \varepsilon^2 P(u^{(n)}(\phi_\sigma^{(n)}, \sigma, \varepsilon), \phi_\sigma^{(n)}, \sigma, \varepsilon)]d\sigma \\ &+ \varepsilon^2 \sum_{s<\tau_i<t} H_i(u^{(n)}(\phi_{\tau_i}^{(n)}, \tau_i, \varepsilon), \phi_{\tau_i}^{(n)}, \varepsilon).\end{aligned} \tag{5.30}$$

The assumptions of continuity of the functions P, Q, H_i, and G_i in h and ϕ guarantee the validity of all permutations of the limits required to derive the identities

$$\begin{aligned}\phi_t(s,\phi) &= \phi + \int_s^t [\omega(\varepsilon) + \varepsilon^2 Q(u(\phi_\sigma,\sigma,\varepsilon),\phi_\sigma,\sigma,\varepsilon)]d\sigma \\ &\quad + \varepsilon^2 \sum_{s<\tau_i<t} G_i(u(\phi_{\tau_i},\tau_i,\varepsilon),\phi_{\tau_i},\varepsilon), \qquad (5.31)\\ u(\phi_t,t,\varepsilon) &= u(\phi,s,\varepsilon) + \int_s^t [\varepsilon\lambda u(\phi_\sigma,\sigma,\varepsilon) + \varepsilon^2 P(u(\phi_\sigma,\sigma,\varepsilon),\phi_\sigma,\sigma,\varepsilon)]d\sigma \\ &\quad + \varepsilon^2 \sum_{s<\tau_i<t} H_i(u(\phi_{\tau_i},\tau_i,\varepsilon),\phi_{\tau_i},\varepsilon)\end{aligned}$$

from relations (5.28)–(5.30).

It follows from identities (5.31) that the function $u(\phi,t,\varepsilon)$ continuous in ϕ, piecewise continuous in t, and periodic in ϕ and t with period 2π for all $t \in \mathbb{R}$ satisfies the system of equations (5.17), (5.18). Hence, $h = u(\phi,t,\varepsilon)$ is an invariant set of the system of equations (5.11).

Thus, the following theorem is proved:

Theorem 2 ([90]). *Assume that the system of equations* (5.11) *satisfies inequalities* (5.13) *and* (5.14) *in domain* (5.12). *Then there exists a positive number* $\varepsilon^0 \le \varepsilon$ *such that, for all* $0 < \varepsilon \le \varepsilon^0$, *the system of equations* (5.11) *possesses an invariant set* $\Gamma(\varepsilon)\colon h = u(\phi,t,\varepsilon)$, *where the function* $u(\phi,t,\varepsilon)$ *is periodic in* ϕ *and* t *with period* 2π, *satisfies the Lipschitz condition with respect to* ϕ *with a constant proportional to* ε, *and is piecewise continuous in* t *with discontinuities of the first kind for* $t = \tau_i$.

On the invariant set $\Gamma(\varepsilon)$, *the system of equations* (5.11) *reduces to the system*

$$\begin{aligned}\frac{d\phi}{dt} &= \omega(\varepsilon) + \varepsilon^2 Q(u(\phi,t,\varepsilon),\phi,t,\varepsilon), \quad t \ne \tau_i,\\ \Delta\phi|_{t=\tau_i} &= \varepsilon^2 G_i(u(\phi,\tau_i,\varepsilon),\phi,\varepsilon).\end{aligned}$$

Note that the function $u(\phi,t,\varepsilon)$ continuously depends on the parameter ε and approaches zero as $\varepsilon \to 0$.

In view of Theorems 1 and 2, with regard for the changes of variables (5.4), (5.6), and (5.9) reducing Eq. (5.1) to the system of equations (5.11), we can now formulate the following assertion for the original system of equations (5.1):

Theorem 3 ([90]). *Assume that the system of equations with pulse action* (5.1) *satisfies the following conditions:*

(1) *the function* $f(x,\dot{x})$ *is continuously differentiable in domain* (5.3) *with respect to both variables and its derivatives in this domain satisfy the Lipschitz condition;*

(2) *the functions* $I_i(x, \dot{x})$ *defined as finite polynomials of their variables and the times* τ_i *satisfy equalities* (5.2);

(3) *the equation*

$$F(a) \equiv f_0^{(1)}(a) + \frac{1}{2\pi} \sum_{i=1}^{p} \overline{I}_i^{(1)}(a) = 0,$$

where

$$I_0^{(1)} = \frac{1}{2\pi} \int_0^{2\pi} f(a \sin \psi, a\omega \cos \psi) \cos \psi d\psi,$$

possesses an isolated positive root $a = a_0$ *such that* $F'(a_0) < 0$.

Then there exists a positive number ε^* *such that, for all* $0 < \varepsilon \le \varepsilon^*$,

(1) *the system of equations possesses an invariant set*

$$\sqrt{x^2 + \frac{1}{\omega^2}\dot{x}^2} = u^*(\phi, t, \varepsilon),$$

where the function $u^*(\phi, t, \varepsilon)$ *satisfies the Lipschitz condition with respect to* ϕ, *is piecewise continuous in* t *with discontinuities of the first kind for* $t = \tau_i$, *periodic in* ϕ *and* t *with period* 2π, *and such that* $u^*(\phi, t, \varepsilon) \to a_0$ *as* $\varepsilon \to 0$ *uniformly in* ϕ *and* t;

(2) *there exists a* $\delta_0(\varepsilon)$*-neighborhood* ($\delta_0(\varepsilon) \to 0$ *as* $\varepsilon \to 0$) *of the cylinder*

$$x^2 + \frac{1}{\omega^2}\dot{x}^2 = a_0^2$$

such that any solution of Eq. (5.1) *starting at* $t = t_0$ *from this neighborhood remains bounded together with its derivative for all* $t \ge t_0$.

Resonance Case. In the case where ω is close to one of the numbers of the form $\frac{r}{s}$, where r and s are natural numbers, system (5.1) is investigated in [139].

In this case, the first approximations to the solutions of Eqs. (5.1) should be taken in the form $x = a \sin(\frac{rt}{s} + \theta)$, where a and θ are solutions of the equations

$$\dot{a} = \varepsilon s \frac{A(a) + I^{(1)}(a, \theta)}{r}, \quad \dot{\theta} = -\varepsilon s \frac{\frac{\Delta a}{2} + B(a) + I^{(2)}(a, \theta)}{ra},$$

where, in turn,

$$A(a) = \frac{1}{2\pi} \int_0^{2\pi} f\left(a \sin \phi, a\frac{r}{s} \cos \phi\right) \cos \phi d\phi,$$

$$B(a) = \frac{1}{2\pi} \int_0^{2\pi} f\left(a \sin \phi, a\frac{r}{s} \cos \phi\right) \sin \phi d\phi,$$

$$I^{(1)}(a,\theta) = \frac{1}{2\pi s}\sum_{i=1}^{ps} I_i\left(a\sin\left(\frac{r}{s}\tau_i+\theta\right), a\frac{r}{s}\cos\left(\frac{r}{s}\tau_i+\theta\right)\right)\cos\left(\frac{r}{s}\tau_i+\theta\right),$$

$$I^{(2)}(a,\theta) = \frac{1}{2\pi s}\sum_{i=1}^{ps} I_i\left(a\sin\left(\frac{r}{s}\tau_i+\theta\right), a\frac{r}{s}\cos\left(\frac{r}{s}\tau_i+\theta\right)\right)\sin\left(\frac{r}{s}\tau_i+\theta\right),$$

$$\varepsilon\overline{\Delta} = \frac{r^2}{s^2} - \omega^2.$$

As an example, we study the influence of pulse action on a linear oscillator under the assumption that this action leads to an increase in the kinetic energy of the oscillator by a constant value εI. In other words, we study the system of equations

$$\ddot{x} + \omega^2 x = -\frac{\varepsilon\lambda}{x}, \quad t \neq \tau_i, \quad \Delta\dot{x}^2 = 2\varepsilon I, \quad \lambda > 0,\ I > 0,$$
$$\omega\tau_i = 2\pi i, \quad i = 0, \pm 1, \ldots. \tag{5.32}$$

In these equations, we perform the change of variables

$$x = a\sin(\omega t+\theta), \quad x = a\cos(\omega t+\theta)$$

and average the equations obtained as a result over the explicit time. As a result, in the first approximation, we get the following equations:

$$\dot{a} = -\varepsilon\frac{\lambda a - \frac{I}{\pi a\omega}}{2}, \quad \dot{\theta} = -\varepsilon I\frac{\tanh\theta}{2\pi a^2\omega}. \tag{5.33}$$

The first equation has the stationary solution

$$a = a_0 = \left(\frac{I}{\lambda\pi\omega}\right)^{\frac{1}{2}},$$

and the second equation has two stationary solutions

$$\theta = 0 (\operatorname{mod} 2\pi) \quad \text{and} \quad \theta = \pi (\operatorname{mod} 2\pi).$$

By using the relations of the improved first approximations, we get

$$x(t) = a\sin(\omega t+\theta) - \varepsilon\lambda a\frac{\cos(\omega t+\theta)}{4} + \frac{\varepsilon I\sin\omega t}{\pi a\omega^2\cos\theta}\sum_{k=1}^{\infty}\frac{\sin k\omega t}{k},$$

where a and θ are the solutions of the averaged equations (5.33).

We now analyze system (5.32) in the first approximation. It is easy to see that all solutions of this system approach, as functions of time, one of two asymptotically stable periodic solutions

$$x(t) = \pm a_0\sin\omega t \pm \varepsilon\lambda a_0\frac{\cos\omega t}{4} \pm \frac{\varepsilon I\sin\omega t}{\pi a_0\omega^2}\sum_{k=1}^{\infty}\frac{\sin k\omega t}{k}. \tag{5.34}$$

Moreover, the solutions corresponding to the stationary solution of the first equation in (5.33) cover an integral set determined (in the extended phase space $(x, \dot{x}, t)$) by the equation

$$(\omega^2 x^2 + \dot{x}^2)^{\frac{1}{2}} = \left(\frac{\omega}{\lambda}\right)^{\frac{1}{2}} \left(1 - \varepsilon\lambda \frac{\sin 2(\omega t + \theta)}{4}\right) + \varepsilon \left(\frac{I\lambda}{\pi\omega}\right)^{\frac{1}{2}} \sum_{k=1}^{\infty} \frac{\sin k\omega t}{k}.$$

This set is asymptotically stable and the solutions lying in this set are attracted, in the course of time, by one of the two periodic solutions (5.34).

The possibility of application of the method of averaging to Eqs. (5.1) in the resonance case is justified by the following assertion:

Theorem 4 ([139]). *Assume that the functions $f(x, \dot{x})$ and $I_i(x, \dot{x})$, $i = \overline{1, p}$ in the system of equations* (5.1) *are defined and satisfy the Lipschitz condition with respect to their variables in a certain domain*

$$0 < \alpha^2 \le x^2 + \omega^{-2}\dot{x}^2 \le \beta^2.$$

Suppose that the system of equations

$$A(a) + I^{(1)}(a, \theta) = 0, \quad B(a) + \frac{\overline{\Delta} a}{2} + I^{(2)}(a, \theta) = 0 \tag{5.35}$$

has an isolated solution $a = a^0$, $\theta = \theta^0$ from the strip $\alpha < a^0 < \beta$ together with a certain its ρ-neighborhood and is such that the index of the point (a^0, θ^0) under the mapping specified by the left-hand sides of Eqs. (5.35) *differs from zero. Then there exists a positive number ε_0 such that, for all $0 < \varepsilon \le \varepsilon_0$, system* (5.1) *possesses, in the resonance case, a solution 2π-periodic in $\frac{rt}{s}$.*

An oscillating system subjected to a pulse action when the phase point $(x, \dot{x})$ passes through the position $x = x_0$, namely,

$$\ddot{x} + \omega^2 x = \varepsilon f(x, \dot{x}), \quad x \ne x_0, \tag{5.36}$$

$$\Delta\dot{x}|_{x=x_0} = \dot{x}_+ - \dot{x}_- = \begin{cases} \varepsilon I(\dot{x}_-), & x = x_0,\ \dot{x}_- \ge 0, \\ 0, & x = x_0,\ \dot{x}_- < 0, \end{cases} \tag{5.37}$$

where $\dot{x}_-$ and $\dot{x}_+$ are the velocities of the point prior to and after the action of the instantaneous pulse, is analyzed by the method of averaging in [133, 136, 141].

System (5.36), (5.37) can be interpreted as a motion of oscillator (5.36) subjected to the action of instantaneous pulses increasing the velocity of the oscillator by $\varepsilon I(\dot{x}_-)$ every time when the oscillator passes through the position $x = x_0$ (with a nonnegative velocity). As a simple example of this oscillator, we can mention an impact model of a clock studied by many authors [16, 72].

It is clear that the motions of system (5.36), (5.37) may have the following form:

(1) Motions without pulse actions. These motions are described by Eq. (5.36) and satisfy the inequality $x(t) \neq x_0$ for all $t \geq 0$.

(2) Motions subjected to the pulse action for $t \geq 0$ only finitely many times. These are motions for which $x(t) = x_0$ for finitely many values of $t \geq 0$.

(3) Motions subjected to the pulse action for $t \geq 0$ infinitely many times.

Motions of the first two types either approach the infinity as $t \to \infty$ or are attracted to the equilibrium position and cycles of Eq. (5.36) lying in the domain $x < x_0$ of the phase plane $(x, \dot{x})$. In this case, periodic motions may appear only for the first type of motions. They are investigated without using conditions (5.37) according to the standard schemes of asymptotic methods. Motions of the third type either approach the infinity as $t \to \infty$ or are attracted to discontinuous cycles of system (5.36), (5.37). Motions of the last type are not cycles of Eq. (5.36) and generate periodic motions with exactly one pulse action per period. They can be studied by using the scheme of the method of averaging proposed in [132].

Suppose that system (5.36), (5.37) possesses a periodic solution. In view of the autonomy of system (5.36), (5.37), the function obtained from this solution with t replaced by $t + s$ is also a periodic solution of system (5.36), (5.37). Therefore, we can always assume that the analyzed periodic solution takes the value $x(0) = x_0$ at time $t = 0$ and, in addition, $\dot{x}(+0) > 0$.

By ν we denote the frequency of this solution. For this solution, the times τ_k of pulse action are given by the formula

$$\nu\tau_k = 2k\pi, \quad k = 1, 2, \ldots,$$

and the jumps of velocity $\Delta\dot{x}$ at these times are given by the following formula:

$$\Delta\dot{x} = \varepsilon I(\dot{x}_-) = \varepsilon I_0.$$

By virtue of periodicity of the solution, the quantity I_0 is constant for all $\nu\tau_k = 2k\pi$. Hence, the periodic solution of system (5.36), (5.37) satisfies the equation

$$\ddot{x} + \omega^2 x = \varepsilon f(x, \dot{x}), \quad \nu t \neq 2k\pi, \tag{5.38}$$

and the condition

$$\Delta\dot{x} = \varepsilon I_0, \quad \nu t = 2k\pi, \tag{5.39}$$

where k can be set equal to $0, \pm 1, \pm 2, \ldots$. On the other hand, assume that, for some ν and t_0, system (5.38), (5.39) possesses periodic solutions with period 2π with respect to νt: $x = x(t)$. It is clear that these solutions are solutions of the original system (5.36), (5.37) whenever

$$\begin{gathered} x(0) = x_0, \quad I_0 = I(\dot{x}_-), \\ \dot{x}_- = \dot{x}\left(\frac{2\pi}{\nu} - 0\right) \geq 0, \quad \dot{x}_+ \geq 0. \end{gathered} \tag{5.40}$$

Thus, the problem of finding periodic solutions of system (5.36), (5.37) is reduced to the determination of the periodic solutions of system (5.38), (5.39) with period 2π with respect to νt and the choice of the parameters ν and t_0 for these solutions satisfying relations (5.40). In view of the fact that any cycle of system (5.36), (5.37) is formed by a part of the spiral of Eq. (5.36) for which the difference between its two successive crossings of the beam $\dot{x} \leq 0$ by the straight line $x = x_0$ is equal to εI_0, we conclude that the frequency ν of periodic solutions of system (5.36), (5.37) for small ε should be taken close to the frequency of natural oscillations of system (5.36), i.e., close to the value ω.

Therefore, we seek periodic solutions of system (5.38) under the condition $\nu \simeq \omega$. We represent ν in the form

$$\nu^2 = \omega^2 + \varepsilon\overline{\Delta}, \tag{5.41}$$

where $\overline{\Delta}$ is an unknown constant. In view of (5.41), system (5.38), (5.39) can be rewritten in the form

$$\ddot{x} + \nu^2 x = \varepsilon[\overline{\Delta}x + f(x, \dot{x})], \quad \nu t \neq 2k\pi, \tag{5.42}$$

$$\Delta\dot{x} = \varepsilon I_0, \quad \nu t = 2k\pi. \tag{5.43}$$

By using the formalism of generalized functions, we pass from system (5.42), (5.43) to the equation

$$\ddot{x} + \nu^2 x = \varepsilon[\overline{\Delta}x + f(x, \dot{x})] + \varepsilon I_0 \sum_{k=-\infty}^{\infty} \delta\left(t - \frac{2k\pi}{\nu}\right).$$

Further, in view of the properties of the delta function, we get

$$\ddot{x} + \nu^2 x = \varepsilon[\overline{\Delta}x + f(x, \dot{x})] + \varepsilon\frac{I_0\nu}{\pi}\left(\frac{1}{2} + \sum_{k=1}^{\infty} \cos k\nu t\right). \tag{5.44}$$

In (5.44), we perform the change of variables

$$x = \varepsilon\frac{I_0\nu}{\pi\nu}\left(\frac{1}{2} - \sum_{k=2}^{\infty} \frac{\cos k\nu t}{k^2 - 1}\right) + z. \tag{5.45}$$

This yields the following ordinary differential equation for z:

$$\begin{aligned}\ddot{z} + \nu^2 z = {} & \varepsilon[\overline{\Delta}z + f(z + \varepsilon I_0\phi(\nu t), \dot{z} + \varepsilon I_0\dot{\phi}(\nu t)) + \varepsilon\overline{\Delta}I_0\phi(\nu t)] \\ & + \varepsilon\frac{I_0\nu}{\pi}\cos\nu t,\end{aligned} \tag{5.46}$$

where

$$\phi(\nu t) = \frac{1}{\pi\nu}\left(\frac{1}{2} - \sum_{k=2}^{\infty} \frac{\cos k\nu t}{k^2 - 1}\right).$$

Relations (5.40) for the variable z take the form

$$z(0) = \varepsilon I_0 \phi(0) = x_0, \quad I_0 = I\left(\dot{z}(0) - \frac{\varepsilon I_0}{2}\right), \quad \dot{z}(0) \geq \frac{\varepsilon I_0}{2}. \tag{5.47}$$

Hence, it is necessary to find periodic solutions $z(t)$ of Eq. (5.46) with period 2π with respect to νt satisfying relations (5.49). To do this, we perform the change of variables

$$z = a\cos\psi, \quad \dot{z} = -a\nu\sin\psi, \quad \psi = \nu t + \theta. \tag{5.48}$$

As a result, we arrive at the following system of equations:

$$\frac{da}{dt} = -\frac{\varepsilon}{\nu}[\overline{\Delta} a\cos\psi + f(a\cos\phi + \varepsilon I_0\psi, -a\nu\sin\psi + \varepsilon I_0\dot{\phi}) + \varepsilon\overline{\Delta} I_0\phi]\sin\psi$$
$$- \frac{\varepsilon I_0}{\pi}\cos\nu t\sin\psi, \tag{5.49}$$
$$\frac{d\psi}{dt} = -\frac{\varepsilon}{a\nu}[\overline{\Delta} a\cos\psi + f(a\cos\phi + \varepsilon I_0\psi, -a\nu\sin\psi + \varepsilon I_0\dot{\phi}) + \varepsilon\overline{\Delta} I_0\phi]\cos\psi$$
$$- \frac{\varepsilon I_0}{a\pi}\cos\nu t\cos\psi.$$

Then relations (5.47) take the form

$$a_0\cos\theta_0 = x_0 - \varepsilon I_0\phi(0), \quad I_0 = I\left(-a_0\nu\sin\theta_0 - \frac{\varepsilon I_0}{2}\right), \tag{5.50}$$
$$-a_0\nu\sin\theta_0 \geq \frac{\varepsilon I_0}{2},$$

where $a_0 = a(0)$ and $\theta_0 = \theta(0)$. Assume that the function $f(x, \dot{x})$ satisfies the Lipschitz condition with respect to x and $\dot{x}$ in an annular domain

$$\nu^2\alpha^2 \leq \nu^2 x^2 + y^2 \leq \beta^2\nu^2.$$

Then, for small ε, (5.49) is a T-system ($T = \frac{2\pi}{\nu}$) [134] in the band

$$0 < \alpha \leq a \leq \beta. \tag{5.51}$$

Hence, any internal point of band (5.51) has a Δ-constant [134] and this constant is determined, to within the quantities of the second order of smallness, from the system

$$\Delta_1 = -\frac{\varepsilon}{\nu}A(a_0) - \frac{\varepsilon I_0}{2\pi}\sin\theta_0 + \varepsilon^2\cdots,$$
$$\Delta_2 = -\frac{\varepsilon}{2\nu}\overline{\Delta} - \frac{\varepsilon}{a_0\nu}B(a_0) - \frac{\varepsilon I_0}{2\pi a_0}\cos\theta_0 + \varepsilon^2\cdots,$$

where

$$A(a) = A(a,\nu) = \frac{1}{2\pi}\int_0^{2\pi} f(a\cos\phi, -a\nu\sin\phi)\sin\phi d\phi, \tag{5.52}$$

$$B(a) = B(a,\nu) = \frac{1}{2\pi}\int_0^{2\pi} f(a\cos\phi, -a\nu\sin\phi)\cos\phi d\phi.$$

According to [134], the solution of system (5.49) is periodic with period 2π with respect to νt provided that a_0 and θ_0 satisfy the relations

$$-\frac{\varepsilon}{\nu}A(a_0) - \frac{\varepsilon I_0}{2\pi}\sin\theta + \varepsilon^2\cdots = 0, \tag{5.53}$$

$$-\frac{\varepsilon}{2\nu}\overline{\Delta} - \frac{\varepsilon}{a_0\nu}B(a_0) - \frac{\varepsilon I_0}{2\pi a_0}\cos\theta_0 + \varepsilon^2\cdots = 0. \tag{5.54}$$

Moreover, the indicated periodic solution is the limit of the sequence

$$\begin{aligned} a_n(t) = a_0 - \frac{\varepsilon}{\nu}\int_0^t &[F_1(t, a_{n-1}(t), \theta_{n-1}(t), \varepsilon) \\ &- \overline{F_1(t, a_{n-1}(t), \theta_{n-1}(t), \varepsilon)}]dt, \\ \theta_n(t) = a_0 - \frac{\varepsilon}{\nu}\int_0^t &[F_2(t, a_{n-1}(t), \theta_{n-1}(t), \varepsilon) \\ &- \overline{F_2(t, a_{n-1}(t), \theta_{n-1}(t), \varepsilon)}]dt, \end{aligned} \tag{5.55}$$

where $-\frac{\varepsilon}{\nu}F_1(t,a,\theta)$ and $-\frac{\varepsilon}{\nu}F_2(t,a,\theta)$ are the right-hand sides of the first and second equations of system (5.49), respectively, and

$$\overline{F(t)} = \frac{1}{T}\int_0^T F(t)dt, \quad T = \frac{2\pi}{\nu}, \quad n = 1, 2, \ldots.$$

In particular, for $n = 1$, to within the quantities of the order ε^2, relations (5.55) yield the following formulas:

$$\begin{aligned} a(t) &= a_0 - \frac{\varepsilon\overline{\Delta}}{\nu^2}a_0\frac{\cos 2\theta_0 - \cos 2(\nu t + \theta_0)}{4} \\ &\quad - \frac{\varepsilon}{\nu^2}\int_{\theta_0}^{\nu t + \theta_0} f(a\cos\phi, -a\nu\sin\phi)\sin\phi d\phi - \frac{\varepsilon I_0}{\pi\nu}\frac{\cos\theta_0 - \cos(2\nu t + \theta_0)}{4}, \\ \theta(t) &= \theta_0 - \frac{\varepsilon\overline{\Delta}}{\nu^2}\frac{\sin 2(\nu t + \theta_0) - \sin 2\theta_0}{4} \\ &\quad - \frac{\varepsilon}{\nu^2 a_0^2}\int_{\theta_0}^{\nu t + \theta_0} f(a\cos\phi, -a\nu\sin\phi)\cos\phi d\phi \\ &\quad - \frac{\varepsilon I_0}{\pi\nu a_0}\frac{\sin\theta_0 - \sin(2\nu t + \theta_0)}{4}. \end{aligned} \tag{5.56}$$

We now study the problem of solvability of Eqs. (5.53) and (5.54). Since the Δ-constant is a continuous function of the point a_0, θ_0 and the parameters $I_0, \overline{\Delta}$, and ε and satisfies the same Lipschitz condition with respect to a_0, θ_0, $I_0, \overline{\Delta}$, and ε [134], Eq. (5.54) is always solvable in the form

$$\overline{\Delta} = \overline{\Delta}(a_0, \theta_0, I_0, \varepsilon). \tag{5.57}$$

Moreover, the function $\overline{\Delta}$ is continuous in the collection of variables a_0, θ_0, I_0, and ε in the domain of their variation and satisfies, in this domain, the Lipschitz condition with respect to a_0, θ_0, I_0, and ε. Equation (5.54) also implies that, to within the quantities of order ε, the function $\overline{\Delta}(a_0, \theta_0, I_0, \varepsilon)$ is given by the formula

$$\overline{\Delta}(a_0, \theta_0, I_0, \varepsilon) = \overline{\Delta}(a_0, \theta_0, I_0, 0) = -2\frac{B(a_0)}{a_0} - \frac{I_0\omega}{\pi a_0}\cos\theta_0. \tag{5.58}$$

Substituting $\overline{\Delta}(a_0, \theta_0, I_0, \varepsilon)$ in the first equation in (5.53), we arrive at the relation for the initial values of periodic solutions of system (5.49). To within the quantities of order ε, this relation takes the form

$$A(a_0) + \frac{I_0\omega}{2\pi}\sin\theta_0 = 0. \tag{5.59}$$

It is clear that periodic solutions of system (5.49) are also periodic solutions of the original system (5.36), (5.37) provided that their initial values satisfy relations (5.50). Hence, to determine these values, it suffices to solve the system of relations (5.50), (5.53) by substituting relation (5.57) for the function $\overline{\Delta}$ in (5.53).

Since a_0 is taken from interval (5.51), relations (5.50) imply that

$$\cos\theta_0 = \frac{x_0 - \varepsilon I_0\phi(0)}{a_0}, \quad \sin\theta_0 = -\sqrt{1 - \left(\frac{x_0 - \varepsilon I_0\phi(0)}{a_0}\right)^2}. \tag{5.60}$$

Both relations (5.60) and the inequality in (5.50) taking, in view of of relation (5.60), the form

$$\nu\sqrt{a_0^2 - (x_0 - \varepsilon I_0\phi(0))^2} \geq \frac{\varepsilon I_0}{2},$$

are always true for small ε if a_0 varies within the range

$$\beta \geq a_0 \geq \alpha > |x_0|. \tag{5.61}$$

Thus, we assume that $a > |x_0|$ and substitute (5.60) in the remaining relation in (5.50). As a result, instead of (5.50), we get relations (5.60) and the equation

$$I_0 = I\left(\nu\sqrt{a_0^2 - (x_0 - \varepsilon I_0\phi(0))^2} - \frac{\varepsilon I_0}{2}\right). \tag{5.62}$$

We now assume that the function $I(\dot{x})$ satisfies the Lipschitz condition in the segment $\nu\beta \geq \dot{x} \geq 0$.

Then Eq. (5.62) is solvable for small ε in the form

$$I_0 = I_0(a_0, \varepsilon). \tag{5.63}$$

Moreover, the function $I_0(a_0, \varepsilon)$ satisfies the Lipschitz condition with respect to a_0 and ε in the domain of their variation. Relation (5.62) also implies that, to within the quantities of order ε, the function $I_0(a_0, \varepsilon)$ is given by the formula

$$I_0(a_0, \varepsilon) = I_0(a_0, 0) = I\left(\omega\sqrt{a_0^2 - x_0^2}\right). \tag{5.64}$$

Thus, the system of relations (5.50), (5.53), (5.57) is reduced to system (5.53), (5.57), (5.60), (5.63).

Substituting relations (5.63) in (5.60), we express θ_0 in terms of a_0. Then we substitute (5.63) and (5.60) in (5.57) and express $\overline{\Delta}$ via a_0. Similarly, we arrive at the expression

$$A(a_0) - \frac{I_0(a_0, \varepsilon)\nu}{2\pi}\sqrt{1 - \left(\frac{x_0 - \varepsilon I_0 \phi(0)}{a_0}\right)^2} + \varepsilon \cdots .$$

By virtue of relations (5.53), (5.57), (5.60), and (5.63), this expression satisfies the original system (5.50), (5.53), (5.54) or, in view of (5.41) and (5.64), the relation

$$A(a_0) - \frac{I(\omega\sqrt{a_0^2 - x_0^2})}{2\pi}\,\omega\sqrt{1 - \frac{x_0^2}{a_0^2}} + \varepsilon \cdots = 0. \tag{5.65}$$

Since the left-hand side of Eq. (5.63) is continuous in a_0 and ε, Eq. (5.65) is solvable for small ε provided that the function

$$A(a_0) - \frac{I(\omega\sqrt{a_0^2 - x_0^2})}{2\pi a_0}\,\omega\sqrt{a_0^2 - x_0^2}, \tag{5.66}$$

takes values of different signs in the interval $\alpha < a_0 < \beta$ [135]. Moreover, the solutions of Eq. (5.65) are continuous in ε and, to within the quantities of a certain order $\delta(\varepsilon)$, where $\delta(\varepsilon) \to 0$ as $\varepsilon \to 0$, are determined by the solutions of the equation

$$A(a_0) = \frac{I(\omega\sqrt{a_0^2 - x_0^2})\omega}{2\pi a_0}\sqrt{a_0^2 - x_0^2}. \tag{5.67}$$

Thus, to satisfy the system of relations guaranteeing the existence of periodic solutions of system (5.36), (5.37), it is sufficient that Eq. (5.67) have a solution lying in segment (5.61) together with its certain neighborhood and that this solution be neither a point of maximum nor a point of minimum of function (5.66). By using relations (5.41), (5.58), (5.60), and (5.64), we get the following formula for the frequency of the periodic solution:

$$\nu = \omega + \frac{1}{2}\frac{\varepsilon\overline{\Delta}}{\omega} + \varepsilon^2 \cdots \simeq \omega - \varepsilon\left(\frac{B(a_0)}{a_0\omega} + \frac{I(\omega\sqrt{a_0^2 - x_0^2})x_0}{2\pi a_0^2}\right). \tag{5.68}$$

According to the changes of variables (5.45) and (5.48), the periodic solutions of system (5.36), (5.37) are determined (to within the quantities of order $\delta(\varepsilon)$) by the formula

$$x(t) = \frac{\varepsilon I(\omega\sqrt{a_0^2 - x_0^2})}{\pi\omega}\left(\frac{1}{2} - \sum_{k=2}^{\infty}\frac{\cos k\nu(t+s)}{k^2 - 1}\right)$$
$$+ a(t+s)\cos(\nu(t+s) + \theta(t+s)),$$

where $a(t), \theta(t)$ is the function given by relation (5.56), ν is given by relation (5.68), a_0 is the root of Eq. (5.67), and s is an arbitrary constant. This proves the following theorem:

Theorem 5 ([136]). *Assume that the functions $f(x, \dot{x})$ and $I(\dot{x})$ characterizing system* (5.36), (5.37) *are defined and continuous and that they satisfy the Lipschitz condition with respect to x and $\dot{x}$ for x and $\dot{x}$ from the domain*

$$\alpha^2 \leq x^2 + \frac{\dot{x}^2}{\omega^2} \leq \beta^2,$$

where β and α are positive constants satisfying the inequality $\beta > \alpha > |x_0|$.

Suppose that the equation

$$A(a) - \frac{I(\omega\sqrt{a^2 - x_0^2})\omega}{2\pi a}\sqrt{a^2 - x_0^2} = 0, \tag{5.69}$$

where $A(a)$ is given by relation (5.52), *possesses a solution $a = a_0$ lying in the segment $\alpha \leq a_0 \leq \beta$ together with its certain ρ-neighborhood and such that the left-hand side of Eq.* (5.69) *turns neither into a maximum nor into a minimum for this solution.*

Then there exists $\varepsilon_0 > 0$ such that, for all ε smaller than ε_0, the system of equations (5.36), (5.37) *has a one-parameter family of periodic solutions*

$$x = x_0(\omega_1(t+s), \varepsilon) = a(\omega_1(t+s))\cos\phi(\omega_1(t+s))$$

satisfying the inequality

$$|a(t) - a_0| + |\omega_1 - \omega - \varepsilon\Delta(a_0)| \leq \eta(\varepsilon),$$

where s is an arbitrary constant, $\eta(\varepsilon)$ is a constant approaching zero as $\varepsilon \to 0$,

$$\Delta(a_0) = -\frac{I(\omega\sqrt{a_0^2 - x_0^2})}{2\pi a_0^2} x_0 - \frac{B(a_0)}{a_0\omega},$$

and $B(a)$ is given by relation (5.52).

The case where the right-hand side of the first equation in system (5.36), (5.37) depends on t, i.e., has the form $f(\nu t, x, \dot{x})$, is studied in [141]. In the cited work, it is shown that, in the resonance case, i.e., for

$$\omega^2 = \left(\frac{r\nu}{s}\right)^2 + \varepsilon\overline{\Delta},$$

the isolated solution of the system of equations

$$A(a, \theta) + r\nu\sqrt{a^2 - x_0^2}\frac{I\left(\frac{r\nu\sqrt{a^2-x_0^2}}{s}\right) - I\left(\frac{-r\nu\sqrt{a^2-x_0^2}}{s}\right)}{2\pi sa} = 0,$$

$$B(a, \theta)\frac{\overline{\Delta}a}{2} + r\nu x_0 \frac{I\left(\frac{r\nu\sqrt{a^2-x_0^2}}{s}\right) - I\left(\frac{-r\nu\sqrt{a^2-x_0^2}}{s}\right)}{2\pi sa} = 0,$$

where

$$A(a, \theta) = \frac{1}{2\pi s}\int_0^{2\pi s} f\left(\phi, a\sin\left(\frac{r}{s}\phi + \theta\right),\right.$$
$$\left.\frac{ar\nu}{s}\cos\left(\frac{r}{s}\phi + \theta\right)\right)\cos\left(\frac{r}{s}\phi + \theta\right)d\phi,$$

$$B(a, \theta) = \frac{1}{2\pi s}\int_0^{2\pi s} f\left(\phi, a\sin\left(\frac{r}{s}\phi + \theta\right),\right.$$
$$\left.\frac{ar\nu}{s}\cos\left(\frac{r}{s}\phi + \theta\right)\right)\sin\left(\frac{r}{s}\phi + \theta\right)d\phi,$$

generates a periodic solution (with period 2π in $\frac{r\nu t}{s}$) of the original system (5.36), (5.37) for small values of the parameter ε.

A similar result is obtained in [97] for the integrodifferential equation subjected to the pulse action

$$\ddot{x} + \omega^2 x = \varepsilon f\left(\nu t, x, \dot{x}, \int_t^{t+s} h(\nu t, \nu s, x(s), \dot{x}(s))ds\right), \quad x \neq x_0,$$
$$\Delta\dot{x} = \varepsilon I(\dot{x}),$$

where $f(\nu t, x, \dot{x}, v)$ and $h(\nu t, \nu s, x, \dot{x})$ are functions periodic in νt and νs with period 2π, $\omega \simeq \nu$, and s and x_0 are constants.

In [63], the method of averaging is substantiated for the nonresonance case under an additional assumption concerning the straight line in which the system is subjected to the pulse action. Under fairly general assumptions imposed on the functions $f(t, x, \dot{x})$ and $I(\dot{x})$, it is shown that, for small values of the parameter ε, the isolated positive root $a = a_0$ of the equation

$$F(a) \equiv \frac{1}{4\pi^2}\int_0^{2\pi}\int_0^{2\pi} f(t, a\sin\theta, a\omega\cos\theta)\cos t\cos\theta\, dt\, d\theta + \omega\frac{I(a\omega) - I(-a\omega)}{2\pi},$$

generates an integral set of the system of equations

$$\ddot{x} + \omega^2 x = \varepsilon f(\nu t, x, \dot{x},), \quad x \neq \varepsilon x_0, \ \Delta\dot{x}|_{x=\varepsilon x_0} = \varepsilon I(\dot{x}). \tag{5.70}$$

This set is specified by the equation

$$\sqrt{(x - \varepsilon x_0)^2 + \omega^{-2}\dot{x}^2} = u(\phi, t, \varepsilon),$$

where the function $u(\phi, t, \varepsilon)$ satisfies the Lipschitz condition with respect to t, is piecewise continuous in ϕ with discontinuities of the first kind for $\phi = k\pi$, 2π-periodic in ϕ and t, and such that $u(\phi, t, \varepsilon) \to 0$ as $\varepsilon \to 0$. In addition, if $F'(a_0) < 0$, then there exists a $\delta_0(\varepsilon)$-neighborhood ($\delta_0(\varepsilon) \to 0$ as $\varepsilon \to 0$) of the cylinder $x^2 + \omega^{-2}\dot{x}^2 = a_0^2$ such that all solutions of Eq. (5.70) starting from this neighborhood remain bounded together with their first derivatives for all $t \geq 0$.

The first and improved first approximations and equations of the first approximation for the equations with slowly varying coefficients subjected to the pulse action

$$\frac{d}{dt}(m(s)\dot{x}) + c(s)x = \varepsilon f(s, x, \dot{x}), \quad t \neq iT,$$
$$\Delta(m(s)\dot{x})|_{t=iT} = \varepsilon I(x, \dot{x}),$$

where $m(s)$ and $c(s)$ are positive functions of the "slow" time $s = \varepsilon t$, are constructed in [42].

5.2 Systems with Fixed Times of the Pulse Action

We now consider a system of differential equations with pulse action of the form

$$\frac{dx}{dt} = \varepsilon X(t, x) \quad \text{for } t \neq \tau_i, \tag{5.71}$$
$$\Delta x|_{t=\tau_i} = \varepsilon I_i(x).$$

The following assumptions are made for the functions $X(t, x)$ and $I_i(x)$ and the times τ_i:

(a) $X(t,x)$ and $I_i(x)$ are continuous functions of their arguments with continuous derivatives with respect to x up to the second order inclusively in a certain domain $t \in \mathbb{R}$, $x \in D \subset \mathbb{R}^n$;

(b) the times τ_i at which the state of the system instantaneously changes are enumerated by integer numbers so that $\tau_i \to -\infty$ as $i \to -\infty$ and $\tau_i \to +\infty$ as $i \to +\infty$. Moreover, one can find numbers L and d such that any time interval of length L contains at most d points of the sequence $\{\tau_i\}$;

(c) the finite limits

$$\lim_{T\to\infty} \frac{1}{T} \int_t^{t+T} X(s,x)ds = X_0(x), \quad \lim_{T\to\infty} \frac{1}{T} \sum_{t<\tau_i<t+T} I_i(x) = I^0(x), \tag{5.72}$$

exist uniformly in $t \in \mathbb{R}$ and $x \in D$.

Parallel with system (5.71), we consider an averaged system

$$\frac{dx}{dt} = \varepsilon[X_0(x) + I^0(x)] \equiv \varepsilon F(x) \tag{5.73}$$

and assume that it has an isolated equilibrium position $x = x_0$, $F(x_0) = 0$ lying in the domain D together with its certain ρ-neighborhood and that the real parts of the eigenvalues of the matrix

$$A = \left.\frac{\partial F}{\partial x}\right|_{x=x_0}$$

are not equal to zero.

We now prove the theorem on the behavior and properties of solutions of the system of equations (5.71) in a neighborhood of the equilibrium position $x = x_0$ of the averaged system of equations (an analog of the famous Bogolyubov theorem).

For the sake of convenience, in what follows, we transform the system of equations (5.71) to a certain special form. To this end, we need the following lemma:

Lemma 4 ([140]). *Let the functions $X(t,x)$ and $I_i(x)$ and times τ_i satisfy conditions* (a)–(c). *If the quantities $X_0(x)$ and $I^0(x)$ are defined by relations* (5.72), *then, for $\eta > 0$, there exists a continuous function $\zeta(\eta)$, $\zeta(\eta) \to 0$ as $\eta \to 0$, such that if*

$$f_\eta(t,x) = \int_{-\infty}^t e^{-\eta(t-s)}[X(s,x) - X_0(x) - I^0(x)]ds + \sum_{\tau_i<t} I_i(x)e^{-\eta(t-\tau_i)},$$

then

$$\|f_\eta(t,x)\| \le \eta^{-1}\zeta(\eta) \tag{5.74}$$

for all $t \in \mathbb{R}$ and $x \in D_\rho$.

Furthermore,

$$\left\| \frac{\partial f_\eta}{\partial t} - X(t,x) + F(x) \right\| \le \zeta(\eta) \tag{5.75}$$

for all $t \in \mathbb{R}$, $t \ne \tau_i$, *and* $x \in D_\rho$.

Proof. It follows from conditions (a)–(c) that there exists a continuous decreasing function $\varepsilon(T)$ such that $\varepsilon(T) \to 0$ as $T \to \infty$ and

$$\left\| \frac{1}{T} \int_t^{t+T} [X(s,x) - X_0(x)]ds \right\| + \left\| \frac{1}{T} \sum_{t<\tau_i<t+T} [I_i(x) - I^0(x)] \right\| \le \varepsilon(T)$$

for all $t \in \mathbb{R}$ and $x \in D_\rho$. Moreover, the function $f_\eta(t,x)$ satisfies the inequality

$$\|f_\eta(t,x)\| \le 2T(1-e^{-\eta T})^{-1}\varepsilon(T) + TM\left(1 + \frac{d}{L} + \varepsilon(T) + \frac{M}{\eta}\left(\frac{\eta T}{1-e^{-\eta T}} - 1\right)\right),$$

where

$$M \ge \sup_{t\in\mathbb{R},\ x\in D_\rho} [\|X(t,x) - X_0(x)\| + \|I_i(x) - I^0(x)\|].$$

We choose T, as a function of η, so that $1 - e^{-\eta T} = \varepsilon(T)$. Since $\varepsilon(T) \to 0$ as $T \to \infty$, the solution T_η of this equation satisfies the condition $\eta T_\eta \to 0$ as $\eta \to 0$ and is continuous in η. We set

$$\eta T_\eta \left[2 + M\left(1 + \frac{d}{L} + \varepsilon(T_\eta)\right)\right] + M\left(\frac{\eta T_\eta}{1 - e^{-\eta T_\eta}} - 1\right) = \zeta(\eta).$$

Thus, $\zeta(\eta) \to 0$ as $\eta \to 0$, and the first part of the lemma is proved.

Estimate (5.75) follows from inequality (5.74) in view of the fact that $f_\eta(t,x)$, for $t \ne \tau_i$, is a function differentiable with respect to t. The lemma is proved. □

In the system of equations (5.71), we now set $x = x_0 + b$, where x_0 is the equilibrium position of the averaged system and b are new variables. This enables us to rewrite the system in the form

$$\frac{db}{dt} = \varepsilon Ab + \varepsilon B(t,b) \quad \text{for } t \ne \tau_i, \tag{5.76}$$

$$\Delta b|_{t=\tau_i} = \varepsilon I_i(x_0) + b,$$

where

$$B(t,b) = X(t, x_0 + b) - Ab.$$

In the system of equations (5.76), we pass from the variable b to the variable y as follows:

$$b = y + \varepsilon v(t,y,\varepsilon), \tag{5.77}$$

where the function $v(t, y, \varepsilon)$ coincides with the function $f_\eta(t, y)$ if, in the last function, we choose $\eta = \eta(\varepsilon)$ as a function of ε such that $\eta(\varepsilon) \to 0$ as $\varepsilon \to 0$ and $\varepsilon\eta^{-1}(\varepsilon) \to$ const as $\varepsilon \to 0$.

In view of conditions (a)–(c) and the properties of the function $f_\eta(t, y)$ established by Lemma 4, by the change of variables (5.77), we pass from system (5.76) to a system of the form

$$\frac{dy}{dt} = \varepsilon A y + \varepsilon Q(t, y, \varepsilon), \quad t \neq \tau_i, \qquad (5.78)$$
$$\Delta y|_{t=\tau_i} = \varepsilon I_i^0(y, \varepsilon).$$

Replacing y with εy in system (5.78), we finally get the following system of equations:

$$\frac{dy}{dt} = \varepsilon A y + P(t, y, \varepsilon), \quad t \neq \tau_i, \qquad (5.79)$$
$$\Delta y|_{t=\tau_i} = I_i(y, \varepsilon),$$

where

$$P(t, y, \varepsilon) = Q(t, \varepsilon y, \varepsilon), \quad I_i(y, \varepsilon) = I_i^0(\varepsilon y, \varepsilon).$$

It is easy to see that the system of equations (5.79) satisfies the following conditions:

(1) the functions $P(t, x, \varepsilon)$ and $I_i(y, \varepsilon)$ are defined in the domain

$$t \in \mathbb{R}, \quad y \in D_{\rho_1} = \left\{ y \in \mathbb{R}^n : \|y\| \leq \frac{\rho_1}{\varepsilon} < \frac{\rho}{\varepsilon} \right\}, \quad 0 < \varepsilon \leq \varepsilon_0;$$

(2) the following inequalities hold for $t \in \mathbb{R}$, $0 < \varepsilon \leq \varepsilon_0$, and $i \in \mathbb{Z}$:

$$\|P(t, 0, \varepsilon)\| + \|I_i(0, \varepsilon)\| \leq M(\varepsilon),$$

where $M(\varepsilon) \to 0$ as $\varepsilon \to 0$;

(3) for any positive $\sigma < \rho_1$ and all $i \in \mathbb{Z}$, the inequality

$$\|P(t, y', \varepsilon) - P(t, y'', \varepsilon)\| + \|I_i(y', \varepsilon) - I_i(y'', \varepsilon)\| \leq \lambda(\varepsilon, \sigma)\|y', y''\|,$$

where $\varepsilon^{-1}\lambda(\varepsilon, \sigma) \to 0$ as $\varepsilon \to 0$ and $\sigma \to 0$, holds in the domain $t \in \mathbb{R}$, y', $y'' \in D_\sigma$, $0 < \varepsilon \leq \varepsilon_0$;

(4) the real parts of the eigenvalues of the matrix A are not equal to zero.

Note that if the original system of equations (5.71) is periodic in t with period ω, then the system of equations (5.79) is also periodic with the same period. Indeed, the periodicity of system (5.71) in t with period ω means that $X(t + \omega, x) = X(t, x)$ and there exists a natural number p such that

$$\tau_{i+p} - \tau_i = \omega, \quad I_{i+p}(x) = I_i(x)$$

for all $i \in \mathbb{Z}$. To prove the periodicity of the system of equations (5.79), it suffices to establish the periodicity of the function $v(t, y, \varepsilon)$ in t with period ω. This follows from the definition of the function $v(t, y, \varepsilon)$ in view of the identity

$$\sum_{\tau_i < t+\omega} I_i(x) e^{-\eta_\varepsilon (t+\omega-\tau_i)} = \sum_{\tau_i < t} I_{i+p}(x) e^{-\eta_\varepsilon (t-\tau_i)} = \sum_{\tau_i < t} I_i(x) e^{-\eta_\varepsilon (t-\tau_i)}.$$

The system of equations (5.71) is called almost periodic in t uniformly with respect to $x \in D$ if $X(t, x)$ is continuous in $t \in \mathbb{R}$ and $x \in D$ except, possibly, the points $t = \tau_i$ at which it has discontinuities of the first kind and, for any $\delta > 0$, one can find an integer p and a positive number $l = l(\delta)$ for which any interval of length l contains a point s such that

$$\begin{gathered} \|X(t+s, x) - X(t, x)\| < \delta, \\ \|I_{i+p}(x) - I_i(x)\| < \delta, \quad |\tau_{i+p} - \tau_i - s| < \delta, \end{gathered} \tag{5.80}$$

for all $t \in \mathbb{R}$ and $x \in D$.

If the system of equations (5.71) is almost periodic, then we can seek its solutions almost periodic in a generalized sense, namely: a solution of the system of equations (5.71) is called almost periodic if, for any $\delta > 0$, one can find $l = l(\delta) > 0$ for which any time interval of length l contains a point s such that $\|x(t+s) - x(t)\| < \delta$ for all $t \in \mathbb{R}$, $|t - \tau_i| > \delta$, and $i \in \mathbb{Z}$.

We now show that if system (5.71) is almost periodic in t, then the function $v(t, y, \varepsilon)$ is almost periodic in the above-mentioned sense. Indeed, let the system of equations (5.71) be almost periodic. We fix $\delta > 0$ and choose l, s, and p such that inequalities (5.80) hold. Then

$$\begin{aligned} v(t+s, x\varepsilon) - v(t, x, \varepsilon) = & \int_{-\infty}^{t} e^{-\eta_\varepsilon (t-\sigma)} [X(\sigma+s, x) - X(\sigma, x)] d\sigma \\ & + \sum_{\tau_i < t+s} I_i(x) e^{-\eta_\varepsilon (t+s-\tau_i)} - \sum_{\tau_i < t} I_i(x) e^{-\eta_\varepsilon (t-\tau_i)}. \end{aligned}$$

Further, we have

$$\begin{aligned} \|v(t+s, x, \varepsilon) - v(t, x, \varepsilon)\| \le & \frac{\delta}{\eta_\varepsilon} + M |e^{-\eta_\varepsilon (\tau_i + s - \tau_{i+1})} - 1| \sum_{\tau_i < t} e^{-\eta_\varepsilon (t-\tau_i)} \\ & + \delta \sum_{\tau_i < t} e^{-\eta_\varepsilon (t-\tau_i)} \le \delta \left[\frac{1}{\eta_\varepsilon} + \left(M \frac{e^{\eta_\varepsilon \delta} - 1}{\delta} + 1 \right) s \right] \end{aligned}$$

for all $|t - \tau_i| > \delta$, $i \in \mathbb{Z}$, where

$$s = e^{\eta_\varepsilon L (1 - \frac{1}{d})} \sum_{m=1}^{\infty} e^{-\eta_\varepsilon \frac{Lm}{d}}.$$

This means that $v(t,x,\varepsilon)$ is an almost periodic function for any fixed $0 < \varepsilon \le \varepsilon_0$. Hence, the transformed system of equations (5.79) is also almost periodic for any fixed ε, $0 < \varepsilon \le \varepsilon_0$.

Lemma 5 ([140]). *Assume that the system of equations*

$$\frac{dx}{dt} = \varepsilon Ax + P(t), \quad t \ne \tau_i, \quad \Delta|_{t=\tau_i} = I_i, \tag{5.81}$$

satisfies the following conditions:

(1) *the real parts of the eigenvalues of the matrix A are not equal to zero;*

(2) *the function $P(t)$ is piecewise continuous with points of discontinuity of the first kind at $t = \tau_i$ and the functions I_i are bounded for all $t \in \mathbb{R}$ and $i \in \mathbb{Z}$;*

(3) *the sequence $\{\tau_i\}_{i=-\infty}^{\infty}$ satisfies condition* (b).

Then

(1) *the system of equations* (5.81) *possesses a unique solution $x^*(t)$ bounded in the entire axis;*

(2) *there exists a positive constant $c = c(\varepsilon, L, d)$ such that*

$$\|x^*(t)\| \le c \max\left\{ \sup_{t\in\mathbb{R}} \|P(t)\|, \sup_{i\in\mathbb{Z}} \|I_i\| \right\} \tag{5.82}$$

for any fixed $0 < \varepsilon \le \varepsilon_0$;

(3) *if the system of equations* (5.81) *is periodic in t with period T or almost periodic in t, then the solution $x^*(t)$ is also periodic with the same period or almost periodic, respectively.*

Proof. Without loss of generality, we can assume that $A = \operatorname{diag}(A_+, A_-)$, where A_+ is a matrix whose eigenvalues have positive real parts and A_- is a matrix whose eigenvalues have negative real parts. We define a matrix $J_A(t)$ by the formula

$$J_A(t) = \begin{cases} \operatorname{diag}(-e^{-\varepsilon A_+ t}, 0), & t > 0, \\ \operatorname{diag}(0, e^{-\varepsilon A_- t}), & t < 0. \end{cases}$$

Since the real parts of the eigenvalues of the matrix A are not equal to zero, there exist positive constants K and α independent of ε such that

$$\|J_A(t)\| \le K e^{-\varepsilon\alpha|t|} \tag{5.83}$$

for all $t \in \mathbb{R}$. By using the matrix $J_A(t)$, we now define the function $x^*(t)$ as follows:

$$x^*(t) = \int_{-\infty}^{\infty} J_A(\sigma) P(t+\sigma)\, d\sigma + \sum_{-\infty<\tau_i<\infty} J_A(\tau_i - t) I_i. \tag{5.84}$$

The right-hand side of the last equality is defined because, according to inequality (5.83), both the integral and the sum are uniformly convergent. Indeed, the right-hand side of equality (5.84) can be estimated as follows:

$$\|x^*(t)\| \le \frac{2K}{\varepsilon\alpha} \sup_{t\in\mathbb{R}} \|P(t)\| + 2Ke^{\varepsilon\alpha L(1-\frac{1}{d})} \sum_{m=0}^{\infty} e^{-\frac{\alpha L m\varepsilon}{d}} \sup_{i\in\mathbb{Z}} \|I_i\|$$
$$\le c(\varepsilon) \max\left\{ \sup_{t\in\mathbb{R}} \|P(t)\|, \sup_{i\in\mathbb{Z}} \|I_i\| \right\}. \tag{5.85}$$

By direct verification, we show that the function $x^*(t)$ is a bounded solution of the system of equations (5.81). The uniqueness of this solution follows from the fact that the values of pulses for any $t = \tau_i$ are constant for all solutions and, hence, the difference of two bounded solutions of the system of equations (5.81) is a bounded solution of the system of equations $\frac{dx}{dt} = Ax$. However, under the assumptions made above, this system possesses solely the trivial bounded solution.

Estimate (5.82) follows from (5.85) if we set

$$c(\varepsilon) = 4K \max\left(\frac{1}{\varepsilon\alpha}, e^{\varepsilon\alpha L(1-\frac{1}{d})} \sum_{m=0}^{\infty} e^{-\frac{\alpha L m\varepsilon}{d}} \right).$$

If the system of equations (5.81) is periodic in t with period ω, then representation (5.84) readily implies that the function $x^*(t)$ is also periodic with period ω.

Now let the system of equations (5.81) be almost periodic. It is necessary show that, in this case, the function $x^*(t)$ is also almost periodic. Let δ be an arbitrarily small positive number and let constants $s > 0$ and $l = l(\delta) > 0$ and an integer number p be such that

$$\|P(t+s) - P(t)\| < \delta, \quad \|I_{i+p} - I_i\| < \delta, \quad |\tau_{i+p} - \tau_i - s| < \delta$$

for all $t \in \mathbb{R}$, $|t - \tau_i| > \delta$, and $i \in \mathbb{Z}$.

We estimate the difference $x^*(t+s) - x^*(t)$ for $|t - \tau_i| > \delta$. Indeed,

$$\|x^*(t+s) - x^*(t)\| \le \frac{2K}{\varepsilon\alpha}\delta + \left\| \sum_{i\in\mathbb{Z}} J_A(\tau_i - t - s) I_i - \sum_{i\in\mathbb{Z}} J_A(\tau_i - t) I_i \right\|$$
$$= \frac{2K}{\varepsilon\alpha}\delta + \Bigg\| \sum_{i\in\mathbb{Z}} [J_A(\tau_{i+p} - \tau_i - s) - J_A(\tau_i - s)] I_{i+p}$$
$$+ \sum_{i\in\mathbb{Z}} J_A(\tau_i - t)[I_{i+p} - I_i] \Bigg\|$$
$$\le \frac{2K}{\varepsilon\alpha}\delta + 2Ke^{\varepsilon\alpha L(1-\frac{1}{d})} \sum_{m=0}^{\infty} e^{-\frac{\alpha L m\varepsilon}{d}} \left[\frac{e^{\varepsilon\|A\|\delta} - 1}{\delta} + 1 \right] < \delta.$$

This means that $x^*(t)$ is an almost periodic function for any fixed $0 < \varepsilon \le \varepsilon_0$.

We now prove the existence of a bounded solution of the system of equations (5.78). To this end, we construct a sequence of functions $x_n(t)$ each of which is a bounded solution of the system of equations

$$\frac{dx}{dt} = \varepsilon A x + P(t, x_{n-1}(t), \varepsilon), \quad t \ne \tau_i, \qquad (5.86)$$
$$\Delta x|_{t=\tau_i} = I_i(x_{n-1}(\tau_i), \varepsilon),$$

where the matrix A, the functions $P(t, x, \varepsilon)$ and $I_i(x, \varepsilon)$, and the sequence $\{\tau_i\}$ satisfy conditions (1)–(4). As the initial function, we take $x_0 = 0$.

According to Lemma 5, for any $n = 0, 1, 2, \ldots$, a bounded solution of the system of equations (5.86) exists and is given by the formula

$$x_n(t) = \int_{-\infty}^{\infty} J_A(\sigma - t) P(\sigma, x_{n-1}(\sigma), \varepsilon) d\sigma + \sum_{-\infty<\tau_i<\infty} J_A(\tau_i - t) I_i(x_{n-1}(\tau_i), \varepsilon). \qquad (5.87)$$

Moreover, for any fixed

$$0 < \varepsilon \le \varepsilon_0 \quad \text{and} \quad 0 < \frac{\sigma}{\varepsilon} < \frac{\rho_1}{\varepsilon},$$

we have

$$\|x_1(t)\| \le \left(\frac{2K}{\varepsilon\alpha} + 2Ks\right) M(\varepsilon),$$
$$\|x_2(t) - x_1(t)\| \le \lambda(\varepsilon, \sigma) \left(\frac{2K}{\varepsilon\alpha} + 2Ks\right)^2 M(\varepsilon),$$

where

$$s = e^{\varepsilon\alpha L(1-\frac{1}{d})} \sum_{m=0}^{\infty} e^{-\frac{\alpha L m \varepsilon}{d}}.$$

By induction, we can readily show that

$$\|x_{k+1}(t)\| \le \frac{\rho M(\varepsilon)}{1 - \lambda(\varepsilon, \sigma)\rho}, \qquad (5.88)$$

$$\|x_{k+1}(t) - x_{k+1}(t)\| \le \frac{(\lambda(\varepsilon, \sigma)\rho)^k}{1 - \lambda(\varepsilon, \sigma)\rho} \rho M(\varepsilon) \qquad (5.89)$$

for all $k = 0, 1, 2, \ldots$ and

$$\rho = \frac{2K}{\varepsilon\alpha} + 2Ks.$$

Indeed, for $k = 0$, estimates (5.88) and (5.89) are proved. Under the assumption that they are true for $k = \overline{1, n-1}$, we find

$$\begin{aligned}\|x_n(t)\| &\le \|x_n(t) - x_1(t)\| + \|x_1(t)\| \\ &\le \int_{-\infty}^{\infty} Ke^{-\varepsilon\alpha|s|}\lambda(\varepsilon,\sigma)\|x_{n-1}(t+s)\|d\sigma \\ &\quad + \sum_{-\infty<\tau_i<\infty} Ke^{-\alpha\varepsilon|\tau_i - t|}\lambda(\varepsilon,\sigma)\|x_{n-1}(\tau_i)\| + \left(\frac{2K}{\varepsilon\alpha} + 2Ks\right)M(\varepsilon) \\ &\le \frac{\rho^2 M(\varepsilon)}{1-\lambda(\varepsilon,\sigma)\rho}\lambda(\varepsilon,\sigma) + \rho M(\varepsilon) = \frac{\rho M(\varepsilon)}{1-\lambda(\varepsilon,\sigma)\rho}\lambda(\varepsilon,\sigma),\end{aligned}$$

i.e., inequality (5.88) is proved for all $k = 1, 2, 3, \ldots$. Estimate (5.89) is checked similarly. We now choose $0 < \overline{\varepsilon} \le \varepsilon_0$ and $0 < \sigma^0 < \rho_1$ sufficiently small such that the inequalities

$$\frac{\rho M(\varepsilon)}{1-\lambda(\varepsilon,\sigma)\rho} < \frac{\sigma}{\varepsilon}, \quad \rho\lambda(\varepsilon,\sigma) < 1 \tag{5.90}$$

hold for all $0 < \varepsilon \le \overline{\varepsilon}$ and $0 < \sigma \le \sigma^0$.

This choice of $\overline{\varepsilon}$ and σ^0 is always possible because

$$\frac{1}{\varepsilon}\lambda(\varepsilon,\sigma) \to 0 \quad \text{and} \quad M(\varepsilon) \to 0 \quad \text{as } \varepsilon \to 0 \text{ and } \sigma \to 0.$$

Thus, each function $x_k(t)$ takes values from the domain $\|x\| \le \frac{\sigma}{\varepsilon}$ for fixed $\varepsilon \in (0, \varepsilon_0)$ and, hence, the iterative process can be continued. For any fixed $\varepsilon, 0 < \varepsilon \le \overline{\varepsilon}_0$, in view of inequalities (5.90), relations (5.89) imply that the sequence $x_k(t)$ is uniformly convergent (in $t \in \mathbb{R}$). Moreover, the limit function

$$x^*(t) = \lim_{k\to\infty} x_k(t)$$

takes values from the domain

$$\|x\| \le \frac{\sigma^0}{\varepsilon}$$

and, in addition, as follows from (5.88), satisfies the inequality

$$\|x^*(t,\varepsilon)\| \le \frac{D(\varepsilon)}{\varepsilon}, \quad t \in \mathbb{R},\ D(\varepsilon) \to 0,\ \varepsilon \to 0,\ D(\varepsilon) < \sigma^0.$$

Passing to the limit as $n \to \infty$ in (5.87), we conclude that, for $\tau_i < t < \tau_{i+1}$, the limit function $x^*(t)$ satisfies the equation

$$\frac{dx}{dt} = \varepsilon Ax + P(t,x,\varepsilon), \quad t \ne \tau_i.$$

At the same time, for $t = \tau_i$, this function satisfies the condition

$$\Delta x|_{t=\tau_i} = I_i^0(x^*(\tau_i), \varepsilon).$$

Therefore, $x^*(t)$ is a solution of the system of equations (5.78). The function $x^*(t)$, as the uniform limit of periodic (almost periodic) functions is also periodic (almost periodic) for any fixed $0 < \varepsilon \leq \bar{\varepsilon}$ if this is true for system (5.78).

The properties of stability of the solution $x^*(t)$ are established by analogy with the same properties for systems of equations without pulse action [27, 88]. As a result, we conclude that, in the case where the real parts of the eigenvalues of the matrix A are negative, the solution is asymptotically stable. At the same time, if r, $0 \leq r \leq n$, eigenvalues have negative real parts and the other $n - r$ eigenvalues have positive real parts, then, in the neighborhood of the point $x^*(t_0)$, there exists an r-dimensional point manifold $\mathcal{M}_{t_0}$ for which the inclusion $x^*(t_0) \in \mathcal{M}_{t_0}$ implies that the difference $\|x(t) - x^*(t)\|$ exponentially vanishes.

Thus, we arrive at the following analog of the famous Bogolyubov theorem on substantiation of the method of averaging for an infinite time interval. □

Theorem 6 ([140]). *Assume that the system of differential equations with pulse action* (5.71) *satisfies conditions* (a)–(c). *Suppose that the system of averaged Eqs.* (5.73) *possesses a quasistatic equilibrium position $x = x_0$ for which the real parts of all roots of the characteristic equation*

$$\det(\lambda E - F_x'(x_0)) = 0$$

differ from zero.

Then one can find positive constants ε', σ_1, $D(\varepsilon)$, $D(\varepsilon) \leq \sigma_1 < \rho_1$, $D(\varepsilon) \to 0$, $\varepsilon \to 0$, such that the following assertions are true for any positive $\varepsilon < \varepsilon'$:

1. *The system of equations* (5.71) *possesses a unique solution $x = x^*(t)$ defined for all $t \in \mathbb{R}$ and satisfying the inequality*

$$\|x^*(t) - x_0\| < \sigma_0.$$

2. *If the system of equations* (5.71) *is periodic (almost periodic) in t, then the solution $x^*(t)$ is also periodic (almost periodic).*
3. *Let $x(t)$ be an arbitrary solution of the system of equations* (5.71) *satisfying the inequality $\|x(t_0) - x_0\| < \sigma_0$ for some $t = t_0$. In this case, if the real parts of all roots of the characteristic Eq.* (5.73) *are negative, then*

$$\|x(t) - x^*(t)\| < ce^{-\alpha\varepsilon(t-t_0)}, \quad t \geq t_0, \tag{5.91}$$

where c and α are positive constants. At the same time, if the real parts of all roots of the characteristic Eq. (5.73) *are positive, then there exists $t' > t_0$ such that*

$$\|x(t') - x_0\| > \sigma_1. \tag{5.92}$$

Finally, if the real parts of r roots are negative and the real parts of the remaining $n-r$ toots are positive, then the σ_0-neighborhood of the point x_0 contains an r-dimensional manifold $\mathcal{M}_{t_0}$ such that the inclusion $x(t_0) \in \mathcal{M}_{t_0}$ yields inequality (5.91), *whereas the relation $x(t_0) \notin \mathcal{M}_{t_0}$ yields inequality* (5.92).

5.3 Systems with Nonfixed Times of the Pulse Action

The problem of qualitative correspondence between the exact solutions of Eqs. (5.71) and the solutions of the corresponding averaged system (5.73) in an infinite time interval is studied in [137, 138, 140]. We now discuss the results obtained in the cited works.

Theorem 7 ([137]). *Assume that the right-hand side of the system*

$$\frac{dx}{dt} = \varepsilon X(t,x), \quad t \neq \tau_i(x),$$
$$\Delta x|_{t=\tau_i(x)} = \varepsilon I_i(x), \quad i = 1,2,\ldots, \quad \tau_i(x) < \tau_{i+1}(x), \tag{5.93}$$

satisfy the following conditions:

(1) *there exist positive constants M and K such that*

$$\left\| \frac{\partial \tau_i(x)}{\partial x} \right\| + \|X(t,x)\| + \|I_i(x)\| \leq M,$$
$$\|X(t,x') - X(t,x'')\| + \left| \frac{\partial \tau_i(x')}{\partial x_j} - \frac{\partial \tau_i(x'')}{\partial x_j} \right| + \|I_i(x') - I_i(x'')\| \leq K\|x'-x''\|,$$
$$\left\| \frac{\partial \tau_i(x)}{\partial x} \right\| \leq K$$

for $t \in (0,\infty)$, x, x', $x'' \in D$, $j = \overline{1,n}$, $i = 1,2,\ldots$;

(2) *the finite limits*

$$\lim_{T\to\infty} \frac{1}{T}\int_t^{t+T} X(s,x)ds = X_0(x),$$
$$\lim_{T\to\infty} \frac{\sum_{t<\tau_i<t+T} I_i(x)}{T} = I_0(x), \tag{5.94}$$
$$\lim_{T\to\infty} \frac{\sum_{t<\tau_i<t+T} 1}{T} = d$$

exist uniformly in t and x for $t \geq 0$ and $x \in D$.

(3) *the averaged system*

$$\frac{d\overline{x}}{dt} = \varepsilon[X_0(\overline{x}) + I_0(\overline{x})] \tag{5.95}$$

possesses a solution $\overline{x} = \overline{x}(\varepsilon t, x_0)$, $\overline{x}(0, x_0) = x_0$; *for* $\varepsilon = 1$, $t \in [0, L]$, $L < \infty$, *this solution belongs to* D *together with its certain* ρ*-neighborhood and satisfies the inequalities*

$$\frac{\partial \tau_i(\overline{x}(\varepsilon t, x_0))}{\partial x} I_i(\overline{x}(\varepsilon t, x_0)) \leq \beta < 0, \quad t_i' < t < t_i'', \tag{5.96}$$

or

$$\frac{\partial \tau_i(x)}{\partial x} \equiv 0,$$

where $t_i' = \inf_{x\in D} \tau_i(x)$, $t_i'' = \sup_{x\in D} \tau_i(x)$, $i = \overline{1, d}$, *and* $\tau_d < \frac{L}{\varepsilon} < \tau_{d+1}$.

Then, for any $\eta > 0$, *there exists* $\varepsilon_0 > 0$ *such that, for all* $\varepsilon < \varepsilon_0$, *the system of equations* (5.93) *possesses a solution* $x_t(x_0)$, $x_0(x_0) = x_0$ *defined for* $t \in [0, \frac{L}{\varepsilon}]$ *and such that*

$$\|x_t(x_0)) - \overline{x}(\varepsilon t, x_0)\| \leq \eta \quad \text{for } t \in \left[0, \frac{L}{\varepsilon}\right].$$

Theorem 7 is an analog of the well-known Bogolyubov result concerning the substantiation of the method of averaging.

Proof. It follows from relations (5.94) that there exists a function $\varphi(t)$ vanishing as $t \to \infty$ and such that

$$\left\| \int_t^{t+T} [X(t, x) - X_0(x)]dt \right\| \leq \frac{\varphi(T)}{2} T, \tag{5.97}$$

$$\left\| \sum_{t<\tau_i<t+T} I_i(x) - I_0(x)T \right\| \leq \frac{\varphi(T)}{2} T.$$

We choose a sufficiently large but fixed T and consider system (5.93) for $t \in [0, T]$. Assume that exactly d_1 points belong to the half interval $[0, T)$:

$$\tau_1(x_0) = t_1^0, \ldots, \tau_{d_1}(x_0) = t_{d_1}^0$$

and, in addition, $0 < t_1^0, t_i^0 < t_{i+1}^0, t_{d_1}^0 < T, i = \overline{1, d_1 - 1}$.

By $x(t, \tau, c)$, $x(\tau, \tau, c) = c$, we denote a solution of the system

$$x(t, \tau, c) = c + \varepsilon \int_\tau^t X(t, x(t, \tau, c))dt. \tag{5.98}$$

It is clear that

$$x(t, \tau, c) = c + \varepsilon \int_\tau^t X(t, c)dt + R(t, \varepsilon, T),$$

where

$$\|R(t,\varepsilon,T)\| = \varepsilon \left\| \int_{\tau}^{t} [X(t,x(t,\tau,c)) - X(t,c)]dt \right\| \leq \varepsilon^2 KMT^2$$

for $0 < \tau \leq t \leq T$.

The solution $x_t(x_0)$ of system (5.93) is formed by pieces of functions (5.98). Thus, to within the quantities of order ε^2, $x_t(x_0)$ is given by the formula

$$x_1(t,\tau,c) = c + \varepsilon \int_{\tau}^{t} X(t,c)dt \quad (0 \leq \tau \leq t \leq T).$$

In the construction of $x_t(x_0)$, we restrict ourselves to the interval $0 < t < T$. This enables us to establish an approximate expression for $x_t(x_0)$.

If the subsequent relations are understood as true to within the quantities of order ε^2, then we can write

$$x_t(x_0) = x_1(t,0,x_0) \quad \text{for } 0 \leq t \leq t_1,$$

where t_1 is the solution of the equation

$$\begin{aligned} t = \tau_1(x_1(t,0,x_0)) &\equiv \tau_1\left(x_0 + \varepsilon \int_0^t X(t,x_0)dt\right) \\ &\equiv \tau_1(x_0) + \varepsilon \frac{\partial \tau_1(x_0)}{\partial x} \int_0^t X(t,x_0)ds. \end{aligned} \tag{5.99}$$

It follows from Eq. (5.99) that

$$t_1 = t_1^0 + \varepsilon \frac{\partial t_1^0}{\partial x} \int_0^{t_1^0} X(t,x_0)dt = t_1^0 + \varepsilon\theta_1.$$

Thus,

$$x_t(x_0) = x_1(t,0,x_0) \quad \text{for } 0 < t < t_1^0 + \varepsilon\theta_1 = t_1.$$

Further, we have

$$x_{t_1}^{+}(x_0) = x_1(t_1,0,x_0) + \varepsilon I_1^0 = x_0 + \varepsilon \int_0^{t_1} X(t,x_0)dt + \varepsilon I_1^0$$

and, hence,

$$\begin{aligned} x_t(x_0) &= x_0 + \varepsilon \int_0^{t_1} X(t,x_0)dt + \varepsilon I_1^0 + \varepsilon \int_{t_1}^{t} X(t,x_0)dt \\ &= x_0 + \varepsilon \int_0^t X(t,x_0)dt + \varepsilon I_1^0 = x_1(t,0,x_0) + \varepsilon I_1^0 \end{aligned}$$

for $t_1 < t < t_1'$, where t_1' is a solution either of the equation

$$t = \tau_1(x_1(t, 0, x_0) + \varepsilon I_1^0) \tag{5.100}$$

or of the equation

$$t = \tau_2(x_1(t, 0, x_0) + \varepsilon I_1^0) \tag{5.101}$$

if (5.100) has no solutions for $t_1 < t < t_2^0$. As a result of the solution of Eq. (5.100), we conclude that its root t_1' is given by the formula

$$\overline{t_1'} = t_1^0 + \varepsilon \frac{\partial t_1^0}{\partial x}\left[\int_0^{t_1^0} X(t, x_0)dt + I_1^0\right] = t_1 + \varepsilon W_1$$

and, therefore, $t_1' = \overline{t_1'}$ provided that

$$W_1 = \frac{\partial t_1^0}{\partial x} I_1^0 \geq \alpha > 0. \tag{5.102}$$

Thus, if inequality (5.102) is true, then

$$x_t(x_0) = x_0 + \varepsilon\left[\int_0^{t_1^0} X(t, x_0)dt + I_1^0\right]$$

for $t_1 < t < t_1 + \varepsilon W_1 = t_1'$.

It is clear that

$$x'_{t_1+\varepsilon W_1}(x_0) = x_0 + \varepsilon\left[\int_0^{t_1^0} X(t, x_0)dt + 2I_1^0\right],$$

whence it follows that

$$\begin{aligned} x_t(x_0) &= x_0 + \varepsilon\left[\int_0^{t_1^0} X(t, x_0)dt + 2I_1^0\right] + \varepsilon \int_{t_1^0}^{t} X(t, x_0)dt \\ &= x_1(t, 0, x_0) + 2\varepsilon I_1^0 \quad \text{for } t_1' < t < t_1'', \end{aligned}$$

where t_1'' is a solution either of the equation

$$t = \tau_1(x_1(t, 0, x_0) + 2\varepsilon I_1^0) \tag{5.103}$$

or of the equation

$$t = \tau_2(x_1(t, 0, x_0) + 2\varepsilon I_1^0)$$

if (5.103) has no solutions for $t_1' < t < t_2^0$. However, the root $\overline{t_1}$ of Eq. (5.103) is equal to

$$\overline{t_1''} = t_1^0 + \varepsilon \frac{\partial t_1^0}{\partial x}\left[\int_0^{t_1^0} X(t, x_0)dt + 2I_1^0\right] = t_1 + 2\varepsilon W_1 = t_1' + \varepsilon W_1.$$

Therefore, $t_1'' = \overline{t_1''}$ and, hence,

$$x_t(x_0) = x_1(t, 0, x_0) + 2\varepsilon I_1^0$$

for $t_1 + \varepsilon W_1 < t < t_1 + 2\varepsilon W_1$.

Similarly, we conclude that, for any fixed k, the function $x_t(x_0)$ undergoes exactly $k - 1$ instant changes in the time interval from t_1 to $t_1 + \varepsilon k W_1$ and takes the values

$$x_1(x_0) = x_0 + \varepsilon(\Phi(t_1^0, x_0) + k_1 I_0)$$

for

$$t_1 + (k_1 - 1)\varepsilon W_1 < t < t_1 + k_1 \varepsilon W_1, \quad k_1 = \overline{1, k}.$$

Thus, the system is characterized by the presence of beating upon the surface $t = \tau_1(x)$ in the vicinity of the point t_1^0, x_0. Hence, to cross the surface $t = \tau_1(x)$, i.e., pass through the beating, the curve $x_t(x_0)$ must undergo about $\frac{1}{\varepsilon}$ instant changes for a finite period of time. In this case, the solution $x_t(x_0)$ naturally "moves" along the surface $t = \tau_1(x)$ and strongly differs from the solution of the averaged system.

Therefore, it remains to assume that the point t_1^0, x_0 is not a point of beating upon the surface $t = \tau_1(x)$. Hence, one must set

$$W_1 = \frac{\partial t_1^0}{\partial x} I_1^0 \le \beta < 0 \quad \text{or} \quad \frac{\partial \tau_1(x)}{\partial x} \equiv 0. \tag{5.104}$$

In this case, the root $\overline{t_1'}$ becomes smaller than t_1 and, thus, t_1' should be found from Eq. (5.101). By using this equation, we find

$$t_1' = t_2^0 + \varepsilon \frac{\partial t_2^0}{\partial x} \left[\int_0^{t_2^0} X(t, x_0) dt + I_1^0 \right] = t_2^0 + \varepsilon \theta_2.$$

Since $t_2^0 > t_1^0$, we have $t_2^0 + \varepsilon\theta_2 > t_1^0$ and

$$x_t(x_0) = x_1(t, 0, x_0) + \varepsilon I_1^0 \quad \text{for } t_1 < t < t_2^0 + \varepsilon\theta_2 = t_2.$$

Further,

$$x_{t_2}^+(x_0) = x_1(t_2, 0, x_0) + \varepsilon(I_1^0 + I_2^0) = x_1(t_2^0, 0, x_0) + \varepsilon(I_1^0 + I_2^0)$$

and, hence,

$$\begin{aligned} x_t(x_0) &= x_1(t_2, 0, x_0) + \varepsilon(I_1^0 + I_2^0) + \varepsilon \int_{t_2}^{t} X(t, x_0) dt \\ &= x_1(t, 0, x_0) + \varepsilon(I_1^0 + I_2^0) \quad \text{for } t_2 < t < t_3. \end{aligned}$$

To avoid beating upon the surface $t = t_2(x)$, we assume that

$$\frac{\partial t_2^0}{\partial x} I_2^0 \le \beta < 0 \quad \text{or} \quad \frac{\partial \tau_2(x)}{\partial x} \equiv 0.$$

Thus, the quantity t_3 is determined from the equation

$$\begin{aligned} t &= \tau_2(x_1(t,0,x_0) + \varepsilon(I_1^0 + I_2^0)) \\ &\equiv t_3^0 + \varepsilon \frac{\partial t_3^0}{\partial x}\left[\int_0^{t_3^0} X(t,x_0)dt + I_1^0 + I_2^0\right] \equiv t_3^0 + \varepsilon\theta_3, \end{aligned}$$

etc. If we now assume that

$$\frac{\partial t_i^0}{\partial x} I_2^0 \le \beta < 0 \quad \text{or} \quad \frac{\partial \tau_i(x)}{\partial x} \equiv 0 \tag{5.105}$$

for $i = \overline{1, d_1}$, then we get the following relation for $x_t(x_0)$:

$$x_t(x_0) = x_1(t,0,x_0) + \varepsilon \sum_{i=0}^{k} I_i^0 \quad \text{for } t_k^0 + \varepsilon\theta_k < t < t_{k+1}^0 + \varepsilon\theta_{k+1}, \tag{5.106}$$

where

$$\theta_k = \frac{\partial t_k^0}{\partial x}\left[\int_0^{t_k^0} X(t,x_0)dt + \sum_{i=0}^{k-1} I_k^0\right], \tag{5.107}$$

$$k = \overline{1, d_1}, \quad t_0 = \theta_0 = \theta_{d_1+1} = I_0^0 = 0, \quad t_{d_1+1}^0 = T.$$

Thus, if conditions (5.105) are satisfied, then $x_t(x_0)$ exists for $t \in [0, T]$ and is determined, to within the quantities of order ε^2, by relations (5.106) and (5.107). This enables us to write

$$\begin{aligned} x_T(x_0) &= x_1(T,0,x_0) + \varepsilon \sum_{0<t_i^0<T} I_i^0 + \varepsilon^2 \cdots \\ &= x_0 + \varepsilon \int_0^T X(t,x_0)dt + \varepsilon \sum_{0<t_i^0<T} I_i^0 + \varepsilon^2 \cdots \\ &= x_0 + \varepsilon[X_0(x_0) + I_0(x_0)]T + \varepsilon \int_0^T [X(t,x_0) - X_0(x_0)]dt \\ &\quad + \varepsilon\left[\sum_{0<t_i^0<T} I_i^0 - I_0(x_0)T\right] + \varepsilon^2 \cdots . \end{aligned} \tag{5.108}$$

We set

$$A_0 x_0 = x_0 + \varepsilon T[X_0(x_0) + I_0(x_0)]. \tag{5.109}$$

Thus, in view of (5.97), inequality (5.108) yields the following estimate:

$$\|x_T(x_0) - A_0 x_0\| \le \varepsilon\varphi(T)T + \varepsilon^2 M_1,$$

where $M_1 = M(T, d_1)$ is a constant depending on T and d_1.

Let $\overline{x}(\varepsilon t, x_0)$ be a solution of the averaged system (5.95). We set

$$\overline{A}x_0 = \overline{x}(\varepsilon T, x_0).$$

Thus, it is clear that

$$\|\overline{A}x_0 - A_0 x_0\| \le \varepsilon \int_0^T \|X_0(\overline{x}) + I_0(\overline{x}) - X_0(x_0) - I_0(x_0)\| dt \le \varepsilon^2 KMT^2 \tag{5.110}$$

and, hence,

$$\|x_T(x_0) - \overline{A}x_0\| \le \varepsilon\varphi(T)T + \varepsilon^2(M_1 + KMT^2). \tag{5.111}$$

Inequalities (5.110) and (5.111) show that $x_T(x_0)$ belongs to the domain D together with its ρ_1-neighborhood, where

$$\rho_1 = \rho - \varepsilon[\varphi(T)T + \varepsilon(M_1 + KMT^2)],$$

and $A_0 x_0$ belongs to the same domain together with its ρ_1'-neighborhood, where

$$\rho_1' = \rho - \varepsilon^2 RMT^2.$$

Assume that the half interval $[T, 2T)$ contains exactly d_2 points, i.e.,

$$T < \tau_{d_1+1}(\overline{A}x_0), \ldots, \tau_{d_1+d_2}(\overline{A}x_0) < 2T.$$

Hence, it follows from estimate (5.111) and the continuity of the functions $\tau_i(x)$ that the segment $[T, 2T]$ contains the following d_2 points:

$$T < \tau_{d_1+1}(x_T(x_0)) = t_1^{(1)}, \ldots, \tau_{d_1+d_2}(x_T(x_0)) = t_{d_2}^{(1)} < 2T.$$

Thus, condition (5.96), estimate (5.111), and the fact that the functions $\frac{\partial \tau_i}{\partial x}$ and $I_i(x)$ are continuous imply that

$$\frac{\partial t_i^{(1)}}{\partial x} I_{i+d_1}(x_T(x_0)) = \frac{\partial t_i^{(1)}}{\partial x} I_i^{(1)} \le \beta_1 < 0$$

for $i = \overline{1, d_2}$, $\beta_1 = \beta - \varepsilon^2$.

We extend the solution $x_t(x_0)$ of system (5.93) constructed for $0 \le t < T$ to the segment $[T, 2T]$. This yields

$$x_t(x_0) = x(t, T, x_T) = x_T(x_0) + \varepsilon \int_T^t X(t, x(t, T, x_T))dt$$
$$= x_T + \varepsilon \int_T^t X(t, x_T)dt + \varepsilon^2 \cdots = x_1(t, T, x_T) + \varepsilon^2 \cdots$$

for $T \le t < t_{d_1+1}$, where t_{d_1+1} is the solution of the equation

$$t = \tau_{d_1+1}\left(x_T + \varepsilon \int_T^t X(t, x_T)dt + \varepsilon^2 \cdots\right)$$
$$\equiv t_1^{(1)} + \varepsilon \frac{\partial \tau_{d_1+1}(x_T)}{\partial x} \int_T^t X(t, x_T)dt + \varepsilon^2 \cdots . \tag{5.112}$$

It follows from Eq. (5.112) that

$$t_{d_1+1} = t_1^{(1)} + \varepsilon \frac{\partial t_1^{(1)}}{\partial x} \int_T^{t_1^{(1)}} X(t, x_1)dt + \varepsilon^2 \cdots$$
$$= t_1^{(1)} + \varepsilon \theta_1^{(1)} + \varepsilon^2 \cdots .$$

Hence, to within the quantities of order ε^2,

$$x_t(x_0) = x_1(t, T, x_T) \quad \text{for } T \le t < t_1^{(1)} + \varepsilon \theta_1^{(1)}.$$

Further,

$$x^+_{t_1^{(1)}}(x_0) = x_1(t_1^{(1)}, T, x_T) + \varepsilon I_1^{(1)}$$

and, hence,

$$x_t(x_0) = x_1(t_1^{(1)}, T, x_T) + \varepsilon I_1^{(1)} + \varepsilon \int_{t_1^{(1)}}^t X(t, x_T)dt$$
$$= x_T + \varepsilon \int_T^t X(t, x_T)dt + \varepsilon I_1^{(1)} = x_1(t, T, x_T) + \varepsilon I_1^{(1)}$$

for $t_1^{(1)} < t < t_2^{(1)}$, etc. In general, we can write

$$x_t(x_0) = x_1(t, T, x_T) + \varepsilon \sum_{i=0}^{k} I_i^{(1)} \tag{5.113}$$

for $t_k^{(1)} + \varepsilon \theta_k^{(1)} < t < t_{k+1}^{(1)} + \varepsilon \theta_{k+1}^{(1)}$, where

$$\theta_k^{(1)} = \frac{\partial t_k^{(1)}}{\partial x}\left[\int_T^{t_k^{(1)}} X(t, x_T)dt + \sum_{i=0}^{k} I_i^{(1)}\right], \tag{5.114}$$

$$k = \overline{1, d_2}, \quad t_0^{(1)} = \theta_0^{(1)} = \theta_{d_2+1}^{(1)} = I_0^{(1)} = 0, \quad t_{d_2+1}^{(1)} = 2T.$$

Thus, under conditions (5.110), $x_t(x_0)$ is determined for all $t \in [T, 2T]$, to within the quantities of order ε^2, by relations (5.113) and (5.114). This enables us to conclude that

$$x_{2T}(x_0) = x_T + \varepsilon \int_T^{2T} X(t, x_T)dt + \varepsilon \sum_{T<t_i^{(1)}<2T} I_i^{(1)} + \varepsilon^2 \cdots$$
$$= x_T + \varepsilon[X_0(x_T) + I_0(x_T)]T + \varepsilon \int_T^{2T} [X(t, x_T) - X_0(x_T)]dt$$
$$+ \varepsilon \sum_{T<t_i^{(1)}<2T} [I_i^{(1)} - I_0(x_T)T] + \varepsilon^2 \cdots .$$

This yields

$$\|x_{2T}(x_0) - A_0 x_T\| \leq \varepsilon\varphi(T)T + \varepsilon^2 M(T, d_2). \tag{5.115}$$

Further, in view of (5.109), (5.110), and (5.115), we get

$$\|x_{2T}(x_0) - \overline{x}(2\varepsilon T, x_0)\| = \|x_{2T} - \overline{A}^2 x_0\|$$
$$\leq \|x_{2T} - A_0 x_T\| + \|A_0 x_T - A_0(\overline{A}x_0)\| + \|A_0(\overline{A}x_0) - \overline{A}(\overline{A}x_0)\|$$
$$\leq \varepsilon\varphi(T)T + \varepsilon^2 M(t, d_2) + (1 + \varepsilon KT)\|x_T - \overline{A}x_0\| + \varepsilon^2 KMT^2$$
$$\leq \varepsilon[1 + (1 + \varepsilon KT)][\varphi(T)T + \varepsilon\overline{M}], \tag{5.116}$$

where $\overline{M} = KMT^2 + \max_{i=1,2} M(T, d_i)$.

It follows from (5.116) that $x_{2T}(x_0)$ belongs to D together with its ρ_2-neighborhood, where

$$\rho_2 = \rho - \varepsilon \sum_{i=0}^{1} (1 + \varepsilon KT)^i (\varphi(T)T + \varepsilon\overline{M}).$$

By using $x_{2T}(x_0)$, we can construct the solution $x_t(x_0)$ for $t \in [2T, 3T]$. Thus, we obtain

$$\|x_{3T}(x_0) - A_0 x_{2T}\| \leq \varepsilon[\varphi(T)T + \varepsilon M(T, d_3)], \tag{5.117}$$

where d_3 is the number of points $\tau_i(\overline{x}(2\varepsilon T, x_0))$ lying in $[2T, 3T]$:

$$2T < \tau_{d_1+d_2+1}(\overline{x}(2\varepsilon T, x_0)), \ldots, \tau_{d_1+d_2+d_3}(\overline{x}(2\varepsilon T, x_0)) < 3T.$$

Inequality (5.117) implies that

$$\|x_{3T}(x_0) - \overline{x}(3\varepsilon T, x_0)\| = \|x_{3T} - \overline{A}^3 x_0\|$$
$$\leq \|x_{3T} - A_0 x_{2T}\| + \|A_0 x_{2T} - A_0(\overline{A}^2 x_0)\| + \|A_0(\overline{A}^2 x_0) - \overline{A}(\overline{A}^2 x_0)\|$$

$$\leq \varepsilon[\varphi(T)T + \varepsilon M(t, d_3)] + (1 + \varepsilon KT)\|x_{2T} - \overline{A}^2 x_0\| + \varepsilon^2 KMT^2$$

$$\leq \varepsilon \sum_{i=0}^{2} (1 + \varepsilon KT)^i [\varphi(T)T + \varepsilon \overline{M}], \tag{5.118}$$

where $\overline{M} = KMT^2 + \max_{i=\overline{1,3}} M(T, d_i)$.

It follows from inequality (5.118) that $x_{3T}(x_0)$ belongs to D together with its ρ_3-neighborhood, where

$$\rho_3 = \rho - \varepsilon \sum_{i=0}^{2} (1 + \varepsilon KT)^i (\varphi(T)T + \varepsilon \overline{M}).$$

The quantity $x_{3T}(x_0)$ can be used to construct the solution $x_t(x_0)$ for $t \in [3T, 4T]$, etc. In the kth stage, we construct the solution $x_t(x_0)$ for $t \in [(k-1)T, kT]$, $kT \leq \frac{L}{\varepsilon}$, and conclude that

$$\|x_{kT}(x_0) - \overline{x}(\varepsilon kT, x_0)\| = \|x_{kT} - \overline{A}^k x_0\| \leq \varepsilon \sum_{i=0}^{k-1} (1 + \varepsilon KT)^i [\varphi(T)T + \varepsilon \overline{M}],$$

where $\overline{M} = KMT^2 + \max_{i=\overline{1,k}} M(T, d_i)$.

According to conditions (5.94), we have $d_i \leq c < \infty$. Therefore,

$$\max_i M(T, d_i) \leq M_0(T) < \infty$$

and, hence,

$$\|x_{kT}(x_0) - \overline{x}(\varepsilon kT, x_0)\| \leq \varepsilon \sum_{i=0}^{k-1} (1 + \varepsilon KT)^i [\varphi(T)T + \varepsilon M_0(T)].$$

The last inequality yields

$$\begin{aligned} \|x_{kT}(x_0) - \overline{x}(\varepsilon kT, x_0)\| &\leq \varepsilon \frac{(1 + \varepsilon KT)^k - 1}{\varepsilon KT} [\varphi(T)T + \varepsilon M_0(T)] \\ &\leq (1 + \varepsilon KT)^{\frac{L}{\varepsilon T}} \left[\frac{\varphi(T)}{K} + \varepsilon \frac{M_0(T)}{KT} \right] \\ &\leq (e^{kL} + 0(\varepsilon)) \left[\frac{\varphi(T)}{K} + \varepsilon \frac{M_0(T)}{KT} \right]. \end{aligned} \tag{5.119}$$

We now fix T and ε_0 from the conditions

$$e^{kL} \frac{\varphi(T)}{K} \leq \frac{\eta}{4}, \quad \varepsilon(e^{kL} + 0(\varepsilon)) \frac{M_0(T)}{KT} + 0(\varepsilon) \frac{\varphi(T)}{K} \leq \frac{\eta}{4}.$$

Thus, it follows from (5.119) that

$$\|x_{kT}(x_0) - \overline{x}(\varepsilon kT, x_0)\| \le \frac{\eta}{2} \tag{5.120}$$

for $k = 0, 1, \dots, [\frac{L}{\varepsilon T}]$, where $[\frac{L}{\varepsilon T}]$ is the integer part of the number $\frac{L}{\varepsilon T}$.

In the half interval $[(k-1)T, kT)$, the difference

$$\overline{x}(\varepsilon t, x_0) - \overline{x}(\varepsilon(k-1)T, x_0)$$

satisfies the following estimate:

$$\|\overline{x}(\varepsilon t, x_0) - \overline{x}(\varepsilon(k-1)T, x_0)\| \le \int_{(k-1)T}^{kT} [X_0(\overline{x}) + I_0(\overline{x})]dt \le \varepsilon MT. \tag{5.121}$$

At the same time, the difference $x_t(x_0) - x_{(k-1)T}(x_0)$ satisfies the estimate

$$\begin{aligned} \|x_t(x_0) - x_{(k-1)T}(x_0)\| &\le \varepsilon \int_{(k-1)T}^{kT} \|X(t, x(t, \tau, c))\| dt \\ &\quad + \varepsilon \sum_{(k-1)t < \tau_i(x_{(k-1)T}T) < kT} \|I_i(x)\| \\ &\le \varepsilon MT + \varepsilon M d_k \le \varepsilon M(T + c). \end{aligned} \tag{5.122}$$

For

$$\varepsilon MT < \frac{\eta}{4} \quad \text{and} \quad \varepsilon M(T+c) < \frac{\eta}{4},$$

inequalities (5.121) and (5.122), together with (5.120), yield the required estimate

$$\|x_t(x_0) - \overline{x}(\varepsilon t, x_0)\| \le \eta \quad \text{for } t \in \left[0, \frac{L}{\varepsilon}\right].$$

We now study the problem of qualitative correspondence between the exact solutions of system (5.93) and its approximations. Assume that the averaged system (5.95) possesses an isolated equilibrium position $\overline{x} = x^0$:

$$X_0(x^0) + I_0(x^0) = 0.$$

Thus, we want to establish the correspondence between the existence of this equilibrium position and certain properties of the exact system (5.93). To do this, we formulate several assertions. □

Theorem 8 ([137]). *Let conditions* (1) *and* (2) *of the previous theorem be satisfied. If the equilibrium position* $\overline{x} = x^0$ *of the averaged system is asymptotically stable and*

$$\frac{\partial \tau_i(x)}{\partial x} I_i(x) \le \beta < 0 \quad \text{or} \quad \frac{\partial \tau_i(x)}{\partial x} \equiv 0 \tag{5.123}$$

for all $i = 1, 2, \dots$ *and all* x *from a certain* ρ_0*-neighborhood of the position* x^0*, then there exist a* ρ*-neighborhood* $D_\rho(\rho \le \rho_0)$ *of the point* x^0 *and a number* $\varepsilon^0 > 0$ *such that, for all* $\varepsilon < \varepsilon^0$ *and* $x \in D_\rho$*, the solutions* $x_t(x)$*,* $x_0(x) = x$ *of system* (5.93) *are uniformly bounded for* $t \in (0, \infty)$.

Proof. Let $\overline{x}(t, x)$, $\overline{x}(0, x) = x$ be a solution of the averaged system for $\varepsilon = 1$. Since $\overline{x}(t, x^0) = x^0$ is an asymptotically stable solution of the averaged system, the fact that the dependence on the initial conditions is continuous implies the existence of ρ' such that

$$\|\overline{x}(t, x) - x^0\| \le \rho_0 \quad \text{for } \|x - x^0\| \le \rho',\ t \ge 0. \tag{5.124}$$

In view of (5.123), it follows from inequality (5.124) that, for $x \in T_{\rho'} = \{x\colon \|x - x^0\| \le \rho'\}$, the solutions $\overline{x}(\varepsilon t, x)$ satisfy conditions (3) of Theorem 7 for all $t > 0$.

By using this theorem, one can find $\varepsilon^0 = \varepsilon^0(L, \rho)$ such that, for all $\varepsilon < \varepsilon^0$ and $t \in [0, \frac{L}{\varepsilon})$ and some ρ $(\rho \le \rho_0, \rho \le \rho')$, we have

$$\|x_{t+\tau}(x, \tau) - \overline{x}(\varepsilon t, x)\| \le \frac{\rho}{2}, \tag{5.125}$$

where $x_{t+\tau}(x, \tau)$ is a solution of system (5.93) that passes through the point $x \in T_\rho$ for $t = \tau$.

We choose ρ such that T_ρ belongs to the domain of asymptotic stability of the solution $\overline{x} = x^0$ and L such that

$$\|\overline{x}(t, x \in T_\rho) - x^0\| \le \frac{\rho}{2} \quad \text{for } t \ge L. \tag{5.126}$$

In this case, inequalities (5.124)–(5.126) yield the estimates

$$\begin{aligned}
\|x_{t+\tau}(x \in T_\rho, \tau) - x^0\| &\le \|x_{t+\tau}(x \in T_\rho, \tau) - \overline{x}(\varepsilon t, x \in T_\rho)\| \\
&\quad + \|\overline{x}(\varepsilon t, x \in T_\rho) - x^0\| \\
&\le \frac{\rho}{2} + \rho_0 \quad \text{for } t \in \left[0, \frac{L}{\varepsilon}\right),
\end{aligned} \tag{5.127}$$

$$\|x_{\frac{L}{\varepsilon}}(x \in T_\rho, 0) - x^0\| \le \rho.$$

The last of these estimates means that

$$x_{\frac{L}{\varepsilon}}(x \in T_\rho, 0) \in T_\rho. \tag{5.128}$$

By using (5.128) and the fact that

$$x_{t+\tau}(x, 0) = x_{t+\tau}(x_\tau(x, 0), \tau),$$

we get

$$x_{t+\frac{L}{\varepsilon}}(x \in T_\rho, 0) = x_{t+\frac{L}{\varepsilon}}\left(x_{\frac{L}{\varepsilon}}(x \in T_\rho, 0), \frac{L}{\varepsilon}\right) = x_{t+\frac{L}{\varepsilon}}\left(x' \in T_\rho, \frac{L}{\varepsilon}\right).$$

In view of (5.127), this yields

$$\|x_{t+\frac{L}{\varepsilon}}(x \in T_\rho, 0) - x^0\| \leq \frac{\rho}{2} + \rho_0 \quad \text{for } t \in \left[0, \frac{L}{\varepsilon}\right),$$
$$\|x_{\frac{2L}{\varepsilon}}(x \in T_\rho, 0) - x^0\| \leq \rho. \tag{5.129}$$

The last inequality in (5.129) means that

$$x_{\frac{2L}{\varepsilon}}(x \in T_\rho, 0) \in T_\rho$$

and leads to the estimates

$$\|x_{t+k\frac{L}{\varepsilon}}(x \in T_\rho, 0) - x^0\| \leq \frac{\rho}{2} + \rho_0, \tag{5.130}$$
$$\|x_{\frac{(k+1)L}{\varepsilon}}(x \in T_\rho, 0) - x^0\| \leq \rho$$

for $t \in [0, \frac{L}{\varepsilon})$, $k = 0, 1, 2, \ldots$. Inequalities (5.130) mean that

$$\|x_t(x \in T_\rho) - x^0\| \leq \frac{\rho}{2} + \rho_0$$

for all $t \in [0, \infty)$, □

Theorem 8 guarantees the existence of solutions bounded for $t \geq 0$. The following theorem solves the problem of solutions bounded for $t \in (-\infty, \infty)$:

Theorem 9 ([137]). *Assume that system* (5.93) *satisfies conditions* (i) *and* (ii) *of Theorem 7 both for $t > 0$ and for $t < 0$. We set*

$$I_0(x) = \lim_{T\to\infty} \frac{\sum_{t<\tau_i<t+T} I_i(x)}{T} \quad \textit{for } t \geq 0,$$
$$I^0(x) = \lim_{T\to\infty} \frac{\sum_{t-T<\tau_i<t} I_i(x)}{T} \quad \textit{for } t \leq 0.$$

Suppose that the system

$$\frac{d\overline{x}}{dt} = \varepsilon[X_0(\overline{x}) + I_0(\overline{x})] \tag{5.131}$$

averaged for $t \geq 0$ possesses an asymptotically stable equilibrium position $x = x^0$ satisfying inequalities (5.123) *for x from a certain ρ_0-neighborhood of this position. Assume that, in the ρ-neighborhood D_ρ of the solution x^0 mentioned in Theorem* 8, *the system*

$$\frac{d\overline{x}_1}{dt} = \varepsilon[X_0(\overline{x}_1) + I^0(\overline{x}_1)] \tag{5.132}$$

averaged over $t \leq 0$ possesses an equilibrium position $x_1 = x_1^0$ for which either

$$\frac{\partial \tau_i(x)}{\partial x} I_i(x) \leq \beta < 0 \quad or \quad \frac{\partial \tau_i(x)}{\partial x} \equiv 0$$

for all $i = -1, -2, \dots$ and all x from a certain ρ_0'-neighborhood of the position x_1^0.

Under these conditions, the following assertions are true:

1. *If the equilibrium position x_1^0 of system* (5.132) *is asymptotically unstable (asymptotically stable for $t < 0$), then there exist $\varepsilon^0 > 0$ and a domain D_{ρ_1} containing x^0 and x_1^0 such that, for $\varepsilon < \varepsilon^0$, all solutions $x_t(x)$ of system* (5.93) *for which $x \in D_\rho$ are uniformly bounded for $t \in (-\infty, \infty)$.*
2. *If the equilibrium positions x_1^0 of system* (5.132) *are asymptotically stable, then one can find $\varepsilon_0 > 0$ and x^* such that the solution $x_t(x^*)$ of system* (5.93) *is uniformly bounded for $t \in (-\infty, \infty)$.*

Proof. Assume that the equilibrium position x_1^0 of system (5.132) is asymptotically unstable. Applying Theorem 8 to the intervals $t \geq 0$ and $t \leq 0$, we conclude that there exist ρ- and $\overline{\rho}$-neighborhoods of the points x^0 and x_1^0 such that

$$\|x_t(x \in T_\rho)\| \leq c_1 \quad \text{for } t \in (0, \infty),$$
$$\|x_t(x \in T_{\overline{\rho}})\| \leq c_2 \quad \text{for } t \in (-\infty, 0).$$

Since $x_1^0 \in D_\rho$, the set $D_\rho \bigcap T_{\overline{\rho}} = D_{\rho_1}$ is nonempty. Therefore,

$$\|x_t(x \in D_{\rho_1})\| \leq c = \max(c_1, c_2) \quad \text{for } t \in (-\infty, \infty).$$

Now let x_1^0 be an asymptotically stable solution of system (5.132). Applying Theorem 7 to (5.93), we show that estimates of the form (5.127) are true for $x_t(x \in T_{\rho_1'}, \tau)$, i.e.,

$$\|x_{t+\tau}(x \in T_{\rho_1'}, \tau) - x_1^0\| \leq \frac{\rho_1'}{2} + \rho_1^0$$

for $t \in [0, \frac{L}{\varepsilon})$ and $\tau \geq -\frac{L}{\varepsilon}$ and

$$\left\| x_0\left(x \in T_{\rho_1'}, -\frac{L}{\varepsilon}\right) - x_1^0 \right\| \leq \rho_1', \tag{5.133}$$

where $T_{\rho_1'}$ is a ρ_1'-neighborhood of the point x_1^0. Inequalities (5.133) yield the following estimates:

$$\left\| x_{t-k\frac{L}{\varepsilon}}\left(x \in T_{\rho_1'}, -k\frac{L}{\varepsilon}\right) - x_1^0 \right\| \leq \frac{\rho_1'}{2} + \rho_1^0,$$
$$\left\| x_0\left(x \in T_{\rho_1'}, -k\frac{L}{\varepsilon}\right) - x_1^0 \right\| \leq \rho_1'$$

for $t \in (-k\frac{L}{\varepsilon}, 0]$, $k = 1, 2, \dots$.

In the set $D_{\rho_2} = T_{\rho_1'} \cap D_\rho$, we choose a convergent subsequence of points

$$y_k = x_0\left(x_k \in D_{\rho_2}, -k\frac{L}{\varepsilon}\right), \quad k = 1, 2, \ldots, \quad \lim_{k\to\infty} y_k = y_0$$

and consider the solution $x_t(y_0)$. We show that $x_t(y_0)$ is defined for $t \in (-\infty, 0)$ and bounded. Indeed, the sequence of solutions $x_t(y_k)$ defined for $-k\frac{L}{\varepsilon} \le t \le 0$ is uniformly bounded for $k = 1, 2, \ldots$ because it coincides with the sequence $x_{\bar{t}-k\frac{L}{\varepsilon}}(x_k, -k\frac{L}{\varepsilon})$ for $\bar{t} \in [-k\frac{L}{\varepsilon}, 0)$. Assume that $\|x_t(y_0)\| > D$ for some $t = t_0 < 0$, $D > \frac{\rho_1'}{2} + \rho_1^0 + \|x_1^0\|$. Since $x_t(y_0)$ is a piecewise continuous function of t in any interval, we have

$$\|x_t(y_0)\| > D_1 > \frac{\rho_1'}{2} + \rho_1^0 + \|x_1^0\|$$

for an interval (t_0, t_0'), which is the interval of continuity for $x_t(y_0)$. Since the points of discontinuity of the solution $x_t(x)$ continuously depend on x in a finite time interval, we can choose $\tau_1 \in (t_0, t_0')$ such that

$$x_{\tau_1}(y_k) \to x_{\tau_1}(y_0) \quad \text{as } k \to \infty. \tag{5.134}$$

Relation (5.134) implies the inequality

$$x_{\tau_1}(y_k) > \frac{\rho_1'}{2} + \rho_1^0 + \|x_1^0\| \tag{5.135}$$

for all sufficiently large k.

However, (5.135) contradicts (5.133) because

$$x_{\tau_1}(y_k) = x_{\overline{\tau}-k\frac{L}{\varepsilon}}\left(x_k, -k\frac{L}{\varepsilon}\right).$$

This contradiction proves that $x_t(y_0)$ is a solution bounded on $(-\infty, 0)$. Since $y_0 \in D_\rho$, the solution $x_t(y_0)$ is also bounded for $t \ge 0$ and, therefore,

$$\|x_t(y_0)\| \le c < \infty,$$

which completes the proof of Theorem 9. □

Theorems 8 and 9 essentially use the asymptotic stability of equilibrium positions. However, in the problem of qualitative correspondence between the exact and approximate solutions, parallel with stability, one can also use some other properties of the equilibrium positions. Thus, in view of the fact that, for systems periodic for $t \ge 0$, the right-hand side of the averaged system (5.131) coincides, to within $\varepsilon\mu(\varepsilon)$, where $\mu(\varepsilon) \to 0$ as $\varepsilon \to 0$, with the operator of shift along the trajectories of system (5.93), we can formulate the following theorem:

Theorem 10 ([137]). *Assume that conditions* (1) *of Theorem* 7 *are satisfied for system* (5.93) *T-periodic for $t \geq 0$. If, in addition, the averaged system* (5.131) *possesses an isolated singular point $\overline{x} = x^0$:*

$$X_0(x^0) + I(x^0) = 0,$$

lying, together with a certain ρ-neighborhood, in the domain D, its index under the mapping $x \to X_0(x) + I(x)$ is not equal to zero, and

$$\frac{\partial \tau_i(x^0)}{\partial x^0} I_i(x) \leq \beta < 0 \quad \text{or} \quad \frac{\partial \tau_i(x)}{\partial x} \equiv 0, \quad i = \overline{1, d},$$

then there exists $\varepsilon_0 > 0$ such that, for all $\varepsilon < \varepsilon_0$, system (5.93) *possesses a periodic solution $x = x_t^0$, $t > t_1^0$, with period T for which* $\lim_{\varepsilon \to 0} x_t^0 = x^0$.

Moreover, if the functions $X_0(x)$ and $I(x)$ are continuously differentiable in a ρ-neighborhood of the equilibrium position $x = x^0$ and the real parts of the eigenvalues of the matrix

$$\frac{\partial [X_0(x^0) + I(x^0)]}{\partial x}$$

are negative, then the solution x_t^0 is asymptotically stable.

In [138], it is shown that if the averaged system (5.73) possesses an asymptotically orbitally stable periodic solution, then, for $t \geq 0$, the solutions of Eqs. (5.71) starting from a sufficiently small vicinity of the trajectory of this solution are uniformly bounded. The sufficient conditions of generation of the integral set of the original Eqs. (5.71) by an asymptotically orbitally stable periodic solution of the averaged system (5.73) are also established in the cited work.

An analog of the second Bogolyubov's theorem on substantiation of the method of averaging for systems with pulse action is proved in [140]. In this work, a system of equations in the standard form (5.71) is investigated under the assumption that the hypersurfaces $t = \tau_i(x)$ are hyperplanes $t = \tau_i$. Under the condition that the corresponding averaged system possesses an isolated equilibrium position $x = x_0$ and the real parts of the eigenvalues of the matrix

$$\left. \frac{\partial (X_0(x) + I_0(x))}{\partial x} \right|_{x = x_0}$$

are not equal to zero, it is shown that the unique solution of the original system bounded on the entire axis exists in the vicinity of the solution $x = x_0$. The properties of this solution and solutions starting in a sufficiently small vicinity of the bounded solution are also investigated.

The presented results concerning the applicability of the method of averaging to systems of the form (5.71) were generalized to systems of differential equations subjected to pulse actions with "slow" and "fast" variables in [12] and to a class of functional-differential equations with pulses in [13].

Chapter 6

Averaging of Differential Inclusions

6.1 Averaging of Inclusions with Pulses at Fixed Times

In the present section, we consider the problem of substantiation of the method of complete and partial averaging on finite and infinite intervals for differential inclusions subjected to pulse actions at fixed times.

Method of Averaging for a Finite Interval. Consider a differential inclusion with multivalued pulses

$$\begin{aligned} \dot{x} &\in \varepsilon X(t,x), \quad t \neq \tau_i, \ x(0) \in X_0, \\ \Delta x|_{t=\tau_i} &\in I_i(x). \end{aligned} \tag{6.1}$$

If, for any $x \in D$, there exists a limit

$$Y(x) = \lim_{T\to\infty}\left(\frac{1}{T}\int_t^{t+T} X(t,x)dt + \frac{1}{T}\sum_{t\le\tau_i<t+T} I_i(x)\right), \tag{6.2}$$

then inclusion (6.1) can be associated with the averaged inclusion

$$\dot{y} \in \varepsilon Y(y), \quad y(0) \in X_0. \tag{6.3}$$

Theorem 1. *Assume that the following conditions are satisfied in the domain $Q\{t \ge 0, x \in D \subset \mathbb{R}^n\}$:*

(1) *the set-valued mappings $X\colon Q \to \operatorname{conv}(\mathbb{R}^n)$ and $I_i\colon D \to \operatorname{conv}(\mathbb{R}^n)$ are continuous, uniformly bounded by a constant M, and satisfy the Lipschitz condition with respect to x with constant λ:*

(2) *limit (6.2) exists uniformly in $t \ge 0$ and $x \in D$ and, in addition,*

$$\frac{1}{T}\, i(t,t+T) \le d < \infty,$$

where $i(t,t+T)$ is the number of points of the sequence τ_i in the interval $(t,t+T]$;

(3) *for all $x_0 \in D' \subset D$ and $t \in [0, L\varepsilon^{-1}]$, the solutions of inclusion (6.3) belong to the domain D together with a certain ρ-neighborhood.*

Then, for any $\eta > 0$ *and* $L > 0$*, there exists* $\varepsilon_0(\eta, L) > 0$ *such that the following assertions are true for* $\varepsilon \in (0, \varepsilon_0]$ *and* $t \in [0, L\varepsilon^{-1}]$*:*

(1) *for any solution* $y(t)$ *of inclusion* (6.3)*, there exists a solution* $x(t)$ *of inclusion* (6.1) *such that*

$$\|x(t) - y(t)\| < \eta; \tag{6.4}$$

(2) *for any solution* $x(t)$ *of inclusion* (6.1)*, there exits a solution* $y(t)$ *of inclusion* (6.3) *such that inequality* (6.4) *is true.*

Proof. By conditions (1) and (2) of the theorem, the set-valued mapping $Y\colon D \to \operatorname{conv}(\mathbb{R}^n)$ is bounded by a constant $M_1 = M(1 + d)$ and satisfies the Lipschitz condition with constant $\lambda_1 = \lambda(1 + d)$.

Let $y(t)$ be a solution of inclusion (6.3). We split the segment $[0, L\varepsilon^{-1}]$ into subsegments with steps $\gamma(\varepsilon)$ such that $\gamma(\varepsilon) \to \infty$ and $\varepsilon\gamma(\varepsilon) \to 0$ as $\varepsilon \to 0$. Then there exists a measurable branch $v(t)$ of the mapping $Y(y(t))$ such that

$$y(t) = y(t_j) + \varepsilon \int_{t_j}^{t} v(s)ds, \quad y(0) = x_0, \; t \in [t_j, t_{j+1}], \tag{6.5}$$

where $t_j = j\gamma(\varepsilon)$ and $m\gamma(\varepsilon) \le L\varepsilon^{-1} < (m + 1)\gamma(\varepsilon)$, $j = \overline{0, m}$.

Consider a function

$$y^1(t) = y^1(t_j) + \varepsilon v_j(t - t_j), \quad t \in [t_j, t_{j+1}], \; y^1(0) = x_0, \tag{6.6}$$

where the vectors v_j are such that

$$\left\| \gamma(\varepsilon)v_j - \int_{t_j}^{t_{j+1}} v(s)ds \right\| = \min_{v \in Y(y^1(t_j))} \left\| \gamma(\varepsilon)v - \int_{t_j}^{t_{j+1}} v(s)ds \right\|. \tag{6.7}$$

The vector v_j exists and is unique by virtue of compactness and convexity of the set $Y(y^1(t_j))$ and strict convexity of the minimized function.

Denote $\delta_j = \|y(t_j) - y^1(t_j)\|$. For $t \in [t_j, t_{j+1}]$, by virtue of (6.5) and (6.6), we find

$$\|y(t) - y(t_j)\| \le M_1\varepsilon\gamma(\varepsilon), \quad \|y^1(t) - y^1(t_j)\| \le M_1\varepsilon\gamma(\varepsilon). \tag{6.8}$$

Hence, for $t \in [t_j, t_{j+1}]$, we get

$$\begin{aligned} &\|y(t) - y^1(t_j)\| \le \|y(t_j) - y^1(t_j)\| + \|y(t) - y(t_j)\| \le \delta_j + \varepsilon M_1(t - t_j), \\ &h(Y(y(t)), Y(y^1(t_j)) \le \lambda_1\|y(t) - y^1(t_j)\| \le \lambda_1(\delta_j + \varepsilon M_1(t - t_j)). \end{aligned} \tag{6.9}$$

It follows from relations (6.7) and (6.9) that

$$\begin{aligned} \left\| \int_{t_j}^{t_{j+1}} (v(s) - v_j)ds \right\| &\le \int_{t_j}^{t_{j+1}} h(Y(y(s)), Y(y^1(t_j)))ds \\ &\le \lambda_1\left(\delta_j\gamma(\varepsilon) + \varepsilon M_1\frac{\gamma^2(\varepsilon)}{2}\right). \end{aligned} \tag{6.10}$$

In view of (6.5) and (6.6), we can write

$$\delta_{j+1} \le \delta_j + \varepsilon\lambda_1\left(\delta_j\gamma(\varepsilon) + \varepsilon M_1\frac{\gamma^2(\varepsilon)}{2}\right) = (1+\lambda_1\varepsilon\gamma(\varepsilon))\delta_j + \lambda_1 M_1\frac{\varepsilon^2\gamma^2(\varepsilon)}{2}. \tag{6.11}$$

Since $\delta_0 = 0$, inequality (6.11) implies that

$$\delta_1 \le \lambda_1 M_1\frac{\varepsilon^2\gamma^2(\varepsilon)}{2},$$

$$\delta_2 \le (1+\lambda_1\varepsilon\gamma(\varepsilon))\delta_1 + \lambda_1 M_1\frac{\varepsilon^2\gamma^2(\varepsilon)}{2}$$

$$\le \lambda_1 M_1\frac{\varepsilon^2\gamma^2(\varepsilon)}{2}((1+\lambda_1\varepsilon\gamma(\varepsilon))+1)$$

etc.,

$$\begin{aligned}\delta_{j+1} &\le \lambda_1 M_1\frac{\varepsilon^2\gamma^2(\varepsilon)}{2}((1+\lambda_1\varepsilon\gamma(\varepsilon))^i + (1+\lambda_1\varepsilon\gamma(\varepsilon))^{i-1} + \dots + 1)\\ &= \frac{M_1\varepsilon\gamma(\varepsilon)}{2}((1+\lambda_1\varepsilon\gamma(\varepsilon))^{i+1} - 1)\\ &\le \frac{M_1\varepsilon\gamma(\varepsilon)}{2}((1+\lambda_1\varepsilon\gamma(\varepsilon))^{\frac{L}{\varepsilon\gamma(\varepsilon)}} - 1)\\ &\le \frac{M_1\varepsilon\gamma(\varepsilon)}{2}(e^{\lambda_1 L} - 1).\end{aligned} \tag{6.12}$$

Thus, by virtue of inequalities (6.8), we arrive at the following estimate:

$$\begin{aligned}\|y(t) - y^1(t)\| &\le \|y(t) - y(t_j)\| + \|y(t_j) - y^1(t_j)\| + \|y^1(t_j) - y^1(t)\|\\ &\le 2M_1\varepsilon\gamma(\varepsilon) + \frac{M_1\varepsilon\gamma(\varepsilon)}{2}(e^{\lambda_1 L} - 1)\\ &\le \frac{M_1\varepsilon\gamma(\varepsilon)}{2}(e^{\lambda_1 L} + 3).\end{aligned} \tag{6.13}$$

It follows from condition (2) of the theorem that, for any $\eta_1 > 0$, there exists $\varepsilon_1(\eta_1) > 0$ such that the following inequality is true for $\varepsilon \le \varepsilon_1(\eta_1)$:

$$h\left(Y(y^1(t_j)), \frac{1}{\gamma(\varepsilon)}\int_{t_j}^{t_{j+1}} X(s, y^1(t_j))ds + \frac{1}{\gamma(\varepsilon)}\sum_{t_j \le \tau_i < t_{j+1}} I_i(y^1(t_j))\right) < \eta_1. \tag{6.14}$$

Hence, one can find vectors $u_j(t) \in X(t, y^1(t_j))$ and $p_{ij} \in I_i(y^1(t_j))$ such that

$$\left\| v_i - \frac{1}{\gamma(\varepsilon)}\left(\int_{t_j}^{t_{j+1}} u_j(s)ds + \sum_{t_j \le \tau_i < t_{j+1}} p_{ij}\right)\right\| < \eta_1. \tag{6.15}$$

Consider a family of functions

$$x^1(t) = x^1(t_j) + \varepsilon \int_{t_j}^{t} u_j(s)ds + \varepsilon \sum_{t_j \le \tau_i < t} p_{ij},\ t \in (t_j, t_{j+1}]. \tag{6.16}$$

By using relations (6.6), (6.16), and (6.15), in view of the fact that $x^1(0) = y^1(0)$, for $j = \overline{1,m}$, we obtain

$$\begin{aligned} \|x^1(t_j) - y^1(t_j)\| &\le \|x^1(t_{j-1}) - y^1(t_{j-1})\| + \eta_1 \varepsilon \gamma(\varepsilon) \\ &\le \cdots \le j\eta_1 \varepsilon \gamma(\varepsilon) \le L\eta_1. \end{aligned} \tag{6.17}$$

Note that, for $t \in (t_j, t_{j+1}]$, we have

$$\|x^1(t) - x^1(t_j)\| \le M(1+d)\varepsilon\gamma(\varepsilon) = M_1 \varepsilon \gamma(\varepsilon).$$

Thus, in view of inequality (6.8), we find

$$\begin{aligned} \|x^1(t) - y^1(t)\| &\le L\eta_1 + 2M_1\varepsilon\gamma(\varepsilon), \\ \|x^1(t) - y^1(t_j)\| &\le L\eta_1 + M_1\varepsilon\gamma(\varepsilon). \end{aligned} \tag{6.18}$$

We now show that there exists a solution

$$x(t) = x(t_j) + \varepsilon \int_{t_j}^{t} u(\tau)\,d\tau + \varepsilon \sum_{t_j \le \tau_i < t} q_i,$$

$$x(0) = x_0, \quad t \in (t_j, t_{j+1}],$$

of inclusion (6.1) sufficiently close to $x^1(t)$.

Let $\theta_1, \dots, \theta_p$ be the times of pulses τ_i, from the half interval $(t_j, t_{j+1}]$. For the sake of convenience, we denote $\theta_0 = t_j$ and $\theta_{p+1} = t_{j+1}$. Further, let

$$\mu_k^+ = \|x^1(\theta_k + 0) - x(\theta_k + 0)\| \quad \text{and} \quad \mu_k^- = \|x^1(\theta_k) - x(\theta_k)\|, \quad k = \overline{0,p}.$$

In view of the Lipschitz condition, we conclude that

$$\begin{aligned} (\dot{x}^1(t), \varepsilon X(t, x^1(t))) &\le h(\varepsilon X(t, y^1(t_j)), \varepsilon X(t, x^1(t))) \\ &\le \varepsilon\lambda \|x^1(t) - y^1(t_j)\| \le \varepsilon\lambda(M_1\varepsilon\gamma(\varepsilon) + L\eta_1) = \eta^*, \\ \rho(\Delta x^1|_{t=\theta_k}, \varepsilon I_i(x^1(\theta_k))) &\le h(\varepsilon I_i(y^1(t_j)), \varepsilon I_i(x^1(\theta_k))) \\ &\le \varepsilon\lambda \|y^1(t_j) - x^1(\theta_k)\| \le \varepsilon\lambda(M_1\varepsilon\gamma(\varepsilon) + L\eta_1) = \eta^*. \end{aligned}$$

According to the Filippov theorem, between the points of the pulses, there exists a solution $x(t)$ of inclusion (6.1) such that the following inequality is true for $t \in (\theta_k, \theta_{k+1}]$:

$$\|x(t) - x^1(t)\| \le \mu_k^+ e^{\varepsilon\lambda(t-\theta_k)} + \varepsilon \int_{\theta_k}^{t} e^{\varepsilon\lambda(t-s)} \eta^* ds.$$

Denote $\gamma_k = \theta_{k+1} - \theta_k \le \gamma(\varepsilon)$, $\gamma_0 + \cdots + \gamma_p = \gamma(\varepsilon)$. Then

$$\mu_{k+1}^- \le \mu_k^+ e^{\varepsilon\lambda\gamma_k} + \frac{\eta^*}{\lambda}(e^{\lambda\varepsilon\gamma(\varepsilon)} - 1). \tag{6.19}$$

In passing through the point of the pulse, we find

$$\begin{aligned}
\mu_{k+1}^+ &\le \mu_{k+1}^- + \varepsilon h(I_i(y^1(t_j)), I_i(x(\theta_{k+1}))) \\
&\le \mu_{k+1}^- + \varepsilon h(I_i(x^1(\theta_{k+1})), I_i(x(\theta_{k+1}))) \\
&\quad + \varepsilon h(I_i(y^1(t_j)), I_i(x^1(\theta_{k+1}))) \\
&\le \mu_{k+1}^- + \varepsilon\lambda\mu_{k+1}^- + \varepsilon h(I_i(y^1(t_j)), I_i(x^1(\theta_{k+1}))) \\
&\le (1+\varepsilon\lambda)\mu_{k+1}^- + \eta^*.
\end{aligned} \tag{6.20}$$

Relations (6.19) and (6.20) imply that

$$\begin{aligned}
\mu_{k+1}^+ &\le (1+\varepsilon\lambda)e^{\varepsilon\lambda\gamma_k}\mu_k^+ + \beta, \\
\beta &= \frac{\eta^*}{\lambda}(1+\varepsilon\lambda)(e^{\lambda\varepsilon\gamma(\varepsilon)} - 1) + \eta^*.
\end{aligned}$$

Therefore,

$$\begin{aligned}
\mu_1^+ &\le (1+\varepsilon\lambda)e^{\lambda\varepsilon\gamma_0}\mu_0^+ + \beta \le (1+\varepsilon\lambda)e^{\lambda\varepsilon\gamma(\varepsilon)}\mu_0^+ + \beta, \\
\mu_2^+ &\le (1+\varepsilon\lambda)e^{\varepsilon\lambda\gamma_1}\mu_1^+ + \beta \\
&\le (1+\varepsilon\lambda)^2 e^{\varepsilon\lambda(\gamma_0+\gamma_1)}\mu_0^+ + \beta(1+\varepsilon\lambda)e^{\varepsilon\lambda\gamma_1} + \beta \\
&\le (1+\varepsilon\lambda)^2 e^{\lambda\varepsilon\gamma(\varepsilon)}\mu_0^+ + \beta((1+\varepsilon\lambda)e^{\lambda\varepsilon\gamma(\varepsilon)} + 1),
\end{aligned}$$

etc.,

$$\begin{aligned}
\mu_{k+1}^+ &\le (1+\varepsilon\lambda)^{k+1} e^{\varepsilon\lambda\gamma(\varepsilon)}\mu_0^+ \\
&\quad + \beta(e^{\lambda\varepsilon\gamma(\varepsilon)}((1+\varepsilon\lambda)^k + \cdots + (1+\varepsilon\lambda)) + 1) \\
&= (1+\varepsilon\lambda)^{k+1} e^{\lambda\varepsilon\gamma(\varepsilon)}\mu_0^+ \\
&\quad + \beta\left(e^{\lambda\varepsilon\gamma(\varepsilon)}\frac{(1+\varepsilon\lambda)^k - 1}{\varepsilon\lambda}(1+\varepsilon\lambda) + 1\right) \\
&\le e^{\lambda(1+d)\varepsilon\gamma(\varepsilon)}\mu_0^+ + \eta^*\left(\frac{1+\varepsilon\lambda}{\lambda}(e^{\lambda\varepsilon\gamma(\varepsilon)} - 1) + 1\right) \\
&\quad \times \left(e^{\lambda\varepsilon\gamma(\varepsilon)}\frac{e^{\lambda d\varepsilon\gamma(\varepsilon)} - 1}{\varepsilon\lambda}(1+\varepsilon\lambda) + 1\right) \\
&= \alpha\mu_0^+ + \beta_1,
\end{aligned}$$

where

$$\alpha = e^{\lambda\varepsilon\gamma(\varepsilon)(1+d)},$$

$$\beta_1 = (\varepsilon\gamma(\varepsilon)M_1 + L\eta_1)\left(\frac{1+\varepsilon\lambda}{\lambda}(e^{\lambda\varepsilon\gamma(\varepsilon)} - 1) + 1\right)$$
$$\times (e^{\lambda\varepsilon\gamma(\varepsilon)}(e^{\lambda d\varepsilon\gamma(\varepsilon)} - 1)(1+\varepsilon\lambda) + \varepsilon\lambda).$$

Thus,

$$\delta_{j+1}^+ = \|x(t_{j+1}) - x^1(t_{j+1})\| \le \alpha\delta_j^+ + \beta_1.$$

As a result, we arrive at the chain of inequalities

$$\delta_0^+ = 0, \quad \delta_1^+ \le \beta_1, \quad \delta_2^+ \le \alpha\beta_1 + \beta_1 = (\alpha+1)\beta_1, \dots,$$

$$\begin{aligned}\delta_{j+1}^+ &\le (\alpha^j + \dots + 1)\beta_1 = \frac{\alpha^{j+1}-1}{\alpha-1}\beta_1 \\ &\le \frac{e^{\lambda L(1+d)} - 1}{e^{\lambda(1+d)\varepsilon\gamma(\varepsilon)} - 1}(M_1\varepsilon\gamma(\varepsilon) + L\eta_1) \\ &\quad \times \left(\frac{1+\varepsilon\lambda}{\lambda}(e^{\lambda\varepsilon\gamma(\varepsilon)} - 1) + 1\right) \\ &\quad \times \left(e^{\lambda\varepsilon\gamma(\varepsilon)}\left(e^{\lambda d\varepsilon\gamma(\varepsilon)} - 1\right)(1+\varepsilon\lambda) + \varepsilon\lambda\right).\end{aligned}$$

Since

$$\lim_{\varepsilon\downarrow 0}\left(\frac{1+\varepsilon\lambda}{\lambda}(e^{\lambda\varepsilon\gamma(\varepsilon)} - 1) + 1\right) = 1$$

and

$$\begin{aligned}&\lim_{\varepsilon\downarrow 0}\frac{e^{\lambda\varepsilon\gamma(\varepsilon)}(e^{\lambda d\varepsilon\gamma(\varepsilon)} - 1)(1+\varepsilon\lambda) + \varepsilon\lambda}{e^{\lambda(1+d)\varepsilon\gamma(\varepsilon)} - 1} \\ &= \lim_{\varepsilon\to 0}\frac{e^{\lambda\varepsilon\gamma(\varepsilon)}\frac{e^{\lambda d\varepsilon\gamma(\varepsilon)}-1}{\lambda\varepsilon\gamma(\varepsilon)} + \frac{1}{\gamma(\varepsilon)}}{\frac{e^{\lambda(1+d)\varepsilon\gamma(\varepsilon)}-1}{\lambda\varepsilon\gamma(\varepsilon)}} = \frac{d}{1+d},\end{aligned}$$

we have

$$\delta_{j+1}^+ \le C(M_1\varepsilon\gamma(\varepsilon) + L\eta_1)$$

for $\varepsilon \le \varepsilon_2$.

Hence, for $t \in (t_j, t_{j+1}]$, we get the inequality

$$\begin{aligned}\|x(t) - x^1(t)\| &\le \|x(t) - x(t_j)\| + \|x(t_j) - x^1(t_j)\| + \|x^1(t) - x^1(t_j)\| \\ &\le M(1+d)\varepsilon\gamma(\varepsilon) + M_1\varepsilon\gamma(\varepsilon) + C(M_1\varepsilon\gamma(\varepsilon) + L\eta_1) \\ &= M_1(2+C)\varepsilon\gamma(\varepsilon) + CL\eta_1. \qquad (6.21)\end{aligned}$$

By virtue of inequalities (6.13), (6.18), and (6.21), we conclude that $\|x(t) - y(t)\|$ can be made smaller than any given η by the proper choice of $\varepsilon \le \varepsilon_0$ and η_1.

The second assertion of the theorem is proved similarly. □

The following statement is obtained as a corollary of Theorem 1:

Theorem 2 ([103]). *Assume that the following conditions are satisfied in the domain* $Q\{t \ge 0,\ x \in D \subset \mathbb{R}^n\}$:

(1) *the set-valued mappings* $X\colon Q \to \operatorname{conv}(\mathbb{R}^n)$ *and* $I_i\colon D \to \operatorname{conv}(\mathbb{R}^n)$ *are continuous, uniformly bounded by a constant* M, *and satisfy the Lipschitz condition with constant* λ *with respect to* x;

(2) *limit* (6.2) *exists uniformly in* x *and, in addition,*

$$\frac{1}{T}\, i(t, t+T) \le d < \infty;$$

(3) *the* R*-solutions of inclusion* (6.3) *belong to the domain* D *together with a certain* ρ*-neighborhood for all* $x_0 \in D' \subset D$ *and* $t \in [0, L\varepsilon^{-1}]$.

Then, for any $\eta > 0$ *and* $L > 0$, *there exists* $\varepsilon_0(\eta, L) > 0$ *such that the following assertions are true for* $\varepsilon \in (0, \varepsilon_0]$ *and* $t \in [0, L\varepsilon^{-1}]$:

(1) *for any* R*-solution* $Y(t)$ *of inclusion* (6.3), *there exists an* R*-solution* $X(t)$ *of inclusion* (6.1) *such that*

$$h(X(t), Y(t)) < \eta; \tag{6.22}$$

(2) *for any* R*-solution* $X(t)$ *of inclusion* (6.1), *there exits an* R*-solution* $Y(t)$ *of inclusion* (6.3) *such that inequality* (6.22) *is true.*

We now consider a scheme of partial averaging. Together with impulsive differential inclusion (6.1), we consider an impulsive differential inclusion

$$\begin{aligned} \dot{y} &\in \varepsilon \widetilde{X}(t, y), \quad t \ne \nu_j,\ y(0) \in X_0, \\ \Delta y|_{t=\nu_j} &\in K_j(y), \end{aligned} \tag{6.23}$$

where the limit

$$\lim_{T\to\infty} \frac{1}{T} h\Bigg(\int_t^{t+T} X(t, x)dt + \sum_{t \le \tau_i < t+T} I_i(x), \\ \int_t^{t+T} \widetilde{X}(t, x)dt + \sum_{t \le \nu_j < t+T} K_j(x)\Bigg) = 0 \tag{6.24}$$

exists for any $x \in D$.

Theorem 3 ([112]). *Assume that the following conditions are satisfied in the domain* $Q\{t \geq 0,\ x \in D \subset \mathbb{R}^n\}$*:*

(1) *the set-valued mappings* $X,\ \widetilde{X}\colon Q \to \operatorname{conv}(\mathbb{R}^n)$ $I_j, K_j\colon\ D \to \operatorname{conv}(\mathbb{R}^n)$ *are continuous, uniformly bounded by a constant* M*, and satisfy the Lipschitz condition with constant* λ *with respect to* x*;*

(2) *limit* (6.24) *exists uniformly in* x *and, in addition,*

$$\frac{1}{T}\, i(t, t+T) \leq d < \infty, \qquad \frac{1}{T}\, j(t, t+T) \leq d < \infty,$$

where $i(t,t+T)$ *and* $j(t,t+T)$ *are, respectively, the numbers of points of the sequences* $\{t_j\}$ *and* $\{\nu_j\}$ *in the segment* $[t, t+T]$*;*

(3) *the solutions of inclusion* (6.23) *belong to the domain* D *together with a certain* ρ*-neighborhood for all* $x_0 \in D' \subset D$ *and* $t \in [0, L\varepsilon^{-1}]$.

Then, for any $\eta > 0$ *and* $L \in (0, L^*]$*, there exists* $\varepsilon_0(\eta, L) \in (0, \sigma]$ *such that the following inequality is true for* $\varepsilon \in (0, \varepsilon_0]$ *and* $t \in [0, L\varepsilon^{-1}]$*:*

$$\|x(t,\varepsilon) - y(t,\varepsilon)\| < \eta, \tag{6.25}$$

where $x(t,\varepsilon)$ *and* $y(t,\varepsilon)$ *are solutions of inclusions* (6.1) *and* (6.23)*, respectively, and* $x(0,\varepsilon) = y(0,\varepsilon) = x_0 \in X_0$.

Theorem 4 ([112]). *Assume that conditions* (1) *and* (2) *of Theorem* 3 *are satisfied in the domain* $Q\{t \geq 0,\ x \in D \subset \mathbb{R}^n\}$ *and, in addition, that*

(3) *for all* $X_0 \subset D' \subset D$ *and* $t \in [0, L\varepsilon^{-1}]$*, the* R*-solutions of inclusion* (6.23) *lie in the domain* D *together with a certain* ρ*-neighborhood.*

Then, for any $\eta > 0$ *and* $L \in (0, L^*]$*, there exists* $\varepsilon_0(\eta, L) \in (0, \sigma]$ *such that the following inequality is true for* $\varepsilon \in (0, \varepsilon_0]$ *and* $t \in [0, L\varepsilon^{-1}]$*:*

$$h(X(t,\varepsilon), Y(t,\varepsilon)) < \eta, \tag{6.26}$$

where $X(t,\varepsilon)$ *and* $Y(t,\varepsilon)$ *are* R*-solutions of inclusions* (6.1) *and* (6.23)*, respectively, and* $X(0,\varepsilon) = Y(0,\varepsilon) = X_0$.

Method of Averaging in an Infinite Interval. For an inclusion

$$\dot{x} \in \varepsilon X(t,x), \quad t \neq \tau_i,\ x(0) \in X_0, \qquad \Delta x|_{t=\tau_i} \in I_i(x), \tag{6.27}$$

we consider the averaged differential inclusion

$$\dot{\xi} \in \varepsilon \overline{X}(\xi), \quad \xi(0) = x_0, \tag{6.28}$$

where

$$\overline{X}(x) = \lim_{T\to\infty} \frac{1}{T}\left(\int_t^{t+T} X(t,x)dt + \sum_{t\le t_j<t+T} I_i^j(x) \right). \tag{6.29}$$

Theorem 5 ([112]). *Assume that the following conditions are satisfied in the domain $Q\{t \ge 0,\ x \in D \subset \mathbb{R}^n\}$:*

(1) *the mappings $X(t,x)$ and $I_j(x)$ are nonempty convex compact sets; moreover, they are continuous and uniformly bounded and satisfy the Lipschitz condition with respect to x with constant λ;*

(2) *limit* (6.29) *exists and, in addition, the inequality*

$$\frac{1}{T} i(t, t+T) \le d < \infty$$

is true uniformly in $x \in D$ and $t \ge 0$;

(3) *for any $x_0 \in D' \subset D$ and $t \ge 0$, the R-solutions of inclusion* (6.28) *are uniformly asymptotically stable and belong to the domain D together with a ρ-neighborhood.*

Then, for any $\eta > 0$, there exists $\varepsilon^0(\eta) > 0$ such that the following inequality is true for $\varepsilon \in (0, \varepsilon^0]$ and $t \ge 0$:

$$h(R(t,\varepsilon), \bar{R}(\varepsilon t)) \le \eta,$$

where $R(t,\varepsilon)$ is the R-solution of inclusion (6.27), *$\overline{R}(\varepsilon t)$ is the R-solution of inclusion* (6.28), *and $R(0,\varepsilon) = \bar{R}(0) = x_0$.*

Theorem 6 ([112]). *Assume that the following conditions are satisfied in the domain $Q\{t \ge 0,\ x \in D \subset \mathbb{R}^n\}$:*

(1) *conditions* (1) *and* (2) *of Theorem* 5;

(2) *the averaged inclusion* (6.28) *possesses a periodic R-solution $\bar{R}(\tau)$ whose trajectory C is asymptotically orbitally stable and lies in the domain D together with a certain δ-neighborhood.*

Then, for any $\eta \in (0, \delta]$, there exist $\varepsilon_0 > 0$ and $\eta_0 \in (0, \eta]$ such that the following inequality is true for $\varepsilon \in (0, \varepsilon_0]$, $\eta_1 \in (0, \eta_0]$, and $t \ge 0$:

$$h(R(t), C) < \eta,$$

where $R(t)$ is an R-solution of inclusion (6.27) *satisfying the following initial condition: $h(R(0), C) < \eta_1$.*

Theorem 7 ([112]). *Assume that the following conditions are satisfied in the domain Q:*

(1) *conditions* (1) *and* (2) *of Theorem* 5*;*

(2) *inclusion* (6.28) *has an asymptotically stable equilibrium position $\bar{R}^0$ lying in the set D together with a certain ρ_0-neighborhood.*

Then, for any $\eta \in (0, \rho_0)$, there exist $\varepsilon_0 > 0$ and $\eta_0 \in (0, \eta]$ such that the following inequality is true for $\varepsilon \in (0, \varepsilon_0]$, $\eta_1 \in (0, \eta_0]$, and $t \geq 0$:

$$h(R(t), \bar{R}^0) < \eta,$$

where $R(t)$ is an R-solution of inclusion (6.27) *satisfying the following initial condition: $h(R(0), \bar{R}^0) < \eta_1$.*

The proofs of Theorems 5–7 are similar to the proofs of the corresponding theorems for differential equations in [44, 45, 88, 89, 108]. However, the first Bogolyubov's theorem in these proofs should be replaced by Theorem 1.

6.2 Krasnosel'skii–Krein Theorem for Differential Inclusions

In the previous section, the Lipschitz condition for the original or averaged inclusions has been essentially used for the substantiation of the method of averaging for differential inclusions. For functional-differential equations satisfying the Lipschitz condition, the method of averaging was substantiated in [59]. In [40], the Lipschitz condition was replaced by a unilateral Lipschitz condition. For ordinary differential equations, similar results were obtained without using the Lipschitz condition in [70]. In what follows, we prove an analog of the Krasnosel'skii–Krein theorem for differential inclusions (with and without pulse action) in terms of ordinary solutions and R-solutions.

Theorem 8 ([122]). *Assume that the following conditions are satisfied for a differential inclusion*

$$\dot{x} \in F(t, x, \lambda), \tag{6.30}$$

where $F(t, x, \lambda)$ is a set-valued mapping taking values in $\operatorname{conv}(\mathbb{R}^n)$ *and defined for $0 \leq t \leq T$ and $x \in D$, D is a bounded domain in $\mathbb{R}^n$, $\lambda \in \Lambda$, and Λ is a set of values of the parameter λ with limit point $\lambda_0 \in \Lambda$:*

(a) *the set-valued mapping $F(t, x, \lambda)$ is uniformly bounded, continuous in t, uniformly continuous in x, and uniform in t and λ;*

(b) *the set-valued mapping* $F(t, x, \lambda)$ *is integrally continuous in* λ *at the point* λ_0, *i.e., the following condition is satisfied for any* $0 \le t_1 < t_2 \le T$ *and* $x \in D$:

$$\lim_{\lambda \to \lambda_0} h\left(\int_{t_1}^{t_2} F(s, x, \lambda) ds, \int_{t_1}^{t_2} F(s, x, \lambda_0) ds \right) = 0; \tag{6.31}$$

(c) *the solutions* $x(t, \lambda_0)$ *of the inclusion*

$$\dot{x} \in F(t, x, \lambda_0) \tag{6.32}$$

satisfying the initial condition $x(0, \lambda_0) = x_0 \in D^1 \subset D$ *are defined for* $0 \le t \le T$ *and lie in the domain* D *together with a certain* ρ*-neighborhood.*

Then, for any $\eta > 0$, *there exists a neighborhood* $U(\lambda_0)$ *of the point* λ_0 *such that, for* $\lambda \in U(\lambda_0)$ *and any solution* $x(t, \lambda)$ *of inclusion* (6.30) *defined for* $0 \le t \le T$ *and satisfying the initial condition* $x(0, \lambda) = x_0$, *there exists a solution* $x(t, \lambda_0)$ *of inclusion* (6.32) *satisfying the inequality*

$$\|x(t, \lambda) - x(t, \lambda_0)\| < \eta, \quad 0 \le t \le T.$$

Proof. It follows from conditions (a) and (b) of the theorem and boundedness of the domain D that the convergence in (6.31) is uniform in t_1, t_2, and x.

Let $x(t, \lambda)$ ($\lambda \to \lambda_0$, $\lambda \in \Lambda$) be a uniformly convergent sequence of solutions of (6.30) satisfying the initial condition $x(0, \lambda) = x_0$. Hence, there exists a continuous function $y(t)$ such that

$$\lim_{\lambda \to \lambda_0} \max_{t \in [0,T]} \|x(t, \lambda) - y(t)\| = 0.$$

We now show that the following equality is true for any $0 \le \tau < t \le T$:

$$\lim_{\lambda \to \lambda_0} \int_\tau^t F(s, x(s, \lambda), \lambda) ds = \int_\tau^t F(s, y(s), \lambda_0) ds. \tag{6.33}$$

It is clear that

$$h\left(\int_\tau^t F(s, x(s, \lambda), \lambda) ds, \int_\tau^t F(s, y(s), \lambda_0) ds \right) \le I_1 + I_2 + I_3 + I_4,$$

where

$$I_1 = h\left(\int_\tau^t F(s, x(s, \lambda), \lambda) ds, \int_\tau^t F(s, y(s), \lambda) ds \right),$$

$$I_2 = h\left(\int_\tau^t F(s, y(s), \lambda) ds, \int_\tau^t F(s, \overline{y}(s), \lambda) ds \right),$$

$$I_3 = h\left(\int_\tau^t F(s, \overline{y}(s), \lambda) ds, \int_\tau^t F(s, \overline{y}(s), \lambda_0) ds \right),$$

$$I_4 = h\left(\int_\tau^t F(s, y(s), \lambda_0) ds, \int_\tau^t F(s, \overline{y}(s), \lambda_0) ds \right),$$

and $\overline{y}(t)$ is a piecewise continuous function such that

$$\max_{t\in[0,T]} \|y(t) - \overline{y}(t)\| < \delta,$$

where δ is chosen from the condition of uniform continuity of the right-hand side to guarantee that the inequality

$$h(F(s,x,\lambda), F(s,y,\lambda)) < \frac{\varepsilon}{4T}$$

holds for $\|x - y\| < \delta$.

Each term is estimated separately. The neighborhood $U(\lambda_0)$ is chosen to guarantee the validity of the inequality $\|x(s,\lambda) - y(s)\| < \delta$ for $\lambda \in U(\lambda_0)$ and all $s \in [0,T]$.

Then

$$I_1 \le \int_\tau^t h(F(s,x(s,\lambda),\lambda), F(s,y(s),\lambda))ds \le \frac{\varepsilon}{4},$$

$$I_2 \le \int_\tau^t h(F(s,y(s),\lambda), F(s,\overline{y}(s),\lambda))ds < \frac{\varepsilon}{4},$$

$$I_4 \le \int_\tau^t h(F(s,y(s),\lambda_0), F(s,\overline{y}(s),\lambda_0))ds < \frac{\varepsilon}{4}.$$

By using condition (b) of the theorem, we make the neighborhood $U(\lambda_0)$ somewhat smaller, such that the inequality

$$I_3 = h\Big(\int_\tau^t F(s,\overline{y}(s),\lambda)ds, \int_\tau^t F(s,\overline{y}(s),\lambda_0)ds\Big) < \frac{\varepsilon}{4}$$

holds for $\lambda \in U(\lambda_0)$.

Thus, we have proved the limit equality (6.33).

The set of ordinary solutions of the differential inclusion (6.30) coincides with the set of generalized solutions [34] defined as the set of continuous functions $x(t,\lambda)$ satisfying the inclusion

$$x(t,\lambda) - x(\tau,\lambda) \in \int_\tau^t F(s,x(s,\lambda),\lambda)ds$$

for all t and τ.

Hence, passing to the limit as $\lambda \to \lambda_0$, we get

$$y(t) - y(\tau) \in \int_\tau^t F(s,y(s),\lambda_0)ds,$$

i.e., $y(t)$ is a generalized solution of (6.32) [34]. Therefore, it is also an ordinary solution of this inclusion.

Thus, we proved that the limit of any uniformly convergent sequence of solutions of (6.30) is a solution of inclusion (6.32).

We now show that, for any η, there exists a neighborhood $U(\lambda_0)$ such that, for any solution $x(t,\lambda)$, $\lambda \in U(\lambda_0)$ of inclusion (6.30) satisfying the initial condition $x(0,\lambda) = x_0$, one can find a solution $x(t,\lambda_0)$ of inclusion (6.32) such that

$$\|x(t,\lambda) - x(t,\lambda_0)\| < \eta, \quad 0 \le t \le T.$$

Assume the contrary. Then there exist η_0 and a sequence of solutions $x(t,\lambda_k)$, $\lambda_k \in U(\lambda_0)$, $\lambda_k \to \lambda_0$, $k \to \infty$, of inclusion (6.30) such that

$$\max_{t\in[0,T]} \|x(t,\lambda_k) - x(t,\lambda_0)\| \ge \eta_0 \tag{6.34}$$

for all solutions $x(t,\lambda_0)$ of inclusion (6.32).

The family $x(t,\lambda_k)$ is uniformly bounded and equicontinuous. Hence, by the Arzelà theorem, it contains a uniformly convergent subsequence. As shown above, the limit of this subsequence must be a solution of (6.32), which contradicts (6.34). The theorem is thus proved. □

Remark 1. If $x(t,\lambda_0)$ is a solution of the differential inclusion (6.32), then it is not always possible to construct a sequence of solutions of inclusion (6.30) convergent to $x(t,\lambda_0)$ as $\lambda \to \lambda_0$.

Example 1. Consider an inclusion

$$\dot{x} \in [1,2](\sqrt{x} + \lambda^2), \quad x(0,\lambda) = 0. \tag{6.35}$$

Then, for $\lambda_0 = 0$, inclusion (6.32) takes the form

$$\dot{x} \in [1,2]\sqrt{x}, \quad x(0,0) = 0. \tag{6.36}$$

It is clear that the solutions $x(t,\lambda)$ of inclusion (6.35) converge to solutions of (6.36) that belong to the integral funnel $[t^2/4,\, t^2]$. At the same time, there is no sequence $x(t,\lambda)$ convergent to the trivial solution of inclusion (6.36).

Remark 2. If inclusion (6.32) possesses a unique solution, then any sequence of solutions $x(t,\lambda)$ of inclusion (6.30) converges to this solution as $\lambda \to \lambda_0$. This statement is similar to Theorem 1 in [34].

We now consider the proof of an analog of the Krasnosel'skii–Krein theorem for differential inclusions in terms of R-solutions.

Theorem 9 ([122]). *Assume that the differential inclusion* (6.30) *satisfies conditions* (a) *and* (b) *of Theorem* 8 *and, in addition,*

(c) *the R-solutions $R(t, \lambda_0)$ of inclusion* (6.32) *satisfying the initial condition*

$$R(0, \lambda_0) = x_0 \in D^1 \subset D$$

are defined for $0 \le t \le T$ *and, together with a certain ρ-neighborhood, lie in the domain D.*

Then, for any $\eta > 0$, there exists a neighborhood $U(\lambda_0)$ of the point λ_0 such that, for $\lambda \in U(\lambda_0)$ and any R-solution $R(t, \lambda)$ of inclusion (6.30) *defined for* $0 \le t \le T$ *and satisfying the initial condition $R(0, \lambda) = x_0$, there exists an R-solution $R(t, \lambda_0)$ of inclusion* (6.32) *for which*

$$h(R(t, \lambda), R(t, \lambda_0)) < \eta, \quad 0 \le t \le T.$$

Proof. Let $R(t, \lambda)$ ($\lambda \to \lambda_0$, $\lambda \in \Lambda$) be a uniformly convergent sequence of R-solutions of (6.30) satisfying the initial condition $R(0, \lambda) = x_0$. The limit of this sequence is a continuous function denoted by $R(t) \in \text{comp}(\mathbb{R}^n)$, i.e.,

$$\lim_{\lambda \to \lambda_0} \max_{t \in [0,T]} h(R(t, \lambda), R(t)) = 0. \tag{6.37}$$

By virtue of the definition of an R-solution the following relation is true everywhere on $[0, T]$:

$$h\Big(R(t + \Delta, \lambda), \bigcup_{x \in R(t,\lambda)} \Big\{x + \int_t^{t+\Delta} F(s, x, \lambda) ds\Big\}\Big) = 0(\Delta), \quad \Delta \downarrow 0. \tag{6.38}$$

In this relation, we pass to the limit as $\lambda \to \lambda_0$. To this end, we show that

$$\lim_{\lambda \to \lambda_0} h\Big(\bigcup_{x \in R(t,\lambda)} \Big\{x + \int_t^{t+\Delta} F(s, x, \lambda) ds\Big\}, \bigcup_{x \in R(t)} \Big\{x + \int_t^{t+\Delta} F(s, x, \lambda_0) ds\Big\}\Big) = 0.$$

We fix arbitrary $\varepsilon > 0$. By using conditions (a) and (b) of the theorem and relation (6.37), we choose $U(\lambda_0)$ such that

$$h\Big(\int_t^{t+\Delta} F(s, x, \lambda) ds, \int_t^{t+\Delta} F(s, x, \lambda_0) ds\Big) < \frac{\varepsilon}{3}$$

and $h(R(t, \lambda), R(t)) < \delta$, where δ is determined from the condition of uniform continuity of the function F, namely,

$$h(F(s, x, \lambda), F(s, y, \lambda)) < \frac{\varepsilon}{3\Delta}$$

for $\|x - y\| < \delta < \frac{\varepsilon}{3}$.

Then

$$h\Big(\bigcup_{x\in R(t,\lambda)}\Big\{x+\int_t^{t+\Delta}F(s,x,\lambda)ds\Big\},\bigcup_{x\in R(t)}\Big\{x+\int_t^{t+\Delta}F(s,x,\lambda_0)ds\Big\}\Big)$$

$$\le h\Big(\bigcup_{x\in R(t,\lambda)}\Big\{x+\int_t^{t+\Delta}F(s,x,\lambda)ds\Big\},\bigcup_{x\in R(t)}\Big\{x+\int_t^{t+\Delta}F(s,x,\lambda)ds\Big\}\Big)$$

$$+h\Big(\bigcup_{x\in R(t)}\Big\{x+\int_t^{t+\Delta}F(s,x,\lambda)ds\Big\},\bigcup_{x\in R(t)}\Big\{x+\int_t^{t+\Delta}F(s,x,\lambda_0)ds\Big\}\Big)$$

$$\le h(R(t,\lambda),R(t))+\max_{\substack{x\in R(t,\lambda),y\in R(t),\\ \|x-y\|<\delta}}\int_t^{t+\Delta}h(F(s,x,\lambda),F(s,y,\lambda))ds$$

$$+\max_{x\in R(t)}h\Big(\int_t^{t+\Delta}F(s,x,\lambda)ds,\int_t^{t+\Delta}F(s,x,\lambda_0)ds\Big)<\varepsilon.$$

Hence, passing to the limit as $\lambda\to\lambda_0$ in (6.38), we get

$$h\Big(R(t+\Delta),\bigcup_{x\in R(t)}\Big\{x+\int_t^{t+\Delta}F(s,x,\lambda_0)\Big\}ds\Big)=o(\Delta),\quad \Delta\downarrow 0.$$

Therefore, $R(t)$ is an R-solution of inclusion (6.32).

The final part of the proof is similar to the final part of the proof of Theorem 8 and uses the properties of uniform boundedness and equicontinuity of the set of R-solutions of inclusion (6.30), which follow from condition (a) of the theorem. □

Remark 3. If $R(t,\lambda_0)$ is an R-solution of the differential inclusion (6.32), then it is not always possible to construct a sequence of R-solutions of (6.30) that converges to $R(t,\lambda_0)$ as $\lambda\to\lambda_0$.

Thus, for inclusion (6.35), the sequences of R-solutions $R(t,\lambda)$ converge to the R-solution $R(t,0)=[\frac{t^2}{4},t^2]$ but there is no sequence $R(t,\lambda)$ convergent, e.g., to the R-solution $R_1(t,0)=[0,t^2]$.

Remark 4. If inclusion (6.32) possesses a unique R-solution, then any sequence of R-solutions $R(t,\lambda)$ of inclusion (6.30) converges to this R-solution as $\lambda\to\lambda_0$.

Remark 5. If inclusion (6.32) possesses a unique R-solution, then, for any ordinary solution of inclusion (6.32), there exists a sequence of ordinary solutions of (6.30) that converges to this solution as $\lambda\to\lambda_0$.

Example 2. Let

$$\dot{x}\in[1,2](1-e^{-\frac{t}{\lambda}})x+(1-e^{-\frac{t}{\lambda}}),\quad x(0,\lambda)=0. \tag{6.39}$$

Then the R-solutions have the form

$$R(t,\lambda) = \left[e^{\tau} - 1, \frac{1}{2}(e^{2\tau} - 1)\right], \qquad \tau = t + \lambda(e^{-\frac{t}{\lambda}} - 1)$$

and converge as $\lambda \downarrow 0$ to the unique R-solution

$$R(t,0) = \left[e^{t} - 1, \frac{1}{2}(e^{2t} - 1)\right]$$

of the inclusion

$$\dot{x} \in [1,2]x + 1, \quad x(0,0) = 0. \tag{6.40}$$

Moreover, despite the nonuniqueness of an ordinary solution of inclusion (6.40), for any solution $x(t,0)$, there exists a sequence of solutions $x(t,\lambda)$ of inclusion (6.39) that converges to this solution as $\lambda \downarrow 0$.

Just as in the case of ordinary differential equations, the Bogolyubov theorem on substantiation of the method of averaging in a finite interval for differential inclusions is directly connected with the theorems on continuous dependence of the solutions of differential inclusions with right-hand sides integrally continuous as functions of the parameter.

We now formulate the Bogolyubov theorems for differential inclusions corresponding to Theorems 8 and 9.

Theorem 10. *Assume that, in a domain $Q = \{t \geq 0,\ x \in D \subset \mathbb{R}^n$, D is a bounded domain$\}$, the differential inclusion*

$$\dot{x} \in \varepsilon F(t,x) \tag{6.41}$$

satisfies the following conditions:

(a) *the set-valued mapping $F\colon \mathbb{R}_+ \times \mathbb{R}^n \to \operatorname{conv}(\mathbb{R}^n)$ is uniformly bounded, continuous in t, and uniformly continuous in x uniformly with respect to t;*

(b) *for all $x \in D$, the following limit exists:*

$$\lim_{T\to\infty} \frac{1}{T}\int_0^T F(s,x)ds = F_0(x);$$

(c) *the solutions $y(\tau)$ of the inclusion*

$$\frac{dy}{d\tau} \in F_0(y), \quad y(0) = x_0 \in D^1 \subset D, \quad \tau = \varepsilon t, \tag{6.42}$$

are defined for $0 \leq \tau \leq L$ and lie in the domain D together with a ρ-neighborhood.

Then, for any $\eta > 0$, *there exists* $\varepsilon_0 > 0$ *such that, for* $0 < \varepsilon \le \varepsilon_0$ *and any solution* $x(t, \varepsilon)$ *of inclusion* (6.41) *satisfying the condition* $x(0, \varepsilon) = x_0$, *there exists a solution of inclusion* (6.42) *for which the inequality*

$$\|x(t, \varepsilon) - y(\varepsilon t)\| < \eta$$

holds on the segment $[0, L\varepsilon^{-1}]$.

Theorem 11. *Assume that the differential inclusion* (6.41) *satisfies conditions* (a) *and* (b) *of Theorem* 10 *and, in addition,*

(c) *the R-solutions* $Y(\tau)$ *of inclusion* (6.42), $Y(0) = x_0 \in D^1 \subset D$, *are defined for* $0 \le \tau \le L$ *and belong to the domain D together with a certain* ρ*-neighborhood.*

Then, for any $\eta > 0$, *there exists* $\varepsilon_0 > 0$ *such that, for* $0 < \varepsilon \le \varepsilon_0$ *and any R-solution* $X(t, \varepsilon)$, $X(0, \varepsilon) = x_0$, *of inclusion* (6.41), *there exists an R-solution of inclusion* (6.42) *for which the inequality*

$$h(X(t, \varepsilon), Y(\varepsilon t)) < \eta$$

holds on the segment $[0, L\varepsilon^{-1}]$.

We now substantiate the method of averaging in the form of Bogolyubov's theorem for impulsive differential inclusions

$$\begin{gathered} \dot{x} \in \varepsilon X(t, x), \quad t \neq \tau_i, \\ \Delta x|_{t=\tau_i} \in \varepsilon I_i(x), \quad x(0) = x_0, \end{gathered} \tag{6.43}$$

where $t \ge 0$ is time, $x \in \mathbb{R}^n$ is the phase vector, $\varepsilon > 0$ is a small parameter, $X\colon \mathbb{R}_+ \times \mathbb{R}^n \to \operatorname{conv}(\mathbb{R}^n)$ and $I_i\colon \mathbb{R}^n \to \operatorname{conv}(\mathbb{R}^n)$ are set-valued mappings, and τ_i are the times of pulses such that $0 \le \tau_i < \tau_{i+1}$.

If the limit exists

$$\lim_{T\to\infty} h\biggl(\frac{1}{T}\int_t^{t+T} X(s, x)ds + \frac{1}{T}\sum_{t\le\tau_i<t+T} I_i(x), \overline{X}(x)\biggr) = 0, \tag{6.44}$$

then inclusion (6.43) is associated with the inclusion

$$\dot{y} \in \varepsilon\overline{X}(y), \quad y(0) = x_0. \tag{6.45}$$

Inclusion (6.45) is called the averaged inclusion.

The proposed averaging scheme was substantiated in [112] and [114] under the assumption that the functions $X(t, x)$ and $I_i(x)$ satisfy the Lipschitz condition with respect to x. In what follows, we study the problem of closeness of solutions of systems (6.43) and (6.45) under weaker conditions.

Theorem 12 ([123]). *Assume that the following conditions are satisfied in the domain* $Q\{t \geq 0, x \in D \subset \mathbb{R}^n\}$*:*

(a) *the set-valued mappings* $X(t,x)$ *and* $I_i(x)$ *are uniformly bounded by a constant* M*,* $X(t,x)$ *is continuous in* t *and uniformly continuous in* x *uniformly with respect to* t*, and* $I_i(x)$ *are equicontinuous;*

(b) *the limit of* (6.44) *exists uniformly in* $x \in D$ *and* $t \geq 0$ *and, in addition,*

$$\frac{1}{T} i(t, t+T) \leq d < \infty,$$

where $i(t, t+T)$ *is the number of points of the sequence* $\{\tau_i\}$ *in the interval* $(t, t+T]$*;*

(c) *the solutions of inclusion* (6.45) *with* $\varepsilon = 1$ *are defined for* $t \in [0, L^*]$ *and lie in the domain* D *together with a certain* ρ*-neighborhood;*

(d) *the modulus of continuity of the set-valued mapping* $\overline{X}(x)$ *is the Kamke function* [23].

Then, for any $\eta > 0$ *and any* $L \in (0, L^*]$*, there exists* $\varepsilon_0 = \varepsilon_0(\eta, L) > 0$ *such that, for* $0 < \varepsilon \leq \varepsilon_0$ *and any solution* $x(t,\varepsilon)$ *of inclusion* (6.43) *satisfying the condition* $x(0,\varepsilon) = x_0$*, one can find a solution of inclusion* (6.45) *for which the inequality*

$$\|x(t,\varepsilon) - y(t,\varepsilon)\| < \eta$$

is true on the segment $[0, L\varepsilon^{-1}]$.

Proof. We fix $L \in (0, L^*]$. Let $x(t) = x(t,\varepsilon)$ be an arbitrary solution of inclusion (6.43). We split the segment $[0, \frac{L}{\varepsilon}]$ into subsegments by points $t_k = kh(\varepsilon)$, $k = \overline{0,m}$, where $m = [\frac{L}{\varepsilon h(\varepsilon)}]$ and $t_{m+1} = \frac{L}{\varepsilon}$. Further, we choose steps $h(\varepsilon)$ such that $h(\varepsilon) \to \infty$ and $\varepsilon h(\varepsilon) \to 0$ as $\varepsilon \to 0$. Since $x(t)$ is a solution of inclusion (6.43), there exist a measurable selector $v(t) \in X(t, x(t))$ [23] and pulse vectors $p_i \in I_i(x(\tau_i))$ such that

$$x(t) = x(t_k) + \varepsilon \int_{t_k}^{t} v(s)ds + \varepsilon \sum_{t_k \leq \tau_i < t} p_i, \quad x(0) = x_0, \tag{6.46}$$

$$t \in [t_k, t_{k+1}], \quad k = \overline{0,m}.$$

For each $k = \overline{0,m}$, for $t \in [t_k, t_{k+1}]$, we define an Euler quasibroken line as follows:

$$\begin{aligned} x_1(t) &= x_1(t_k) + \varepsilon \int_{t_k}^{t} w(s)ds \\ &\quad + \varepsilon \sum_{t_k \leq \tau_i < t} r_i + \varepsilon \alpha_k (t - t_k), \quad x_1(0) = x_0, \end{aligned} \tag{6.47}$$

where

$$w(t) \in X(t, x_1(t_k)) \colon \|w(t) - v(t)\| = \min_{u \in X(t, x_1(t_k))} \|u - v(t)\|,$$
$$r_i \in I_i(x_1(t_k)) \colon \|r_i - p_i\| = \min_{r \in I_i(x_1(t_k))} \|r - p_i\|,$$

$w(t)$ is a measurable function [23], and $\alpha_k \in \mathbb{R}^n$ are constants chosen from the condition

$$x(t_{k+1}) = x_1(t_{k+1}), \quad k = \overline{0, m}.$$

We now show that these α_k exist. Indeed, by using equalities (6.46) and (6.47), for $t = t_{k+1}$, we get

$$\varepsilon \alpha_k (t_{k+1} - t_k) = \varepsilon \int_{t_k}^{t_{k+1}} (v(s) - w(s)) ds + \varepsilon \sum_{t_k \le \tau_i < t_{k+1}} (p_i - r_i).$$

Therefore,

$$\begin{aligned} \varepsilon \|\alpha_k\| (t_{k+1} - t_k) &\le \varepsilon \int_{t_k}^{t_{k+1}} \|v(s) - w(s)\| ds + \varepsilon \sum_{t_k \le \tau_i < t_{k+1}} \|p_i - r_i\| \\ &\le \varepsilon \int_{t_k}^{t_{k+1}} h(X(s, x(s)), X(s, x(t_k))) ds \\ &\quad + \varepsilon \sum_{t_k \le \tau_i < t_{k+1}} h(I_i(x(\tau_i)), I_i(x(t_k))). \end{aligned}$$

In view of the uniform continuity of the function $X(t, x)$ and equicontinuity of the functions $I_i(x)$, for any $\eta_1 > 0$, one can find $\delta > 0$ such that the inequalities $h(X(t, x), X(t, y)) < \eta_1$ and $h(I_i(x), I_i(y)) < \eta_1$ are true for $\|x - y\| < \delta$.

We have

$$\|x(t) - x(t_k)\| \le \varepsilon M h(\varepsilon) + \varepsilon d M h(\varepsilon) = M \varepsilon h(\varepsilon)(1 + d).$$

Further, we choose ε_0 such that the inequality $M\varepsilon h(\varepsilon)(1 + d) < \delta$ is true for $\varepsilon \le \varepsilon_0$. This yields

$$\begin{aligned} \varepsilon \|\alpha_k\| (t_{k+1} - t_k) &\le \varepsilon \eta_1 (t_{k+1} - t_k) + \varepsilon d \eta_1 (t_{k+1} - t_k) \\ &= \eta_1 \varepsilon (1 + d)(t_{k+1} - t_k). \end{aligned}$$

Then $\|\alpha_k\| \le \eta_1 (1 + d)$.

We now show that, for any $\eta > 0$, there exists ε_0 such that the estimate $\|x(t) - x_1(t)\| < \frac{\eta}{3}$ is true for $\varepsilon \le \varepsilon_0$ and $t \in [0, \frac{L}{\varepsilon}]$.

For $t \in [t_k, t_{k+1}]$, we find

$$\begin{aligned}\|x(t) - x_1(t)\| &\le \|x(t) - x(t_k)\| + \|x(t_k) - x_1(t_k)\| \\ &\quad + \|x_1(t_k) - x_1(t)\| \le (M\varepsilon h(\varepsilon) + dM\varepsilon h(\varepsilon)) \\ &\quad + (M\varepsilon h(\varepsilon) + dM\varepsilon h(\varepsilon) + \|\alpha_k\|\varepsilon h(\varepsilon)) \\ &= (1+d)(2M + \eta_1)\varepsilon h(\varepsilon).\end{aligned}$$

We choose ε_0 such that the inequality

$$(1+d)(2M + \eta_1)\varepsilon h(\varepsilon) < \frac{\eta}{3}$$

is true for $\varepsilon \le \varepsilon_0$.

For each $k = \overline{0, m}$, we define

$$y_1(t) = y_1(t_k) + \varepsilon \int_{t_k}^{t} z(s)ds + \varepsilon\alpha_k(t - t_k),$$
$$y_1(0) = x_0, \quad t \in [t_k, t_{k+1}],$$

where $z(s) \equiv z_k$ for $t \in [t_k, t_{k+1}]$ and

$$\left\| \int_{t_k}^{t_{k+1}} w(s)ds + \varepsilon \sum_{t_k \le \tau_i < t_{k+1}} r_i - z_k(t_{k+1} - t_k) \right\|$$
$$= \min_{z \in \overline{X}(x_1(t_k))} \left\| \int_{t_k}^{t_{k+1}} w(s)ds + \varepsilon \sum_{t_k \le \tau_i < t_{k+1}} r_i - z(t_{k+1} - t_k) \right\|.$$

It is necessary to show that, for any $\eta > 0$, there exists ε_0 such that the estimate $\|y_1(t) - x_1(t)\| < \frac{\eta}{3}$ is true for $\varepsilon \le \varepsilon_0$ and $t \in [0, \frac{L}{\varepsilon}]$.

Denote $\delta_k = \|y_1(t_k) - x_1(t_k)\|$. In view of the uniform convergence of (6.44), there exists a monotonically decreasing function $F(t) \to 0$ as $t \to \infty$ such that

$$\begin{aligned}\delta_{k+1} &\le \delta_k + \varepsilon \left\| \int_{t_k}^{t_{k+1}} w(s)ds + \varepsilon \sum_{t_k \le \tau_i < t_{k+1}} r_i - \int_{t_k}^{t_{k+1}} z(s)ds \right\| \\ &\le \delta_k + \varepsilon h(\varepsilon) h\Bigg(\frac{1}{h(\varepsilon)} \int_{t_k}^{t_{k+1}} X(s, x_1(t_k))ds \\ &\qquad + \frac{1}{h(\varepsilon)} \sum_{t_k \le \tau_i < t_{k+1}} I_i(x_1(t_k)), \overline{X}(x_1(t_k)) \Bigg) \\ &\le \delta_k + \varepsilon h(\varepsilon) F(h(\varepsilon)).\end{aligned}$$

Thus, we get the following sequence of estimates:

$$\delta_0 = 0, \quad \ldots, \quad \delta_{k+1} \le \delta_k + \varepsilon h(\varepsilon) F(h(\varepsilon)),$$

whence it follows that

$$\delta_{k+1} \le \sum_{i=0}^{k+1} \varepsilon h(\varepsilon) F(h(\varepsilon)) \le LF(h(\varepsilon)).$$

Hence, for $t \in [t_k, t_{k+1}]$, we obtain

$$\begin{aligned}\|y_1(t) - x_1(t)\| &\le \|y_1(t) - y_1(t_k)\| + \|y_1(t_k) - x_1(t_k)\| + \|x_1(t_k) - x_1(t)\| \\ &\le 2[M\varepsilon h(\varepsilon) + \|\alpha_k\|\varepsilon h(\varepsilon)] + Md\varepsilon h(\varepsilon) + LF(h(\varepsilon)) \\ &= \varepsilon h(\varepsilon)\big[M(2+d) + 2\eta_1(1+d)\big] + LF(h(\varepsilon)).\end{aligned}$$

We choose ε_0 such that the estimate

$$\varepsilon h(\varepsilon)[M(2+d) + 2\eta_1(1+d)] + LF(h(\varepsilon)) < \frac{\eta}{3}$$

is true for $\varepsilon \le \varepsilon_0$.

For $t \in [t_k, t_{k+1}]$, we get

$$\begin{aligned}\|x_1(t_k) - y_1(t)\| &\le \|y_1(t) - y_1(t_k)\| + \|y_1(t_k) - x_1(t_k)\| \\ &\le \varepsilon h(\varepsilon)[M + \eta_1(1+d)] + LF(h(\varepsilon)) \equiv \gamma(\varepsilon).\end{aligned}$$

Then

$$\begin{aligned}\rho(\dot y_1(t), \varepsilon\overline{X}(y_1(t))) &\le \rho(\dot y_1(t), \varepsilon\overline{X}(x_1(t_k))) + \varepsilon h(\overline{X}(x_1(t_k)), \overline{X}(y_1(t))) \\ &\le \varepsilon\|\alpha_k\| + \varepsilon\omega(\gamma(\varepsilon)) \le \varepsilon\eta_1(1+d) + \varepsilon\omega(\gamma(\varepsilon)),\end{aligned}$$

where ω is the modulus of continuity of the function $\overline{X}(x)$.

By the Plis theorem [105, 106], there exists a solution $y(t) = y(t, \varepsilon)$ of inclusion (6.45) such that

$$\|y_1(t) - y(t)\| \le r(\varepsilon t),$$

where $r(\tau)$ is the upper solution of the Cauchy problem

$$\frac{dr}{d\tau} = \omega(r) + \beta(\varepsilon), \quad r(0) = 0,\ 0 \le \tau \le L, \tag{6.48}$$

$\beta(\varepsilon) = \eta_1(1+d) + \omega(\gamma(\varepsilon))$.

Since $\beta(\varepsilon)$ can be made arbitrarily small and the equation

$$\dot r = \omega(r), \quad r(0) = 0$$

possesses solely the trivial solution, by the Hartman theorem [57], we conclude that all solutions of (6.48) uniformly converge to the solution of the limit problem, i.e.,

$r(\tau)$ can be made smaller than $\eta/3$ by the choice of ε_0. Thus, for any solution $x(t)$, there exists a solution $y(t)$ such that the estimate

$$\|x(t) - y(t)\| \leq \|x(t) - x_1(t)\| + \|x_1(t) - y_1(t)\| + \|y_1(t) - y(t)\| < \eta$$

holds for $\varepsilon \leq \varepsilon_0$.

The theorem is proved. □

Remark 6. Let $X(t, \varepsilon)$ be the set of solutions of the impulsive differential inclusion (6.43) and let $X(t, 0)$ be the set of solutions of the differential inclusion (6.45). Then the theorem establishes the upper semicontinuity (by inclusion) of the set-valued mapping $X(t, \varepsilon)$ at the point $\varepsilon = 0$.

6.3 Averaging of Inclusions with Pulses at Nonfixed Times

Consider differential inclusions with pulse actions

$$\dot{x} \in F^1(t, x, \varepsilon), \quad x(0) = x_0,\ t \neq \varepsilon\tau_i^1(x),\ t \neq \sigma_i^1(x), \tag{6.49}$$

$$\Delta x|_{t=\varepsilon\tau_i^1(x)} \in \varepsilon I_i^1(x), \tag{6.50}$$

$$\Delta x|_{t=\sigma_p^1(x)} \in K_p^1(x). \tag{6.51}$$

Inclusion (6.49)–(6.51) is associated with the following differential inclusion:

$$\dot{y} \in F^2(t, y, \varepsilon), \quad y(0) = x_0,\ t \neq \varepsilon\tau_i^2(x),\ t \neq \sigma_i^2(y), \tag{6.52}$$

$$\Delta y|_{t=\varepsilon\tau_i^2(y)} \in \varepsilon I_i^2(y), \tag{6.53}$$

$$\Delta y|_{t=\sigma_p^2(y)} \in K_p^2(y), \tag{6.54}$$

where $t \in [0, L]$ is time, $x \in D \subset \mathbb{R}^n$ is the phase vector, ε is a small parameter, $\tau_i^j, \sigma_i^j : D \to \mathbb{R}$ are pulse surfaces, and $F^j : [0, L] \times D \times \mathbb{R}_+ \to \operatorname{conv}(\mathbb{R}^n)$ and I_i^j, $K_p^j : D \to \operatorname{conv}(\mathbb{R}^n)$, $i = \overline{1, k}$, $p = \overline{1, r}$, $j = 1, 2$, are set-valued mappings.

First, we consider the differential inclusions (6.49) and (6.52) solely with asymptotically small pulse actions (6.50) and (6.53).

Assume that

$$\lim_{\varepsilon \to 0} h\Bigg(\frac{1}{\Delta}\Bigg(\int_t^{t+\Delta} F^1(t, x, \varepsilon)\, dt + \varepsilon \sum_{t < \varepsilon\tau_i(x) < t+\Delta} I_i^1(x)\Bigg),$$
$$\frac{1}{\Delta}\Bigg(\int_t^{t+\Delta} F^2(t, x, \varepsilon)\, dt + \varepsilon \sum_{t < \varepsilon\tau_i^2(x) < t+\Delta} I_i^2(x)\Bigg)\Bigg) = 0. \tag{6.55}$$

Let $J_j(t, t + \Delta)$, $j = 1, 2$, be the number of asymptotically stable pulses of the solutions of inclusions (6.49), (6.50) and (6.52), (6.53), respectively, in the interval $(t, t + \Delta]$ $(0 \leq t,\ t + \Delta \leq L)$.

Theorem 13. *Assume that the following conditions are satisfied in the domain* $Q\{t \in [0, L],\ x \in D\}$*:*

(1) *the set-valued mappings* $F^j(t, x, \varepsilon)$ *and* $I_i^j(x)$ *and the functions* $\tau_i^j(x)$ *satisfy the Lipschitz condition with respect to* x *with constant* λ*, they are continuous in* t*, and, in addition,* $F^j(t, x, \varepsilon) \subset T_D(x)$ *and* $x + I_i^j(x) \subset D$*;*

(2) *the set-valued mappings* $F^j(t, x, \varepsilon)$ *and* $I_i^j(x)$ *are uniformly bounded by a constant* M*;*

(3) *limit* (6.55) *exists uniformly in* $(t, x) \in Q$*;*

(4) *the numbers* $J_j(t, t + \Delta)$*,* $j = 1, 2$*, satisfy the inequalities*

$$\frac{1}{\Delta} J_j(t, t + \Delta) \leq \frac{A}{\varepsilon} < \infty;$$

(5) *the surfaces* $t = \varepsilon\tau_i^j(x)$ *do not intersect and the inequalities*

$$\tau_i^j(x) \geq \tau_i^j(x + z), \quad |\tau_{i+1}^j(x) - \tau_i^j(x)| \leq M$$

hold for any $x \in D$ *and* $z \in I_i^j$*,* $j = 1, 2$*.*

Then, for any $\eta > 0$*, there exists* $\varepsilon_0(\eta) \in (0, \varepsilon_1)$ *such that the following statements are true:*

(1) *if* $y(t) = y(t, \varepsilon)$ *is a solution of* (6.52), (6.53) *with* $\varepsilon \in (0, \varepsilon_0)$*, then there exists a solution* $x(t) = x(t, \varepsilon)$*,* $x(0) = y(0)$ *of inclusion* (6.49), (6.50) *such that*

$$\|x(t) - y(t)\| \leq \eta, \quad t \in [0, L] \setminus \bigcup_i [t_i^2 - \Delta_i, t_i^2 + \Delta_i], \tag{6.56}$$

where $t_i^2 = \tau_i(y((t_i^2))$ *and, in addition,* $\sum_i \Delta_i < C\eta$*;*

(2) *if* $x(t) = x(t, \varepsilon)$ *is a solution of* (6.49), (6.50) *with* $\varepsilon \in (0, \varepsilon_0)$*, then there exists a solution* $y(t) = y(t, \varepsilon)$*,* $x(0) = y(0)$ *of inclusion* (6.52), (6.53) *such that inequality* (6.56) *is true.*

Proof. Let $y(t)$ be an arbitrary solution of inclusion (6.52), (6.53). It follows from conditions (1) and (2) of the theorem that there exist solutions of inclusions (6.49), (6.50) and (6.52), (6.53) that can be extended to $[0, L]$ and all these solutions belong to the domain D.

We now show that, for sufficiently small $\varepsilon \leq \varepsilon_1$, every solution $x(t)$ and $y(t)$ crosses the surfaces $t = \tau_i^j(x)$ exactly once. Assume the contrary, i.e., that the solution $y^j(t)$ of inclusion (6.49), (6.50) or, respectively, of (6.52), (6.53) suffers beating upon the surface $t = \tau_i^j(x)$.

Let $t_0 = \varepsilon\tau_i^j(y^j(t_0))$ and let the solution $y^j(t)$ with the initial condition

$$y^j(t_0) + z \in y^j(t_0) + \varepsilon I_i^j(y^j(t_0))$$

cross the same surface at time t^*, i.e., $t^* = \varepsilon\tau_i^j(y^j(t^*))$. Further, let (t_0, t^*) be the interval of continuity of the function $y^j(t)$. Then we get

$$y^j(t^*) = y^j(t_0) + z + \int_{t_0}^{t^*} u(\tau)\,d\tau, \quad u(t) \in F^j(t, y^j(t), \varepsilon).$$

We choose

$$0 < \varepsilon < \varepsilon_1 < \frac{1}{\lambda M}.$$

By using the Lipschitz condition, we find

$$\begin{aligned} t^* - t_0 &= \varepsilon\tau_i^j(y^j(t^*)) - \varepsilon\tau_i^j(y^j(t_0)) \\ &= \varepsilon\tau_i^j(y^j(t^*)) - \varepsilon\tau_i^j(y^j(t_0) + z) + \varepsilon\tau_i^j(y^j(t_0) + z) - \varepsilon\tau_i^j(y^j(t_0)) \\ &\le \varepsilon\lambda \int_{t_0}^{t^*} u(\tau)\,d\tau + \varepsilon\tau_i^j(y^j(t_0) + z) - \varepsilon\tau_i^j(y^j(t_0)) \\ &\le \varepsilon\lambda M(t^* - t_0) + \varepsilon\tau_i^j(y^j(t_0) + z) - \varepsilon\tau_i^j(y^j(t_0)). \end{aligned}$$

Thus,

$$(1 - \varepsilon\lambda M)(t^* - t_0) \le \varepsilon(\tau_i^j(y^j(t_0) + z) - \tau_i^j(y^j(t_0))),$$

which contradicts condition (5) of the theorem.

We now suppose that the solution $y^j(t)$ crosses another surface $t = \varepsilon\tau_k^j(x)$ at time $t_* \in (t_0, t^*)$. In this case, we get

$$\begin{aligned} &t_0 = \varepsilon\tau_i^j(y^j(t_0)) > \varepsilon\tau_i^j(y^j(t)), \quad t_0 < t \le t_*; \\ &t_* = \varepsilon\tau_k^j(y^j(t_*)) > \varepsilon\tau_k^j(y^j(t)), \quad t_* < t \le t^*; \\ &t^* = \varepsilon\tau_i^j(y^j(t^*)). \end{aligned}$$

For any arbitrarily chosen continuous function $z(t)$ such that $z(t_*) = y^j(t_*)$ and $z(t^*) = y^j(t^*)$, there exists $\bar{t} \in (t_*, t^*]$ for which $\tau_i^j(z(\bar{t})) = \tau_k^j(z(\bar{t}))$. The last equality contradicts the condition that the surfaces $t = \varepsilon\tau_i^j(x)$ and $t = \varepsilon\tau_k^j(x)$ are disjoint. Hence, each solution crosses each surface at most once.

We split the segment $[0, L]$ into subsegments by points $t_k = k\frac{L}{m}$, $k = \overline{0, m}$. Since $y(t)$ is a solution of (6.52), (6.53), there exist a measurable selector $v(t) \in F^2(t, y(t), \varepsilon)$ and pulse vectors $p_i \in I_i^2(y(\varepsilon\tau_i^2))$ such that

$$\begin{aligned} &y(t) = y(t_k) + \int_{t_k}^{t} v(\tau)\,d\tau + \varepsilon \sum_{t_k \le \varepsilon\tau_i^2(y) < t} p_i, \\ &y(0) = x_0, \quad t \in (t_k, t_{k+1}]. \end{aligned}$$

For any $k = \overline{1, m}$, we define a function

$$y^1(t) = y^1(t_k) + \int_{t_k}^{t} \bar{v}(\tau)\, d\tau + \varepsilon \sum_{t_k \le \varepsilon\tau_i^2(y) < t} \bar{p}_i,$$

$$y^1(0) = x_0, \quad t \in (t_k, t_{k+1}],$$

where $\bar{v}(\tau) \in F^2(\tau, y^1(t_k), \varepsilon)$ and $\bar{p}_i \in I_i^2(y^1(t_k))$ are such that

$$\begin{aligned}
&\left\| \int_{t_k}^{t_{k+1}} \bar{v}(\tau) d\tau - \int_{t_k}^{t_{k+1}} v(\tau) d\tau \right\| \\
&\quad = \min_{z(\tau) \in F^2(\tau, y^1(t_k), \varepsilon)} \left\| \int_{t_k}^{t_{k+1}} z(\tau) d\tau - \int_{t_k}^{t_{k+1}} v(\tau) d\tau \right\|, \\
&\|\bar{p}_i - p_i\| = \min_{r \in I_i^2(y^1(t_k))} \|r - p_i\|.
\end{aligned}$$

Denote $\delta_k = \|y(t_k) - y^1(t_k)\|$. Thus, in view of condition (5) of the theorem, we get

$$\begin{aligned}
\|y(t) - y^1(t_k)\| &\le \|y(t) - y(t_k)\| + \|y(t_k) - y^1(t_k)\| \\
&\le \delta_k + M(1 + A)(t - t_k).
\end{aligned} \tag{6.57}$$

For $\delta_{k+1} = \|y(t_{k+1}) - y^1(t_{k+1})\|$, we can write

$$\begin{aligned}
\delta_{k+1} &\le \delta_k + \left\| \int_{t_k}^{t_{k+1}} \bar{v}(\tau)\, d\tau - \int_{t_k}^{t_{k+1}} v(\tau)\, d\tau \right\| + \varepsilon \left\| \sum_{t_k \le \varepsilon\tau_i^2(y) < t_{k+1}} (\bar{p}_i - p_i) \right\| \\
&\le \delta_k + \int_{t_k}^{t_{k+1}} h(F^2(\tau, y(\tau), \varepsilon), F^2(\tau, y^1(t_k), \varepsilon))\, d\tau \\
&\quad + \varepsilon \sum_{t_k \le \varepsilon\tau_i^2(y) < t_{k+1}} h(I_i^2(y^1(t_k)), I_i^2(y(\varepsilon\tau_i^2))) \\
&\le \delta_k + \int_{t_k}^{t_{k+1}} \lambda \|y(t) - y^1(t_k)\|\, dt + \varepsilon \sum_{t_k \le \varepsilon\tau_i^2(y) < t_{k+1}} \lambda \|y(\varepsilon\tau_i^2) - y^1(t_k)\| \\
&\le \delta_k + \lambda \left[\delta_k \frac{L}{m} + M(1 + A) \frac{L^2}{2m^2} \right] + A\lambda \left[\delta_k + M(1 + A) \frac{L}{m} \right] \frac{L}{m}.
\end{aligned}$$

Let $a = \lambda(1 + A)$ and let $b = \lambda M(1 + A)(\frac{1}{2} + A)$. Hence, we have

$$\delta_{k+1} \le \left(1 + a\frac{L}{m}\right) \delta_k + \frac{bL^2}{m^2}.$$

Thus, we get the following sequence of inequalities:

$$\delta_0 = \|y^1(0) - y(0)\| = 0,$$

$$\delta_{k+1} \le \left[\left(1 + a\frac{L}{m}\right)^k + \left(1 + a\frac{L}{m}\right)^{k-1} + \cdots + \left(1 + a\frac{L}{m}\right) + 1\right]\frac{bL^2}{m^2},$$

i.e.,

$$\delta_{k+1} \le \frac{(1 + a\frac{L}{m})^{k+1} - 1}{a\frac{L}{m}} \cdot \frac{bL^2}{m^2} \le \frac{b}{a}\frac{L}{m}\left[\left(1 + a\frac{L}{m}\right)^m - 1\right]$$

$$\le \frac{b}{a}\frac{L}{m}(e^{aL} - 1) = \frac{ML(\frac{1}{2} + A)}{m}(e^{\lambda(1+A)L} - 1). \tag{6.58}$$

It follows from (6.57) and (6.58) that

$$\begin{aligned} \|y(t) - y^1(t)\| &\le \|y(t) - y^1(t_k)\| + \|y^1(t_k) - y^1(t)\| \\ &\le \delta_k + 2M(1 + A)\frac{L}{m} \\ &\le \frac{ML(\frac{1}{2} + A)}{m}(e^{\lambda(1+A)L} - 1) + 2M(1 + A)\frac{L}{m} \\ &\le \frac{ML(1 + A)}{m}(e^{\lambda(1+A)L} + 1), \quad t \in [0, L]. \end{aligned} \tag{6.59}$$

By virtue of condition (3) of the theorem, for any $\eta_1 > 0$, there exist $\varepsilon(\eta_1) > 0$, $\bar{u}(\tau) \in F^1(t, y^1(t_k), \varepsilon)$, and $\bar{q}_i \in I_i^1(y^1(t_k))$ such that the following estimate is true for $0 < \varepsilon < \varepsilon(\eta_1)$:

$$\left\| \left[\int_{t_k}^t \bar{u}(\tau)\, d\tau + \varepsilon \sum_{t_k \le \varepsilon\tau_i^1(y) < t} \bar{q}_i \right] - \left[\int_{t_k}^t \bar{v}(\tau)\, d\tau + \varepsilon \sum_{t_i \le \varepsilon\tau_i^2(y) < t} \bar{p}_i \right] \right\| < \eta_1 \frac{L}{m}.$$

In each interval $(t_k, t_{k+1}]$, we define a function

$$x^1(t) = x^1(t_k) + \int_{t_k}^t \bar{u}(\tau)\, d\tau + \varepsilon \sum_{t_k \le \varepsilon\tau_i^1(y) < t} \bar{q}_i, \quad x^1(0) = x_0.$$

For $t \in (t_k, t_{k+1}]$, $k = \overline{0, m-1}$, we find

$$\|x^1(t) - y^1(t)\| \le \|x^1(t_k) - y^1(t_k)\| + \eta_1 \frac{L}{m} = \delta_k + \eta_1 \frac{L}{m}.$$

Since $\delta_0 = \|x^1(0) - y^1(0)\| = 0$, we have

$$\|x^1(t) - y^1(t)\| \le \eta_1 L, \quad t \in [0, L]. \tag{6.60}$$

We now show that there exists a solution

$$x(t) = x(t_k) + \int_{t_k}^{t} u(\tau)\,d\tau + \varepsilon \sum_{t_k \le \varepsilon\tau_i^1(x) < t} q_i,$$

$$x(0) = x_0, \quad t \in [t_k, t_{k+1}),$$

of inclusion (6.49), (6.50) sufficiently close to $x^1(t)$.

By using the Lipschitz condition, we get

$$\begin{aligned}\rho(\dot{x}^1(t), F^1(t, x^1(t), \varepsilon)) &\le h(F^1(t, x^1(t_k), \varepsilon), F^1(t, x^1(t), \varepsilon)) \\ &\le \lambda \|x^1(t_k) - x^1(t)\| \le \lambda M \frac{L}{m} + \lambda\varepsilon \Bigg\| \sum_{t_k \le \varepsilon\tau_i^1(y) < t} \bar{q}_i \Bigg\| \\ &\le \lambda M \frac{L}{m} + \lambda\varepsilon M \frac{AL}{\varepsilon m} = \frac{\lambda M(1 + A)L}{m} \equiv \gamma.\end{aligned}$$

Let $t \in (t_k, t_{k+1}]$ and let s_i^- and s_i^+ be times at which $x(t)$ and $x^1(t)$ reach the surfaces $t = \tau_i^1(x)$.

Denote

$$\theta_i^- = \|x^1(s_i^-) - x(s_i^-)\| \quad \text{and} \quad \theta_i^+ = \|x^1(s_i^+ + 0) - x(s_i^+ + 0)\|.$$

Consider the intervals $(s_{i-1}^+, s_i^-]$. By the Filippov theorem [23, 49], there exists a solution $x(t)$ of inclusion (6.52), (6.53) such that

$$\|x(t) - x^1(t)\| \le e^{\lambda(t - s_{i-1}^+)} \theta_{i-1}^+ + \int_{s_{i-1}^+}^{t} e^{\lambda(s_{i-1}^+ - s)} \gamma ds.$$

Therefore,

$$\theta_i^- \le e^{\lambda(s_i^- s_{i-1}^+)} \theta_{i-1}^+ + \frac{\gamma}{\lambda}(e^{\lambda(s_i^- - s_{i-1}^+)} - 1).$$

By using conditions (1) and (5) of the theorem, we estimate $s_i^- - s_{i-1}^+$. Without loss of generality, we can assume that $s_i^- = \tau_i^1(x^1(s_i^-))$ and $s_{i-1}^+ = \tau_{i-1}^1(x(s_{i-1}^+))$:

$$\begin{aligned}s_i^- - s_{i-1}^+ &= \varepsilon[\tau_i^1(x^1(s_i^-)) - \tau_{i-1}^1(x(s_{i-1}^+))] \\ &\le \varepsilon[\tau_i^1(x^1(s_i^-)) - \tau_{i-1}^1(x^1(s_i^-)) + \tau_{i-1}^1(x^1(s_i^-)) - \tau_{i-1}^1(x(s_{i-1}^+))] \\ &\le \varepsilon[M + \lambda(\theta_{i-1}^+ + M(s_i^- s_{i-1}^+))].\end{aligned}$$

Thus, we get

$$s_i^- - s_{i-1}^+ \le \frac{\varepsilon(M + \lambda\theta_{i-1}^+)}{1 - \varepsilon\lambda M}.$$

Since the right-hand sides of inclusions (6.49), (6.50), (6.52), and (6.53) are bounded, there exits a constant $K_1 > 0$ such that $\theta_{i-1}^+ \le K_1$. Hence,

$$s_i^- - s_{i-1}^+ \le K\varepsilon, \quad \text{where } K = \frac{M + \lambda K_1}{1 - \varepsilon\lambda M}.$$

This yields

$$\theta_i^- \le e^{\lambda(s_i^- s_{i-1}^+)}\theta_{i-1}^+ + \frac{\gamma}{\lambda}(e^{K\varepsilon} - 1). \tag{6.61}$$

We estimate the difference between $x^1(t)$ and $x(t)$ at the times of pulses. Without loss of generality, we can assume that s_i^- is the time of crossing the surface $t = \varepsilon\tau_i^1(x)$ by $x^1(t)$ and s_i^+ is the time of crossing this surface by $x(t)$.

Further, we get

$$\begin{aligned}
\|x^1(s_i^-) - x(s_i^+)\| &\le \|x^1(s_i^-) - x(s_i^-)\| + \|x(s_i^-) - x(s_i^+)\|\} \\
&\le \theta_i^- + M(s_i^+ - s_i^-), \\
s_i^+ - s_i^- &= \varepsilon[\tau_i^1(x(s_i^+)) - \tau_i^1(x^1(s_i^-))] \le \varepsilon\lambda(\theta_i^- + M(s_i^+ - s_i^-)),
\end{aligned}$$

i.e.,

$$s_i^+ - s_i^- \le \frac{\varepsilon\lambda\theta_i^-}{1 - \varepsilon\lambda M}. \tag{6.62}$$

We now choose $x(s_i^+ + 0)$ such that

$$\begin{aligned}
\theta_i^+ &= \|x^1(s_i^+) - x(s_i^+ + 0)\| = \min_{x \in x(s_i^+) + I_i^1(x(s_i^+))} \|x^1(s_i^+) - x\| \\
&\le h\Bigg(x^1(s_1^-) + \varepsilon I_1^i(x^1(s_i^-)) + \int_{s_1^-}^{s_1^+} \bar{u}(s)ds, \\
&\qquad x(s_i^-) + \varepsilon I_i^1(x(s_i^+)) + \int_{s_i^-}^{s_i^+} u(s)\, ds\Bigg) \\
&\le \|x^1(s_i^-) - x(s_i^-)\| + \left\| \int_{s_1^-}^{s_1^+} (\bar{u}(s) - u(s))\, ds \right\| \\
&\quad + \varepsilon h(I_i^1(x^1(s_i^-)), I_i^1(x(s_i^+))) \\
&\le \theta_i^- + 2M(s_i^+ - s_i^-) + \varepsilon\lambda(\theta_i^- + M(s_i^+ - s_i^-)) \\
&= \frac{1 + \varepsilon\lambda + \varepsilon\lambda M}{1 - \varepsilon\lambda M}\theta_i^-.
\end{aligned}$$

Thus, we get

$$\theta_i^+ \le a_1 \theta_i^-, \quad \text{where } a_1 = \frac{1 + \varepsilon\lambda + \varepsilon\lambda M}{1 - \varepsilon\lambda M}. \tag{6.63}$$

By virtue of inequalities (6.61) and (6.63), we obtain

$$\theta_i^+ \le a_1 e^{\lambda(s_i^- s_{i-1}^+)} \theta_{i-1}^+ + \beta, \quad \beta = \frac{a_1\gamma}{\lambda}(e^{K\varepsilon} - 1).$$

Denote $\delta_k = \|x(t_k+0) - x^1(t_k+0)\|$. Thus, we arrive at the following sequence of estimates (here, we set $s_0^+ = t_k$):

$$\begin{aligned}
&\theta_0^+ = \delta_k, \quad \theta_1^+ \le a_1 e^{\lambda(s_1^- - s_0^+)} \theta_0^+ + \beta, \\
&\theta_2^+ \le a_1 e^{\lambda(s_2^- - s_1^+)} (a_1 e^{\lambda(s_1^- - s_0^+)} \theta_0^+ + \beta) + \beta \\
&\quad \le a_1 e^{\lambda\frac{L}{m}} \theta_0^+ + \beta(a_1 e^{\lambda\frac{L}{m} - 1} + 1), \quad \ldots, \\
&\theta_i^+ \le a_1^i e^{\lambda\frac{L}{m}} \theta_0^+ + \beta(e^{\lambda\frac{L}{m}}(a_1^{i-1} + \cdots + a_1) + 1) \\
&\quad = a_1^i e^{\lambda\frac{L}{m}} \theta_0^+ + \beta\left(e^{\lambda\frac{L}{m}} a_1 \frac{a_1^{i-1} - 1}{a_1 - 1} + 1\right).
\end{aligned}$$

Separately, we perform the following estimates:

$$\begin{aligned}
a_1^i e^{\lambda\frac{L}{m}} &\le e^{i(a_1-1)} e^{\lambda\frac{L}{m}} \le e^{\frac{\lambda L}{m}\left(\frac{A(1+2M)}{1-\varepsilon\lambda M}+1\right)} = K_2 \\
&= 1 + \overline{O}\left(\frac{1}{m}\right), \quad m \to \infty;
\end{aligned}$$

$$\begin{aligned}
\beta\left(e^{\lambda\frac{L}{m}} a_1 \frac{a_1^{i-1}-1}{a_1-1} + 1\right) &\le \beta\left(e^{\lambda\frac{L}{m}} a_1 \frac{e^{\frac{\lambda A L(1+2M)}{m(1-\varepsilon\lambda M)}} - 1}{\frac{\varepsilon\lambda(1+2M)}{1-\varepsilon\lambda M}} + 1\right) \\
&= \frac{a_1\gamma}{\lambda}\frac{e^{K\varepsilon}-1}{\varepsilon}\left(e^{\lambda\frac{L}{m}} a_1 \frac{e^{\frac{\lambda A L(1+2M)}{m(1-\varepsilon\lambda M)}} - 1}{\frac{\lambda(1+2M)}{1-\varepsilon\lambda M}} + \frac{a_1\gamma}{\lambda}(e^{K\varepsilon} - 1)\right) \\
&= \overline{O}\left(\frac{1}{m^2}\right) + \overline{O}\left(\frac{\varepsilon}{m}\right), \quad m \to \infty.
\end{aligned}$$

Thus, there exist positive constants K_3 and K_4 such that the following estimate is true for $m \ge m_1$:

$$\theta_i^+ \le K_2 \theta_0^+ + \frac{K_3}{m^2} + \frac{K_4\varepsilon}{m}.$$

Hence,

$$\delta_{k+1} \le K_2\delta_k + \frac{K_3}{m^2} + \frac{K_4\varepsilon}{m}.$$

As a result, we get the following sequence of estimates:

$$\delta_0 = 0, \quad \delta_1 \le +\frac{K_3}{m^2} + \frac{K_4\varepsilon}{m}, \quad \ldots,$$

$$\begin{aligned} \delta_{k+1} &\le (K_2^k + \cdots + 1)\left(\frac{K_3}{m^2} + \frac{K_4\varepsilon}{m}\right) \\ &= \frac{K_2^{k+1} - 1}{K_2 - 1}\left(\frac{K_3}{m^2} + \frac{K_4\varepsilon}{m}\right) \\ &\le K_5(k+1)\left(\frac{K_3}{m^2} + \frac{K_4\varepsilon}{m}\right) \le K_5\left(\frac{K_3}{m} + K_4\varepsilon\right) \to 0 \end{aligned} \tag{6.64}$$

for $m \ge m_1$ and $\varepsilon \le \varepsilon_1$.

We now choose $0 < \varepsilon \le \varepsilon_0$ and $m \ge m_0$. Thus, inequalities (6.59), (6.60), and (6.64) imply inequality (6.56).

Moreover, by virtue of (6.62), we obtain

$$\sum_i (s_i^+ - s_i^-) \le \frac{A\lambda\eta}{1 - \varepsilon\lambda M}.$$

Similarly, we get the second statement of the theorem. The theorem is proved. □

Remark 7. Let $X(t)$ and $Y(t)$ be the graphs of the solutions $x(t)$ and $y(t)$. At the times of pulses, the vertical segments $[x(t_i^-), x(t_i^+)]$ and $[y(t_i^-), y(t_i^+)]$ belong to these graphs. Then the statement of the theorem can be rewritten in the form $h\big(X(t), Y(t)\big) < \eta$, $t \in [0, L]$.

Remark 8. Assume that the conditions of the theorem are satisfied. We consider inclusion (6.52), (6.53) with $y(0) = y_0$ and denote $\delta = \|x_0 - y_0\|$. Then, for any $\eta > 0$ and each solution $y(t)$ of inclusion (6.52), (6.53), there exist $\varepsilon_0(\eta) > 0$, $\delta_0(\eta) > 0$, a constant C, and a solution $x(t)$ of inclusion (6.49), (6.50) such that

$$\|x(t) - y(t)\| < C\delta + \eta, \quad t \in [0, L], \ 0 < \varepsilon < \varepsilon_0, \ 0 < \delta < \delta_0.$$

We now formulate the theorem of the method of averaging for impulsive differential inclusions (6.49)–(6.51).

Theorem 14. *Assume that all conditions of Theorem* 13 *are satisfied in the domain* Q *and, in addition,*

(5) *the mappings* $K_i^j(x)$ *and the functions* $\sigma_j^i(x)$ *satisfy the Lipschitz condition with constant* μ *and, in addition,* $x + K_i^j(x) \subset D$, $x \in D$*;*

(6) *the surfaces* $t = \sigma_j^i(x)$ *are disjoint and* $\sigma_j^i(x) \geq \sigma_j^i(x+z)$ *for any* $x \in D$ *and* $z \in K_i^j$, $j = 1, 2$*;*

(7) *the following inequalities hold:*

$$\mu M < 1, \quad h(K_i^1(x), K_i^2(x)) \leq \eta,$$

$$|\sigma_i^1(x) - \sigma_i^2(x)| \leq \eta, \quad \|x_0 - y_0\| \leq \delta.$$

Then, for any $\xi > 0$*, there exist* $\eta > 0$ *and* $\delta > 0$ *such that, for each solution* $x(t)$ *of inclusion* (6.49)–(6.51)*, one can find a solution* $y(t)$ *of inclusion* (6.52)–(6.54) *satisfying the estimate*

$$\|x(t) - y(t)\| \leq \eta,$$

$$t \in [0, L] \setminus \left\{ \bigcup_p [s_p^2 - \delta_p, s_p^2 + \delta_p] \bigcup_i [t_i^2 - \Delta_i, t_i^2 + \Delta_i] \right\},$$

where $s_p^2 = \sigma_p(y(s_p^2))$ *and* $t_i^2 = \tau_i(y(t_i^2))$ *and, in addition,* $\sum_i \delta_i + \sum_i \Delta_i < C\eta$.

Remark 9. By analogy with procedure used for differential equations [88], it can be shown that Theorems 13 and 14 lead to theorems on substantiation of the method of averaging for impulsive differential inclusions.

6.4 Averaging of Impulsive Differential Equations with Hukuhara Derivative

In the present section, we consider the method of complete averaging of impulsive differential equations with Hukuhara derivative of the form

$$D_h X(t) = \varepsilon F(t, X), \quad t \neq \tau_i, \; X(0) = X_0, \tag{6.65}$$

$$\Delta X|_{t=\tau_i} = \varepsilon I_i(X). \tag{6.66}$$

Assume that the set-valued mapping $F(t, X)$ is 2π-periodic in t and there exists $p \in N$ such that the equalities $\tau_{i+p} = \tau_i + 2\pi$ and $I_{i+p}(X) \equiv I_i(X)$ are true for all $i \in N$.

Equations (6.65), (6.66) are associated with the following averaged equation:

$$D_h Y(t) = \varepsilon F^0(Y(t)), \quad Y(0) = X_0, \tag{6.67}$$

where

$$F^0(Y) = \frac{1}{2\pi}\int_0^{2\pi} F(s,Y)ds + \frac{1}{2\pi}\sum_{0\le\tau_i<2\pi} I_i(Y). \tag{6.68}$$

Theorem 15. *Assume that the following conditions are satisfied in the domain* $Q = \{t \ge 0,\ X \in D \subset \operatorname{conv}(\mathbb{R}^n)\}$*:*

(1) *the set-valued mappings* $F: \mathbb{R} \times \operatorname{conv}(\mathbb{R}^n) \to \operatorname{conv}(\mathbb{R}^n)$ *and* $I_i: \operatorname{conv}(\mathbb{R}^n) \to \operatorname{conv}(\mathbb{R}^n)$ *are uniformly bounded by a constant* M *and satisfy the Lipschitz condition with constant* λ *with respect to the variable* X*; moreover, the mapping* $F(t,X)$ *is continuous in* t*;*

(2) *there exists a domain* $D' \subset D$ *such that, for any* $X_0 \in D'$*, the solutions of Eq. (6.67) belong to the domain* D *together with a certain* ρ*-neighborhood.*

Then there exist constants $C > 0$ *and* $\varepsilon_0 > 0$ *such that the estimate*

$$h(X(t), Y(t)) \le C\varepsilon \tag{6.69}$$

holds for $\varepsilon \in (0, \varepsilon_0]$ *and* $t \in [0, L\varepsilon^{-1}]$.

Proof. The differential equations with Hukuhara derivative (6.65), (6.66), and (6.67) are equivalent to the integral equations [25]

$$X(t) = X_0 + \varepsilon\int_0^t F(s, X(s))ds + \varepsilon\sum_{0\le\tau_i<t} I_i(X(\tau_i)), \tag{6.70}$$

$$Y(t) = X_0 + \varepsilon\int_0^t F^0(Y(s))ds, \tag{6.71}$$

where the integral is understood in Hukuhara's sense.

We now show that the set-valued mapping $F^0(X)$ is uniformly bounded and satisfies the Lipschitz condition

$$\begin{aligned}
|F^0(Y)| &= h(F^0(Y), \{0\}) \\
&= h\left(\frac{1}{2\pi}\int_0^{2\pi} F(s,Y)ds + \frac{1}{2\pi}\sum_{0\le\tau_i<2\pi} I_i(Y), \{0\}\right) \\
&\le \left|\frac{1}{2\pi}\int_0^{2\pi} F(s,Y)ds\right| + \frac{1}{2\pi}\sum_{0\le\tau_i<2\pi} |I_i(Y)| \\
&\le M + \frac{p}{2\pi}M = M\left(1 + \frac{p}{2\pi}\right),
\end{aligned}$$

$$\begin{aligned} h(F^0(Y'), F(Y'')) &= h\Bigg(\frac{1}{2\pi}\int_0^{2\pi} F(s, Y')ds + \frac{1}{2\pi}\sum_{0\le\tau_i<2\pi} I_i(Y'), \\ &\qquad \frac{1}{2\pi}\int_0^{2\pi} F(s, Y'')ds + \frac{1}{2\pi}\sum_{0\le\tau_i<2\pi} I_i(Y'')\Bigg) \\ &\le h\Bigg(\frac{1}{2\pi}\int_0^{2\pi} F(s, Y')ds, \frac{1}{2\pi}\int_0^{2\pi} F(s, Y'')ds\Bigg) \\ &\quad + \frac{1}{2\pi}\sum_i h(I_i(Y'), I_i(Y'')) \le \lambda h(Y', Y'') + \frac{p}{2\pi}\lambda h(Y', Y'') \\ &= \lambda\left(1 + \frac{p}{2\pi}\right) h(Y', Y''). \end{aligned}$$

The established inequalities mean that the set-valued mapping $F^0(Y)$ is uniformly bounded by a constant $M_1 = M(1 + \frac{p}{2\pi})$ and satisfies the Lipschitz condition with constant $\lambda_1 = \lambda(1 + \frac{p}{2\pi})$.

We now estimate $h(X(t), Y(t))$. By virtue of relations (6.70) and (6.71), we get

$$\begin{aligned} h(X(t), Y(t)) &= h\Bigg(X(t), X_0 + \varepsilon\int_0^t F^0(Y(s))ds\Bigg) \\ &\le h\Bigg(X(t), X_0 + \varepsilon\int_0^t F^0(X(s))ds\Bigg) \\ &\quad + h\Bigg(\varepsilon\int_0^t F^0(X(s))ds, \varepsilon\int_0^t F^0(Y(s))ds\Bigg) \\ &\le \varepsilon\lambda_1\int_0^t h(X(s), Y(s))ds \\ &\quad + h\Bigg(X(t), X_0 + \varepsilon\int_0^t F^0(X(s))ds\Bigg). \end{aligned} \tag{6.72}$$

By using the Gronwall–Bellman lemma and inequality (6.72), we find

$$\begin{aligned} h(X(t), Y(t)) \le e^{\varepsilon\lambda_1 t} h\Bigg(\varepsilon\int_0^t F(s, X(s))ds \\ + \varepsilon\sum_{0\le\tau_i<t} I_i(X(\tau_i)), \varepsilon\int_0^t F^0(X(s))ds\Bigg). \end{aligned} \tag{6.73}$$

Let $t \in [2\pi m, 2\pi(m+1))$. Then

$$
\begin{aligned}
&h\Big(\int_0^t F(s, X(s))ds + \sum_{0\le\tau_i\le t} I_i(X(\tau_i)), \int_0^t F^0(X(s))ds\Big)\\
&= h\Big(\sum_{j=0}^{m-1}\int_{2\pi j}^{2\pi(j+1)} F(s, X(s))ds + \int_{2\pi m}^{t} F(s, X(s))ds\\
&\qquad + \sum_{j=0}^{m-1}\sum_{2\pi j\le\tau_i<2\pi(j+1)} I_i(X(\tau_i)) + \sum_{2\pi m\le\tau_i<t} I_i(X(\tau_i)),\\
&\quad \sum_{i=0}^{m-1}\int_{2\pi j}^{2\pi(j+1)} F^0(X(s))ds + \int_{2\pi m}^{t} F^0(X(s))ds\Big)\\
&= h\Big(\sum_{j=0}^{m-1} F^j + F^m, \sum_{j=0}^{m-1} F^{0j} + F^{0m}\Big),
\end{aligned}
\tag{6.74}
$$

where

$$
\begin{aligned}
F^j &= \int_{2\pi j}^{2\pi(j+1)} F(s, X(s))ds + \sum_{2\pi j\le\tau_i<2\pi(j+1)} I_i(X(\tau_i)), \quad j=\overline{0, m-1},\\
F^m &= \int_{2\pi m}^{t} F(s, X(s))ds + \sum_{2\pi m\le\tau_i<t} I_i(X(\tau_i)),\\
F^{j0} &= \int_{2\pi j}^{2\pi(j+1)} F^0(X(s))ds, \quad j=\overline{0, m-1},\\
F^{0m} &= \int_{2\pi m}^{t} F^0(X(s))ds.
\end{aligned}
$$

We estimate $h(F^j, F^{j0})$:

$$
\begin{aligned}
&h(F^j, F^{j0})\\
&= h\Big(\int_{2\pi j}^{2\pi(j+1)} F(s, X(s))ds + \sum_{2\pi j\le\tau_i<2\pi(j+1)} I_i(X(\tau_i)), \int_{2\pi i}^{2\pi(i+1)} F^0(X(s))ds\Big)\\
&\le h\Big(\int_{2\pi j}^{2\pi(j+1)} F(s, X(s))ds + \sum_{2\pi j\le\tau_i<2\pi(j+1)} I_i(X(\tau_i)),\\
&\qquad \int_{2\pi j}^{2\pi(j+1)} F(s, X(2\pi j))ds + \sum_{2\pi j\le\tau_i<2\pi(j+1)} I_i(X(2\pi j))\Big)
\end{aligned}
$$

$$+ h\left(\int_{2\pi j}^{2\pi(j+1)} F(s, X(2\pi j))ds + \sum_{2\pi j \le \tau_i < 2\pi(j+1)} I_i(X(2\pi j)),\right.$$

$$\left.\int_{2\pi j}^{2\pi(j+1)} F^0(X(2\pi j))ds\right)$$

$$+ h\left(\int_{2\pi j}^{2\pi(j+1)} F^0(X(2\pi j))ds, \int_{2\pi j}^{2\pi(j+1)} F^0(X(s))ds\right)$$

$$\le \int_{2\pi j}^{2\pi(j+1)} h(F(s, X(s)), F(s, X(2\pi j)))ds$$

$$+ \sum_{2\pi j \le \tau_i < 2\pi(j+1)} h(I_i(X(\tau_i)), I_i(X(2\pi j)))$$

$$+ \int_{2\pi j}^{2\pi(j+1)} h(F^0(X(2\pi j)), F^0(X(s)))ds$$

$$\le \lambda \int_{2\pi j}^{2\pi(j+1)} h(X(s), X(2\pi j))ds$$

$$+ \lambda \sum_{2\pi j \le \tau_i < 2\pi(j+1)} h(X(\tau_i), X(2\pi j)) + \lambda_1 \int_{2\pi j}^{2\pi(j+1)} h(X(s), X(2\pi i))ds$$

$$= \lambda \sum_{2\pi j \le \tau_i < 2\pi(j+1)} h(X(\tau_i), X(2\pi j))$$

$$+ \lambda \left(2 + \frac{p}{2\pi}\right) \int_{2\pi j}^{2\pi(j+1)} h(X(s), X(2\pi j))ds. \tag{6.75}$$

On the other hand, we find

$$h(X(s), X(2\pi j))$$

$$= h\left(X(2\pi j) + \varepsilon \int_{2\pi j}^{s} F(\tau, X(\tau))d\tau + \varepsilon \sum_{2\pi j \le \tau_i < s} I_i(X(\tau_i)), X(2\pi j)\right)$$

$$\le \varepsilon \int_{2\pi j}^{s} |F(\tau, X(\tau))|d\tau + \varepsilon \sum_{2\pi j \le \tau_i < s} |I_i(X(\tau_i))|$$

$$\le \varepsilon M(s - 2\pi j) + \varepsilon M p. \tag{6.76}$$

It follows from (6.75) and (6.76) that

$$h(F^j, F^{0j}) \le \varepsilon\lambda \left(2 + \frac{p}{2\pi}\right)$$

$$\times \int_{2\pi j}^{2\pi(j+1)} [M(s - 2\pi j) + Mp]ds + \varepsilon\lambda p\, [2\pi M + Mp]$$

$$
\begin{aligned}
&\leq \varepsilon\lambda\left(2+\frac{p}{2\pi}\right)[2\pi^2 M + 2\pi M p] + \varepsilon\lambda p[2M\pi + Mp] \\
&= 2\varepsilon\lambda M\pi\left[\left(2+\frac{p}{2\pi}\right)(\pi+p) + p\left(1+\frac{p}{2\pi}\right)\right] = 2\varepsilon\pi C_1, \qquad (6.77)
\end{aligned}
$$

where

$$
C_1 = \lambda M\left[\left(2+\frac{p}{2\pi}\right)(\pi+p) + p\left(1+\frac{p}{2\pi}\right)\right].
$$

Thus, we get

$$
\begin{aligned}
&h(F^m, F^{0m}) \\
&= h\Big(\int_{2\pi m}^{t} F(s,X(s))ds + \sum_{2\pi m\leq\tau_i<t} I_i(X(\tau_i)), \int_{2\pi m}^{t} F^0(X(s))ds\Big) \\
&\leq \int_{2\pi m}^{t} |F(s,X(s))|ds + \sum_{2\pi m\leq\tau_i<t} |I_i(X(\tau_i))| + \int_{2\pi m}^{t} |F^0(X(s))|ds \\
&\leq 2\pi M + pM + M\left(1+\frac{p}{2\pi}\right)2\pi = 4M\pi\left(1+\frac{p}{2\pi}\right). \qquad (6.78)
\end{aligned}
$$

By using (6.74), (6.77), and (6.78), we obtain

$$
\begin{aligned}
&h\Big(\int_0^t F(s,X(s))ds + \sum_{0\leq\tau_i\leq t} I_i(X(\tau_i)), \int_0^t F^0(X(s))ds\Big) \\
&\leq \sum_{j=0}^{m-1} h(F^j, F^{j0}) + h(F^m, F^{0m}) \\
&\leq \sum_{j=0}^{m-1} 2\varepsilon\pi C_1 + 4\pi M\left(1+\frac{p}{2\pi}\right) \leq C_1 L + 4\pi M\left(1+\frac{p}{2\pi}\right). \qquad (6.79)
\end{aligned}
$$

It follows from (6.73) and (6.79) that

$$
h(X(t), Y(t)) \leq C\varepsilon,
$$

where

$$
C = \left[C_1 L + 4\pi M\left(1+\frac{p}{2\pi}\right)\right]e^{\lambda_1 L}.
$$

The value of ε_0 is found from the condition

$$
C\varepsilon \leq \rho, \quad \text{i.e., } \varepsilon_0 = \frac{\rho e^{-\lambda_1 L}}{C_1 L + 4\pi M(1+\frac{p}{2\pi})}. \qquad \square
$$

Remark 10. Since the operation of subtraction is not always defined in the space $\text{conv}(\mathbb{R}^n)$, the condition of pulse action (6.66) in the general case should be rewritten in the form

$$X(\tau_i + 0) = \psi_i(X(\tau_i), \varepsilon), \quad \text{where } h(X(\tau_i + 0), \psi_i(X(\tau_i), \varepsilon)) \leq M\varepsilon.$$

In this case, the formula of averaging (6.68) takes the form

$$F^0(Y) = \frac{1}{2\pi} \int_0^{2\pi} F(s, Y)ds + \frac{1}{2\pi} \psi_p(\psi_{p-1}(\cdots \psi_2(\psi_1(Y))\cdots)).$$

Remark 11. If, in Eqs. (6.65), (6.66), the set-valued mappings $F(t, X)$ and $I_i(X)$ are not 2π-periodic, then it is possible to prove an analog of Theorem 15 under the assumption that the following limit exists:

$$F^0(Y) = \lim_{T\to\infty} \frac{1}{T}\left[\int_0^T F(s, Y)ds + \sum_{0\leq\tau_i<T} I_i(Y)\right].$$

Under this assumption, the statement of the theorem takes the form:
Then, for any $\eta > 0$, there exists $\varepsilon_0 > 0$ such that the estimate

$$h(X(t), Y(t)) \leq \eta$$

is true for $\varepsilon \in (0, \varepsilon_0]$ and $t \in [0, L\varepsilon^{-1}]$.

Chapter 7

Differential Equations with Discontinuous Right-Hand Side

7.1 Motions and Quasimotions

In analyzing positional differential games, Krasovskii (see e.g., [71]) assumed that one of the gamblers may use an arbitrary discontinuous function as his strategy, e.g., $v(t,x)$. The differential equation obtained in this case may have no solutions in the ordinary sense.

We generalize the results obtained in [154] to differential equations with discontinuous right-hand side. Note that the solutions $x(t)$ constructed in this case and substituted in the differential equation may violate this equation on a set of nonzero measure.

Consider the dynamical system described by a system of ordinary differential equations

$$\dot{x} = f(t, x, v(t, x)), \tag{7.1}$$

where $x \in \mathbb{R}^n$ is the phase vector, $t \in [t_0, T]$ is time, and $v \in \mathbb{R}^q$.

In what follows, we suppose that the following conditions are satisfied:

Conditions 1. A vector function $f(t, x, v)$ is continuous in the collection of its arguments;

- $v \in Q \in \text{comp}(\mathbb{R}^q)$;
- for any bounded domain $G \subset \mathbb{R}^n$, there exists a constant $\lambda(G) > 0$ such that

  $$\|f(t, x^1, v) - f(t, x^2, v)\| \le \lambda(G)\|x^1 - x^2\|$$

  for any $x^1, x^2 \in G$ uniformly in $t \in [t_0, T]$ and $v \in Q$;
- there exists a constant $\gamma > 0$ such that the inequality

  $$\|f(t, x, v)\| \le \gamma(1 + \|x\|) \tag{7.2}$$

  is true for all $t \in [t_0, T]$, $v \in Q$.

If an arbitrary measurable function $v(t)$ such that $v(t) \in Q$ is taken as $v(t, x)$, then the validity of conditions 1 guarantees the existence, uniqueness, and extendability of the solutions $x(t)$ of system (7.1) to the segment $[t_0, T]$ for $x(t_0) = x_0$.

Assume that the function $v(t, x)$ is such that $f(t, x, v(t, x))$ does not satisfy the conditions of existence theorems.

We now determine the motion [71, 154] of system (7.1). Suppose that the initial point (t_0, x_0) is fixed. We cover the segment $[t_0, T]$ by a system Δ of half intervals $\tau_j \le t < \tau_{j+1}$, $j = \overline{0, m(\Delta) - 1}$, $\tau_0 = t_0$, $\tau_{m(\Delta)} = T$.

Definition 1. A stepwise motion

$$x(\cdot, \Delta) = \{x(t, t_0, x_0, \Delta),\ t_0 \le t \le T\}$$

of system (7.1) is defined as an absolutely continuous solution of the integral equation

$$x(t, \Delta) = x(\tau_j, \Delta) + \int_{\tau_j}^{t} f(s, x(s, \Delta), v(\tau_j, x(\tau_j, \Delta)))\, ds,$$

$$x(\tau_0, \Delta) = x_0, \quad \tau_j \le t \le \tau_{j+1},\ j = \overline{0, m(\Delta) - 1}. \tag{7.3}$$

The existence, uniqueness, and extendability of this stepwise motion $x(\cdot, \Delta)$ to $[t_0, T]$ are established by using the classic theorems of the theory of differential equations. Note that each specific partition Δ of the segment $[t_0, T]$ is associated with "its own" unique stepwise motion $x(\cdot, \Delta)$. We also note that the validity of inequality (7.2) is sufficient for the extendability of the stepwise motion to the left up to the time $t_* \in [0, t_0]$.

Definition 2 ([71]). A function $x(t)$, $t_0 \le t \le T$, is called a motion

$$x(\cdot) = \{x(t, t_0, x_0), t_0 \le t \le T\}$$

of system (7.1) generated from the initial point (t_0, x_0) if, for this function, one can find a sequence of stepwise motions

$$x(t, t_0, x^k, \Delta^k), \quad t_0 \le t \le T, \tag{7.4}$$

uniformly convergent to $x(t)$ on the segment $[t_0, T]$ under the condition

$$\lim_{k\to\infty} \sup_j [\tau_{j+1}^k - \tau_j^k] = 0, \quad \lim_{k\to\infty} \|x^k - x_0\| = 0. \tag{7.5}$$

A fixed initial point $(t_0, x_0) \in [0, T) \times \mathbb{R}^n$ generates not a single motion but, generally speaking, a set of motions called a bundle. Different motions in the bundle (denoted by $X(t_0, x_0)$) are obtained for different sequences of partitions Δ and different initial values of the vector x^k.

In [71], it is shown that a bundle of motions $X(t_0, x_0)$ forms a nonempty compact set in the space of continuous functions $C[t_0, T]$ with standard norm.

Kononenko Counterexample ([69]). As a disadvantage of the presented concept of motion, one can mention the fact that a segment of motion may be not a motion.

Assume that system (7.1) has the form

$$\dot{x} = v, \quad 0 \le t \le 1, \ x(0) = x_0 = 0, \tag{7.6}$$

where

$$v(t,x) = \begin{cases} 1 & \text{for } x > 0, \ t \in [0,1], \\ -1 & \text{for } x < 0, \ t \in [0,1], \\ 0 & \text{for } x = 0, \ t \in [0, \frac{1}{2}) \cup (\frac{1}{2}, 1], \\ 1 & \text{for } x = 0, \ t = \frac{1}{2}. \end{cases}$$

As one of motions of this system from the point $(t_0, x_0) = (0,0)$, we can mention the motion $x(t) = 0$, $0 \le t \le 1$.

However, a part of this motion $x^3(t) \equiv 0$, $\frac{1}{2} \le t \le 1$, is not a motion on the segment $[\frac{1}{2}, 1]$. At the same time, segments of the straight lines $x^1(t) = t - \frac{1}{2}$ and $x^2(t) = -(t - \frac{1}{2})$, $t \in [\frac{1}{2}, 1]$, are motions on this segment. Thus, a part of the motion $x(t) \equiv 0$ is not a motion generated from the "current" initial point $(\frac{1}{2}, 0)$. This is explained by the absence of the limit transition with respect to the initial time in the definition of motion presented above [(7.5)], i.e., initial points of the form (t_0, x^k), where t_0 is a fixed initial time, are considered for a convergent sequence of stepwise motions.

To avoid the "loss of a part of motions," Kononenko proposes to complement the definition of stepwise motions of system (7.1) by the limit transition with respect to the initial time. Namely, every stepwise motion should start at the initial position (t^k, x^k) but not at the initial point (t_0, x^k). Then (for $t^k > t_0$) it is extended to the left up to t_0 [by virtue of condition (7.2), this extension is possible]. Further, in the construction of motions $x(\cdot, t_0, x_0)$, the limit transition is performed either for these extensions to the left (for $t^k > t_0$) or for the "truncated" (for $t^k < t_0$) stepwise motions $x(t, t^k, x^k, \Delta^k)$, $t_0 \le t \le T$, by adding the requirement

$$\lim_{k \to \infty} |t^k - t_0| = 0 \tag{7.7}$$

to conditions (7.5).

Thus, the motion $x^3(t) = 0$, $\frac{1}{2} \le t \le 1$, is the limit of the sequence of stepwise motions originating from

$$(t^k, x^k) = \left(\frac{1}{2} + \frac{1}{10k}, 0\right), \quad k = 1, 2, \ldots.$$

The following assertion is true for (7.5) and (7.7):

Theorem 1 ([69]). *Let $x(t) = x(t, t_0, x_0)$, $t_0 \le t \le T$, be the motion of system* (7.1) *from the initial point (t_0, x_0). Then, for any $t^* \in [t_0, T)$, a part of motion $x(t, t_0, x_0)$, $t^* \le t \le T$, is a motion of the system from the initial point $(t^*, x(t^*, t_0, x_0))$.*

Subbotin Counterexample. The other disadvantage of the definition of motion is connected with the possibility of appearance of new motions absent in the original bundle of motions. This was indicated by Subbotin. In what follows, we present Subbotin's example.

Assume that system (7.1) has the form

$$\dot{x} = v, \quad 0 \le t \le 2,\ x(0) = x_0 = 0, \tag{7.8}$$

where

$$v(t,x) = \begin{cases} 0 & \text{for } x > 0,\ t \in [0,1), \\ 0 & \text{for } x < 0,\ t \in [0,1), \\ 1 & \text{for } x = 0,\ t \in [0,1), \\ 0 & \text{for } x = 0,\ t \in [1,2], \\ 1 & \text{for } x > 0,\ t \in [1,2], \\ -1 & \text{for } x < 0,\ t \in [1,2]. \end{cases}$$

The bundle of motions from the initial point $(t_0, x_0) = (0,0)$ consists of two motions

$$x^1(t) = \begin{cases} 0 & \text{for } t \in [0,1), \\ t-1 & \text{for } t \in [1,2], \end{cases}$$

and

$$x^2(t) = \begin{cases} 0 & \text{for } t \in [0,1), \\ -t+1 & \text{for } t \in [1,2]. \end{cases}$$

Three motions $x^1(t)$, $x^2(t)$, and $x^3(t) \equiv 0$, $1 \le t \le 2$, are generated from the point $(1,0)$. Thus, as the current point $(t, x(t))$ moves along the motions $x^1(t)$ or $x^2(t)$, an additional motion $x^3(t)$ absent in the original bundle $X(t_0, x_0)$ appears in this bundle at time $t = 1$. Thus, in the course of time, we may observe not only the "loss" of parts of the original motions but also the appearance of new motions absent in the original bundle of motions (as in the considered example).

To exclude this possibility, it is necessary to modify the notion of stepwise motions by assuming that stepwise motions with discontinuities (jumps) at the points of partition are possible.

Thus, in the analyzed example, the motion $x(t) \equiv 0$, $0 \le t \le 2$, is the limit of a sequence of stepwise quasimotions

$$x^k(t) = \begin{cases} x^k & \text{for } t \in [0,1), \\ 0 & \text{for } t \in [1,2], \end{cases} \qquad x^k \to 0 \quad \text{as } k \to \infty.$$

These quasimotions have discontinuities of the first kind at the point $t = 1$ and are continuous at the other points of the segment $[0,2]$.

In this case, the convergence of $x^k(t)$ to $x(t) \equiv 0$ should be studied in the space $M[t_0, T]$ of bounded n-dimensional functions with norm

$$\|x(\cdot)\|_{M[t_0,T]} = \sup_{t_0 \le t \le T} \|x(t)\|.$$

For this definition of motion, the appearance of new motions is impossible.

In what follows, we substantiate the proposed approach (the possibility of application of piecewise continuous stepwise motions).

Given an initial point (t_0, x_0) and a number $\alpha \in [0, 1]$, we construct a covering of the segment $[t_0, T]$ with a system Δ of half intervals $\tau_j \le t < \tau_{j+1}, j = \overline{0, m(\Delta) - 1}$, $\tau_0 \ge t_0$, $\tau_{m(\Delta)} = T$.

Definition 3. A stepwise quasimotion of system (7.1) generated from the initial point (t_0, x_0) by a partition of Δ by a number α is defined as an arbitrary function

$$x(\cdot, \Delta, \alpha) = \{x(t, \tau_0, x_0, \overline{x}_0, \Delta, \alpha), \tau_0 \le t \le T\}$$

satisfying the equation

$$x(t, \Delta, \alpha) = \overline{x}_j + \int_{\tau_j}^{t} f(s, x(s, \Delta, \alpha), v(\tau_j, \overline{x}_j))\, ds \tag{7.9}$$

for $\tau_j \le t < \tau_{j+1}$ under the condition

$$\sum_{j=0}^{m(\Delta)-1} \|\overline{x}_j - x_0(\tau_j, \Delta, \alpha)\| \le \alpha, \tag{7.10}$$

where $x_0(\tau_j, \Delta, \alpha)$ is the value of the solution $x(t, \Delta, \alpha)$ of Eq. (7.9) extended to the right on the segment $\tau_{j-1} \le t \le \tau_j$ for $t = \tau_j$, i.e.,

$$x_0(\tau_j, \Delta, \alpha) = \overline{x}_{j-1} + \int_{\tau_{j-1}}^{\tau_j} f(s, x(s, \Delta, \alpha), v(\tau_{j-1}, \overline{x}_{j-1}))\, ds$$

and, in addition,

$$x(\tau_0, \Delta, \alpha) = \overline{x}_0, \quad x_0(\tau_0, \Delta, \alpha) = x_0.$$

The stepwise quasimotions may have finite jumps at points of the partition τ_j. The sum of these jumps should not exceed a given number α. For $\alpha = 0$, a stepwise quasimotion turns into an ordinary stepwise motion.

By $X_*(t_0, x_0)$ we denote the set of stepwise quasimotions of system (7.1) defined as indicated above and generated from the initial point (t_0, x_0) by all possible partitions of Δ and all numbers $\alpha \in [0, 1]$.

We now present several theorems similar to the corresponding theorems in [154]. The proofs of these theorems in terms of the theory of differential equations with discontinuous right-hand sides are somewhat simpler than the corresponding proofs in [154] and illustrate the basic idea of the approach to the investigation of differential equations with discontinuous right-hand sides.

Theorem 2. *The set of all stepwise quasimotions $X_*(t_0, x_0)$ of system (7.1) is bounded in the norm of the space of bounded functions $M[t_0, T]$, i.e., there exists a constant K such that*

$$\|x(\cdot, \Delta, \alpha)\|_{M[t_0,T]} \le K$$

for all $x(\cdot, \Delta, \alpha)$.

Proof. First, we show that the inequality

$$\|x(t, \Delta, \alpha)\| \le (1 + \|x_0\|)e^{\gamma(\tau_{j+1}-t_0)} + \sum_{l=0}^{j} \|\overline{x}_l - x_0(\tau_l, \Delta, \alpha)\|e^{\gamma(\tau_{j+1}-t_0)} - 1 \qquad (7.11)$$

is true for $\tau_j \le t < \tau_{j+1}$, $j = \overline{0, m(\Delta) - 1}$. Moreover, for $x_0(\tau_{j+1}, \Delta, \alpha)$, it is necessary to show that

$$\|x_0(\tau_{j+1}, \Delta, \alpha)\| \le (1 + \|x_0\|)e^{\gamma(\tau_{j+1}-t_0)} + \sum_{l=0}^{j} \|\overline{x_l} - x_0(\tau_l, \Delta, \alpha)\|e^{\gamma(\tau_{j+1}-t_0)} - 1. \qquad (7.12)$$

In the proof, we use the inequality

$$\|\overline{x}_j\| \le \|\overline{x}_j - x_0(\tau_j, \Delta, \alpha)\| + \|x_0(\tau_j, \Delta, \alpha)\|, \qquad (7.13)$$

which follows from the triangle axiom, and the relation

$$\|x(t)\| \le (1 + \|\overline{x}_j\|)e^{\gamma(t-\tau_j)} - 1, \quad t \in [\tau_j, \tau_j + 1], \qquad (7.14)$$

valid for the solution $x(\cdot)$ of system (7.1) under condition (7.2).

Inequalities (7.11) and (7.12) are proved by induction. First, we check the validity of these inequalities for $j = 0$. For $t \in [\tau_0, \tau_1)$, inequality (7.14) implies that

$$\|x(t, \Delta, \alpha)\| \le (1 + \|\overline{x}_0\|)e^{\gamma(t-t_0)} - 1.$$

By using inequality (7.13) and the value $x_0(\tau_0, \Delta, \alpha)$, we get

$$\begin{aligned}\|x(t, \Delta, \alpha)\| &\le (1 + \|\overline{x}_0 - x_0(\tau_0, \Delta, \alpha)\| + \|x_0\|)e^{\gamma(t-t_0)} - 1 \\ &\le (1 + \|x_0\|)e^{\gamma(\tau_1-t_0)} + \|\overline{x}_0 - x_0(\tau_0, \Delta, \alpha)\|e^{\gamma(\tau_1-t_0)} - 1,\end{aligned}$$

i.e., inequality (7.11) is true for $j = 0$. Similarly, we check the validity of inequality (7.12) for $j = 0$.

Assume that inequalities (7.11) and (7.12) are true for $j = k$, i.e., for $\tau_k \le t < \tau_{k+1}$,

$$\begin{aligned}\|x(t,\Delta,\alpha)\| &\le (1+\|x_0\|)e^{\gamma(\tau_{k+1}-t_0)} \\ &\quad + \sum_{l=0}^{k} \|\overline{x}_l - x_0(\tau_l,\Delta,\alpha)\| e^{\gamma(\tau_{k+1}-t_0)} - 1, \\ \|x_0(\tau_{k+1},\Delta,\alpha)\| &\le (1+\|x_0\|)e^{\gamma(\tau_{k+1}-t_0)} \\ &\quad + \sum_{l=0}^{k} \|\overline{x}_l - x_0(\tau_l,\Delta,\alpha)\| e^{\gamma(\tau_{k+1}-t_0)} - 1. \end{aligned} \tag{7.15}$$

It is necessary to prove inequalities (7.11) and (7.12) for $j = k+1$. For $t \in [\tau_{k+1}, \tau_{k+2})$, inequality (7.14) implies that

$$\|x(t,\Delta,\alpha)\| \le (1+\|\overline{x}_{k+1}\|)e^{\gamma(t-\tau_{k+1})} - 1. \tag{7.16}$$

Further, by using (7.13) (for $j = k+1$), we conclude that

$$\|\overline{x}_{k+1}\| \le \|\overline{x}_{k+1} - x_0(\tau_{k+1},\Delta,\alpha)\| + \|x_0(\tau_{k+1},\Delta,\alpha)\|. \tag{7.17}$$

Substituting (7.15) and (7.17) in (7.16), we obtain

$$\begin{aligned}\|x(t,\Delta,\alpha)\| &\le [1 + \|\overline{x}_{k+1} - x_0(\tau_{k+1},\Delta,\alpha)\| \\ &\quad + (1+\|x_0\|)e^{\gamma(\tau_{k+1}-t_0)} \\ &\quad + \sum_{l=0}^{k} \|\overline{x}_l - x_0(\tau_l,\Delta,\alpha)\| e^{\gamma(\tau_{k+1}-t_0)} - 1]e^{\gamma(t-\tau_{k+1})} - 1 \\ &\le (1+\|x_0\|)e^{\gamma(\tau_{k+2}-t_0)} + \sum_{l=0}^{k+1} \|\overline{x}_l - x_0(\tau_l,\Delta,\alpha)\| e^{\gamma(\tau_{k+2}-t_0)} - 1.\end{aligned}$$

Similarly, we prove the corresponding inequalities for $\|x_0(\tau_{k+2},\Delta,\alpha)\|$. Thus, inequalities (7.11) and (7.12) are true.

By using the chain of inequalities

$$\begin{aligned}&(1+\|x_0\|)e^{\gamma(\tau_1-t_0)} + \|\overline{x}_0 - x_0(\tau_0,\Delta,\alpha)\| e^{\gamma(\tau_1-t_0)} - 1 \\ &\le (1+\|x_0\|)e^{\gamma(\tau_2-t_0)} + \sum_{l=0}^{1} \|\overline{x}_l - x_0(\tau_l,\Delta,\alpha)\| e^{\gamma(\tau_2-t_0)} - 1\end{aligned}$$

$$\leq \cdots \leq (1 + \|x_0\|)e^{\gamma(\tau_{m(\Delta)} - t_0)} + \sum_{l=0}^{m(\Delta)-1} \|\overline{x}_l - x_0(\tau_l, \Delta, \alpha)\| e^{\gamma(\tau_{m(\Delta)} - t_0)} - 1,$$

the inequality

$$\|x_0(T, \Delta, \alpha)\| \leq \left[(1 + \|x_0\|) + \sum_{l=0}^{m(\Delta)-1} \|\overline{x}_l - x_0(\tau_l, \Delta, \alpha)\|\right] e^{\gamma(T - t_0)} - 1,$$

and inequalities (7.11), (7.12), and (7.10), we get the following estimates for all $t \in [t_0, T]$:

$$\begin{aligned} \|x(t, \Delta, \alpha)\| &\leq \left[(1 + \|x_0\|) + \sum_{l=0}^{m(\Delta)-1} \|\overline{x}_l - x_0(\tau_l, \Delta, \alpha)\|\right] e^{\gamma(T - t_0)} - 1 \\ &\leq (\alpha + 1 + \|x_0\|)e^{\gamma(T - t_0)} - 1 \leq (2 + \|x_0\|)e^{\gamma(T - t_0)} - 1 = K. \end{aligned}$$

Thus, it is shown that the set of all stepwise quasimotions $X_*(t_0, x_0)$ is bounded. Note that the set $X_*(t_0, x_0)$ is a subset of the space $M[t_0, T]$ of all bounded vector functions $z : [t_0, T] \to \mathbb{R}^n$. □

We now consider the notion of quasimotions [154], i.e., the limits of constructed sequences of stepwise quasimotions in $M[t_0, T]$.

Definition 4. The quasimotion $x(\cdot) = \{x(t, t_0, x_0), t_0 \leq t \leq T\}$ of system (7.1) from the initial point (t_0, x_0) is defined as an arbitrary function continuous on the segment $[t_0, T]$ for which there exists a sequence of stepwise quasimotions (extended to the left up to t_0 if $\tau_0^r > t_0$) convergent (as $r, m \to \infty$) to this function (in the metric of the space $M[t_0, T]$)

$$x(\cdot, \Delta^r, \alpha^m) = \{x(t, \tau_0^r, x_0^r, \overline{x}_0^r, \Delta^r, \alpha^m),\ t_0 \leq t \leq T\}$$

for

$$\operatorname{diam} \Delta^r \to 0$$

and

$$|\tau_0^r - t_0| + \|x_0^r - x_0\| \to 0, \quad \alpha^m \to 0 \quad \text{as } r, m \to \infty. \tag{7.18}$$

Here,

$$\operatorname{diam} \Delta^r = \max_j [\tau_{j+1}^r - \tau_j^r], \quad 0 \leq \alpha^m \leq 1 \quad (r, m = 1, 2, \ldots).$$

Thus, for a given sequence $\{x(\cdot, \Delta^r, \alpha^m)\}$, we have

$$\sup_{t_0 \le t \le T} \|x(t) - x(t, \Delta^r, \alpha^m)\| \to 0$$

provided that condition (7.18) is satisfied.

By $X(t_0, x_0)$ we denote the bundle of constructed quasimotions of system (7.1). Different quasimotions of the bundle are realized for different possible sequences Δ^r, α^m that can be used for the construction of stepwise quasimotions convergent to quasimotions $x(t, t_0, x_0))$.

Consider the properties of the bundles of quasimotions of system (7.1) defined as indicated above.

Theorem 3. *The bundle $X(t_0, x_0)$ is a nonempty bounded (in norm) closed subset of the space $C[t_0, T]$.*

Proof. The set $X(t_0, x_0)$ contains motions [71] (they are obtained for $\alpha = 0$). Hence, by virtue of [71], we conclude that $X(t_0, x_0) \neq \varnothing$.

It follows from Theorem 2 that the set $X_*(t_0, x_0)$ of all stepwise quasimotions $x(\cdot, \Delta, \alpha)$ is bounded in the norm of the space $M[t_0, T]$. Then its closure $\overline{X}_*(t_0, x_0)$ is also bounded in this space. Hence, the bundle of quasimotions $X(t_0, x_0)$ is bounded (in the norm of the space $M[t_0, T]$) as a subset of $\overline{X}_*(t_0, x_0)$. Since the metric of the space $C[t_0, T]$ is induced by the metric of $M[t_0, T]$ and $X(t_0, x_0) \subset C[t_0, T]$, the bundle of quasimotions $X(t_0, x_0)$ is a bounded (in norm) subset of the space $C[t_0, T]$.

We now show that the bundle $X(t_0, x_0)$ is closed in $C[t_0, T]$, i.e., prove that $X(t_0, x_0) \equiv \overline{X}(t_0, x_0)$, where $\overline{X}(t_0, x_0)$ is the closure of the set $X(t_0, x_0)$ in $C[t_0, T]$. Thus, we choose $y(\cdot) \in \overline{X}(t_0, x_0)$ and show that $y(\cdot) \in X(t_0, x_0)$. Since $y(\cdot)$ is the limit point of a sequence of quasimotions $\{x^l(\cdot)\} \subset X(t_0, x_0)$, the function $y(t)$, as the limit of a sequence of continuous functions $x^l(t)$ uniformly convergent on $[t_0, T]$, is continuous on $[t_0, T]$. Further, we choose a sequence of positive numbers ε_l ($l = 1, 2, \ldots$) convergent to zero. For any $l = 1, 2, \ldots$, we construct a ball

$$B(y(\cdot), \varepsilon_l) = \{z(\cdot) \in M[t_0, T] : \|z(\cdot) - y(\cdot)\|_{M[t_0,T]} < \varepsilon_l\}.$$

in the space $M[t_0, T]$.

Since $y(\cdot) \in \overline{X}(t_0, x_0)$, where $\overline{X}(t_0, x_0)$ is the closure of $X(t_0, x_0)$, there exist quasimotions

$$x^l(\cdot) \in X(t_0, x_0) \cap B(y(\cdot), \varepsilon_l).$$

By the definition of quasimotions, for any $l = 1, 2, \ldots$, there exists a sequence of stepwise quasimotions

$$\{x(\cdot, \Delta_r^l, \alpha_m^l)\}$$

convergent to $x^l(\cdot)$ in the space $M[t_0, T]$ [under condition (7.18)]. For any l and "sufficiently large" r and m, we select a single "representative" $x^l(\cdot, \Delta_r^l, \alpha_m^l)$ from each sequence $\{x(\cdot, \Delta_r^l, \alpha_m^l)\}$ such that

(1) $x^l(\cdot, \Delta_r^l, \alpha_m^l) \in B(y(\cdot), \varepsilon_l)$,

(2) diam $\Delta_r^l < \varepsilon_l$, $\alpha_m^l < \varepsilon_l$.

As $l \to \infty$, the sequence

$$\{x(\cdot, \Delta_r^l, \alpha_m^l)\}$$

of stepwise quasimotions constructed from these "representatives" satisfies conditions (7.18) and converges (in the metric of the space $M[t_0, T]$) to a continuous function $y(t), t_0 \le t \le T$. Hence, $y(\cdot)$ is a quasimotion, i.e., $y(\cdot) \in X(t_0, x_0)$. The theorem is proved. □

The following statement directly follows from Theorem 3:

Theorem 4. *For any point (t_0, x_0), the section of the bundle of quasimotions by the hyperplane $t = T$ is a nonempty compact set in $\mathbb{R}^n$.*

Theorem 5. *For any point $(t^*, x(t^*))$, where $t^* \in [t_0, T), x(\cdot)$ is a quasimotion from the bundle $X(t_0, x_0)$, the following inclusion is true:*

$$X(T, t^*, x(t^*)) \subset X(T, t_0, x_0). \tag{7.19}$$

Proof. Let $x(\cdot) = \{x(t), t_0 \le t \le T\}$ be a quasimotion of system (7.1) from the initial point (t_0, x_0) at time $t^* \in [t_0, T)$. We take an arbitrary quasimotion $x^*(\cdot) = \{x^*(t), t^* \le t \le T\}$ but starting from the initial point $(t^*, x(t^*))$, where $x(t^*)$ is the value of $x(t)$ at time t^*.

To prove inclusion (7.19), it is necessary to show that the function

$$\tilde{x}(t) = \begin{cases} x(t) & \text{for } t_0 \le t \le t^*, \\ x^*(t) & \text{for } t^* \le t \le T \end{cases} \tag{7.20}$$

is a quasimotion of system (7.1) from the initial point (t_0, x_0).

To establish this fact, we fix a numerical sequence $\varepsilon_l \in (0, 1)$ ($l = 1, 2, \ldots$) such that

$$\lim_{l \to \infty} \varepsilon_l = 0. \tag{7.21}$$

For any $l = 1, 2, \ldots$, we construct two balls

$$B(x(\cdot), \varepsilon_l) = \{z(\cdot) \in M[t_0, T] : \|z(\cdot) - x(\cdot)\|_{M_n[t_0, T]} < \varepsilon_l\}$$

and

$$B(x^*(\cdot), \varepsilon_l) = \{z(\cdot) \in M[t^*, T] : \|z(\cdot) - x^*(\cdot)\|_{M_n[t^*, T]} < \varepsilon_l\}.$$

According to the definition of quasimotions, there exist two sequences of stepwise quasimotions

$$\{x(\cdot, \Delta^r, \alpha^m)\} \subset M[t_0, T], \quad \{x^*(\cdot, \Delta_*^p, \alpha_*^q)\} \subset M[t^*, T]$$

convergent in the metric of the indicated spaces to $x(\cdot)$ and $x^*(\cdot)$, respectively, under the conditions

$$\begin{aligned} &\text{diam } \Delta^r \to 0, \quad \alpha^m \to 0 \quad \text{as } r, m \to \infty, \\ &\text{diam } \Delta_*^p \to 0, \quad \alpha_*^q \to 0 \quad \text{as } p, q \to \infty. \end{aligned} \tag{7.22}$$

Here,

$$\begin{aligned} &\Delta^r : t_0 \le \tau_0^r < \tau_1^r < \cdots < \tau_{s(\Delta^r)}^r = T, \\ &\Delta_*^p : t^* \le \tilde{\tau}_0^p < \tilde{\tau}_1^p < \cdots < \tilde{\tau}_{s^*(\Delta_*^p)}^p = T. \end{aligned} \tag{7.23}$$

Then, for any $l = 1, 2, \ldots$, the sequences

$$\{x(\cdot, \Delta^r, \alpha^m)\} \quad \text{and} \quad \{x^*(\cdot, \Delta_*^p, \alpha_*^q)\},$$

contain stepwise quasimotions

$$x^l(\cdot, \Delta^l, \alpha^l) \quad \text{and} \quad x^{*l}(\cdot, \Delta_*^l, \alpha_*^l)$$

such that

$$\alpha^l + \alpha_*^l + 2\varepsilon_l \le 1, \tag{7.24}$$

$$x^l(\cdot, \Delta^l, \alpha^l) \in B(x(\cdot), \varepsilon_l), \quad x^{*l}(\cdot, \Delta_*^l, \alpha_*^l) \in B(x^*(\cdot), \varepsilon_l), \tag{7.25}$$

$$\text{diam } \Delta^l < \varepsilon_l, \quad \text{diam } \Delta_*^l < \varepsilon_l. \tag{7.26}$$

Indeed, inequalities (7.24) and (7.26) follow from (7.22) and inclusion (7.25) follows from the convergence of the sequences $\{x(\cdot, \Delta^r, \alpha^m)\}$ and $\{x^*(\cdot, \Delta_*^p, \alpha_*^q)\}$ to $x(\cdot)$ and $x^*(\cdot)$, respectively.

Note that, according to (7.22),

$$\alpha^l \to 0, \quad \alpha_*^l \to 0, \quad \text{diam } \Delta^l \to 0, \quad \text{and} \quad \text{diam } \Delta_*^l \to 0 \quad \text{as } l \to \infty. \tag{7.27}$$

We introduce the functions

$$\tilde{x}^l(t, \tilde{\Delta}^l, \tilde{\alpha}^l) = \begin{cases} x^l(t, \Delta^l, \alpha^l) & \text{for } t_0 \le t \le t^*, \\ x^{*l}(t, \Delta_*^l, \alpha_*^l) & \text{for } t^* \le t \le T. \end{cases} \tag{7.28}$$

By using (7.20), (7.25), and (7.28), we get

$$\tilde{x}^l(\cdot, \tilde{\Delta}^l, \tilde{\alpha}^l) \in B(\tilde{x}(\cdot), \varepsilon_l). \tag{7.29}$$

We now show that $\tilde{x}^l(\cdot, \tilde{\Delta}^l, \tilde{\alpha}^l)$ is a stepwise motion with the following diameter of partition:

$$\text{diam } \tilde{\Delta}^l \le \max\{\text{diam } \Delta^l, \text{diam } \Delta_*^l\}, \tag{7.30}$$

and the number α is equal to

$$\tilde{\alpha}^l \leq \alpha^l + \alpha_*^l + 2\varepsilon_l. \tag{7.31}$$

Indeed, by using (7.28) and (7.23), we conclude that

$$\tilde{\Delta}^l : t_0 \leq \tau_0^l < \tau_1^l < \cdots < \tau_\gamma^l \leq t^* \leq \tilde{\tau}_0^l < \tilde{\tau}_1^l < \cdots < \tilde{\tau}_{s(\Delta_*^l)}^l = T. \tag{7.32}$$

Since the diameter of partition cannot increase as a result of the addition of a point t^*, relation (7.32) yields (7.30).

The number $\tilde{\alpha}^l$ is constructed for $\tilde{x}^l(\cdot, \tilde{\Delta}^l, \tilde{\alpha}^l)$ according to (7.10) and, hence, it follows from (7.32) and (7.28) that the left-hand side of inequality (7.10) for $\tilde{x}^l(\cdot, \tilde{\Delta}^l, \tilde{\alpha}^l)$ does not exceed

$$\alpha^l + \alpha_*^l + \|\overline{x}^l(t, \Delta^l, \alpha^l) - x^{*l}(t_*, \Delta_*^l, \alpha_*^l)\|.$$

By virtue of (7.29), the last term in this sum cannot be larger than $2\varepsilon_l$. This yields inequality (7.31). Note that, according to (7.31) and (7.24), the number $\tilde{\alpha}^l \leq 1$.

Finally, we prove that the function $\tilde{x}(\cdot)$ defined by relation (7.20) is a quasimotion. Let $l \to \infty$. Then, by virtue of (7.21), we have $\varepsilon_l \to 0$. In this case, it follows from (7.30) and (7.27) that

$$\operatorname{diam} \tilde{\Delta}^l \to 0 \quad \text{as } l \to \infty.$$

Moreover, it follows from (7.29) that the stepwise quasimotions $\tilde{x}^l(\cdot, \tilde{\Delta}^l, \tilde{\alpha}^l)$ converge as $l \to \infty$ (in the metric of the space $M[t_0, T]$) to a continuous function $\tilde{x}(\cdot)$. Thus, by definition, the function $\tilde{x}(\cdot)$ is a quasimotion of system (7.1) from the initial point (t_0, x_0). The theorem is proved. □

This means that if the initial point moves along any quasimotion (from a given initial point), then the appearance of new parts of quasimotions absent in the original bundle of quasimotions (from the same initial point) is impossible.

Example 1 ([154]). Consider a system

$$\dot{x} = v, \quad 0 \leq t \leq 2, \ x(0) = x_0 = 0,$$

where

$$v(t,x) = \begin{cases} 1 & \text{for } x > 0, \ t \in [0,2], \\ 1 & \text{for } x = 0, \ t = 1, \\ 0 & \text{for } x = 0, \ t \in [0,1), \\ 0 & \text{for } x = 0, \ t \in (1,2], \\ -1 & \text{for } x < 0, \ t \in [0,2]. \end{cases}$$

The stepwise motions starting at the point $(t_0 = 0,\ x^k)$, where $x^k \to x_0 = 0$, generate the motions

$$x^1(t) \equiv 0, \quad x^2(t) = t, \quad x^3(t) = -t, \quad t \in [0,2],$$

$$x^4(t) = \begin{cases} 0 & \text{for } t \in [0,1], \\ t-1 & \text{for } t \in [1,2]. \end{cases}$$

If motions start at the point $(t^* = 1,\ x(t^*) = 0)$, then a part of the previous motion $x^1(t) = 0,\ \ 0 \le t \le 2$, disappears and a new motion $x^5(t) = 1 - t$, $1 \le t \le 2$, appears.

We now apply the proposed approach to the construction of quasimotions described above.

First, the "new" motion $x^5(t)$ is a part of the quasimotion

$$x(t) = \begin{cases} 0 & \text{for } t \in [0,1), \\ x^5(t) & \text{for } t \in [1,2], \end{cases}$$

because $x(\cdot)$ is the limit of the sequence of stepwise quasimotions

$$x(t) = \begin{cases} 0 & \text{for } t \in [0,1), \\ x^{\delta_k}(t) & \text{for } t \in [1,2], \end{cases}$$

where

$$x^{\delta_k}(t) = 1 - t - \delta_k, \quad 0 < \delta_k < 1, \quad \text{and} \quad \lim_{k\to\infty} \delta_k = 0.$$

Second, the "piece" of motion $x^1(\cdot)$ is the limit of the sequence of stepwise quasimotions $\{x(t, \tau^k, 0, 0, \Delta^k, 0), t \in [1,2]\}$ for $\tau^k \downarrow 1$ as $k \to \infty$.

In this case, the role of quasimotions is played by the functions

$$x_\alpha^+(t) = \begin{cases} 0 & \text{for } t \in [0,\alpha], \\ t-\alpha & \text{for } t \in (\alpha,2], \end{cases}$$

$$x_\alpha^-(t) = \begin{cases} 0 & \text{for } t \in [0,\alpha], \\ \alpha - t & \text{for } t \in (\alpha,2]. \end{cases}$$

We now prove one more theorem necessary in what follows.

Theorem 6. *The set-valued mapping $X(t_0, x)$ is upper semicontinuous with respect to x.*

Proof. Assume the contrary. Let the mapping $X(t,x)$ be not upper semicontinuous at the point x_0. Then there exists $\varepsilon_0 > 0$ such that, for any $\delta > 0$, one can find a quasimotion $x(t,x) \in X(t_0,x)$, where $\|x - x_0\| < \delta$, such that the inequality

$$\|x(t,x) - x(t,x_0)\| \ge \varepsilon_0$$

holds for any quasimotion $x(t,x_0) \in X(t_0,x_0)$.

We fix a sequence $\delta_i \to 0$ as $i \to \infty$ and construct a sequence of quasimotions $x(t, x_i)$, where $x_i \to x_0$ as $i \to \infty$, such that

$$\|x(t, x_i) - x(t, x_0)\| \geq \varepsilon_0 \tag{7.33}$$

for any quasimotion $x(t, x_0) \in X(t_0, x_0)$.

By the definition of quasimotions, for any i, there exists a sequence of stepwise quasimotions $\{x(t, \Delta_r^i, \alpha_m^i)\}$ convergent to $x(t, x_i)$ in the space $M[t_0, T]$ [under conditions (7.18)]. For any i and "sufficiently large" r and m, we select a single "representative" $x^i(t, \Delta_r^i, \alpha_m^i)$ from each sequence such that

$$\operatorname{diam} \Delta_r^i \to 0, \quad \alpha_m^i \to 0 \quad \text{as } i \to \infty$$

and

$$\|x^i(t, \Delta_r^i, \alpha_m^i) - x(t, x_0)\| \geq \frac{\varepsilon_0}{2}$$

for any quasimotion $x(t, x_0) \in X(t_0, x_0)$.

The constructed sequence is a sequence of stepwise quasimotions from the initial point (t_0, x_0). By using the Filippov lemma [50], we select a subsequence of this sequence uniformly convergent to the function $x^*(t, x_0)$ which is continuous and, hence, $x^*(t, x_0) \in X(t_0, x_0)$. This result contradicts inequality (7.33), which completes the proof of the theorem. □

7.2 Impulsive Motions and Quasimotions

Consider an impulsive differential system of the form

$$\dot{x} = f(t, x, v(t, x)), \quad t \neq \tau_i, \tag{7.34}$$

$$\Delta x|_{t=\tau_i} = I_i(x), \quad i = \overline{1, m}, \tag{7.35}$$

where $t \in [t_0, T]$, $t_0 = \tau_0 < \tau_1 < \cdots < \tau_{m+1} = T$, $x \in \mathbb{R}^n$, $v \in \mathbb{R}^q$, and $I_i : \mathbb{R}^n \to \mathbb{R}^n$ are bounded continuous functions.

Definition 5 ([126]). An impulsive quasimotion

$$x(\cdot) = \{x(t, t_0, x_0), t_0 \leq t \leq T\}$$

of system (7.34), (7.35) from an initial point (t_0, x_0) is defined as a piecewise continuous function that coincides with a quasimotion $x(t, \tau_i, x(\tau_i + 0))$ in the intervals $(\tau_i, \tau_{i+1}]$, $i = \overline{0, m}$, and satisfies the condition of jump (7.35) at the points of pulses.

By $X(t_0, x_0)$ we denote a bundle of impulsive quasimotions of system (7.34), (7.35).

Theorem 7 ([126]). *The bundle $X(t_0, x_0)$ is a nonempty bounded (in norm) closed subset of the space $M[t_0, T]$.*

Proof. By virtue of Theorem 3 for quasimotions, the theorem is true for $t \in (\tau_i, \tau_{i+1}]$. Since the functions $I_i(x)$ are continuous and bounded, we conclude that, at the time of a pulse, a nonempty bounded closed subset of the space $M[t_0, T]$ transforms into a nonempty bounded closed subset of the same space. □

Theorem 8 ([126]). *For any points (t_0, x_0), the section of the bundle of impulsive quasimotions by the hyperplane $t = T$ is a nonempty compact set in $\mathbb{R}^n$.*

Theorem 9 ([126]). *For any point $(t^*, x(t^*))$, where $t^* \in [t_0, T)$, $x(\cdot)$ is an impulsive quasimotion from the bundle $X(t_0, x_0)$, the following inclusion is true:*

$$X(T, t^*, x(t^*)) \subset X(T, t_0, x_0).$$

Example 2. Consider an impulsive differential equation

$$\begin{aligned} \dot{x} &= 1 - 2\,\mathrm{sign}\, x, \quad t \neq i,\ i \in N, \\ \Delta x|_{t=i} &= -x + \alpha, \quad 0 < \alpha < 1, \\ x(0) &= x_0 \geq 0. \end{aligned} \tag{7.36}$$

If $x_0 \geq 1$, then $x(t) = x_0 - t$ for $t \in [0, 1]$. At the time $t = 1$, the solution suffers the pulse action and $x(1 + 0) = \alpha$. For $t \in (n, n + \alpha]$, $n \in N$, the solution $x(t) = \alpha + n - t$. For $t \in (n + \alpha, n + 1]$, the solutions do not exist in the ordinary sense and, in these segments, one can consider solutions, e.g., in a sense of the Filippov definition [51] or in a sense of quasimotions. Then $x(t) \equiv 0$ for $t \in (n + \alpha, n + 1]$.

If $0 \leq x_0 < 1$, then, in the interval $(n - 1, n]$, $n \in N$, we have the following solution:

$$x(t) = \begin{cases} A + n - 1 - t, & t \in (n - 1, n - 1 + A], \\ 0, & t \in (n - 1 + A, n], \end{cases}$$

$$A = \begin{cases} x_0, & n = 1, \\ \alpha, & n > 1. \end{cases}$$

Thus, a periodic quasimotion exists for $x_0 = \alpha/$.

Example 3. Consider an impulsive differential equation

$$\begin{aligned} \dot{x} &= v(t, x), \quad t \neq 1, \\ \Delta x|_{t=1} &= -x, \end{aligned}$$

where $v(t,x)$ is the function defined in Example 1. In this case, the role of impulsive quasimotions is played by the functions

$$x^{\pm}_{\alpha,\beta}(t) = \begin{cases} x^{\pm}_{\alpha}(t) & \text{for } t \in [0,1], \\ x^{\pm}_{\beta}(t-1) & \text{for } t \in (1,2]. \end{cases}$$

Theorem 10 ([126]). *The set-valued mapping $X(t_0, x)$ is upper semicontinuous with respect to x.*

The proof of the theorem follows from the definition of impulsive quasimotions and Theorem 6.

We now consider the problem of existence of periodic quasimotions for the impulsive differential equation of the form

$$\begin{aligned} \dot{x} &= A(t)x + f(t,x), \quad t \neq \tau_i, \\ \Delta x|_{t=\tau_i} &= B_i x + I_i(x), \end{aligned} \tag{7.37}$$

where $A : \mathbb{R} \to \mathbb{R}^{n\times n}$ is a continuous T-periodic function, $f : \mathbb{R} \times \mathbb{R}^n \to \mathbb{R}^n$ is a vector function T-periodic in t, $f(t,x) \in Q \in \text{comp}(\mathbb{R}^n)$, and the constant matrices B_i, the continuous functions $I_i(x) \in V \in \text{conv}(\mathbb{R}^n)$, and the times τ_i are such that, for some natural r,

$$B_{i+r} = B_i, \quad I_{i+r}(x) \equiv I_i(x), \quad \tau_{i+r} = \tau_i + T, \quad i \in N.$$

It is also assumed that $0 \leq \tau_1 < \cdots < \tau_r < T$ and $\det(E + B_i) \neq 0$ for all $i = \overline{1,r}$.

Let $\Phi(t,s) = \Phi_{AB_i}(t,s)$ be the matrizant of the homogeneous system corresponding to (7.37).

Theorem 11 ([126]). *If the spectral radius of the matrix $\Phi(T,0)$ is smaller than* 1*, then the impulsive differential Eq.* (7.37) *possesses a T-periodic quasimotion.*

Proof. The existence of a T-periodic quasimotion of Eq. (7.37) is directly connected with the existence of a fixed point of the mapping $\varphi : x \to X_T(x)$, where $X_T(x)$ is the section of the bundle of quasimotions $X(0,x)$ by the hyperplane $t = T$. By virtue of Theorem 8, the set $X_T(x)$ is a nonempty compact set in $\mathbb{R}^n$. In view of Theorem 11, the mapping φ is upper semicontinuous.

We now prove that there exists a set K such that the mapping φ maps the set K onto itself.

Parallel with Eq. (7.37), we consider a linear differential inclusion

$$\begin{aligned} \dot{x} &\in A(t)x + Q, \quad t \neq \tau_i, \\ \Delta x|_{t=\tau_i} &\in B_i x + V. \end{aligned} \tag{7.38}$$

The spectral radius of the matrix $\Phi(T,0)$ is smaller than 1. Thus, inclusion (7.38) has a T-periodic R-solution $R(t)$ whose initial set R_0 is defined as the fixed point of the mapping

$$\chi(R_0) = \Phi(T,0)R_0 + \int_0^T \Phi(T,\tau)Q\,d\tau + \sum_{0\le\tau_i<T} \Phi(T,\tau_i)V.$$

By the definition of impulsive quasimotions, we have

$$X_T(x) \subset R(T) = R_0$$

for any $x \in R_0$. Hence, the mapping φ maps the set R_0 onto itself. By the Kakutani theorem [14], there exists a fixed point $x_0 \in R_0$ of the mapping φ, i.e., there exists a T-periodic quasimotion of Eq. (7.37). The theorem is proved. □

7.3 Euler Quasibroken Lines

We now consider continuous systems of differential equations that do not satisfy the conditions of uniqueness theorems, e.g.,

$$\dot{x} = \sqrt{x}, \quad x(0) = 0. \tag{7.39}$$

It is clear that, for an arbitrarily chosen convergent subsequence of Euler broken lines, the limit of this subsequence is the function $x(t) \equiv 0$, although Eq. (7.39) has infinitely many solutions.

We show that, by using the algorithm of construction of Euler quasibroken lines based on the simplified procedure of construction of quasimotions, it is possible, for any solution of the equation, to construct a subsequence of Euler quasibroken lines convergent to this solution.

Consider a differential equation

$$\dot{x} = f(t,x), \quad x(t_0) = x_0, \tag{7.40}$$

where $f(t,x)$ is continuous in $D:\ t_0 \le t \le t_0 + a_0,\ \ \|x - x_0\| \le b$. We set $a = \min\{a_0, \frac{b}{M+1}\}$, where $M = \max_D \|f(t,x)\|$.

We split the segment $[t_0, t_0 + a]$ into n equal parts by the points $t_k = t_0 + kh_n$, $h_n = \frac{a}{n}$, $k = \overline{0,n}$.

The Euler quasibroken lines are constructed as follows:

$$x^n(t) = x^n(t_i) + (t - t_i)(f(t_i, x^n(t_i)) + \alpha_i), \quad t_i \le t \le t_{i+1}, \tag{7.41}$$
$$i = \overline{0, n-1}, \quad x^n(t_0) = x_0,$$

where $\|\alpha_i\| \le \alpha^m \le 1$, i.e.,

$$x^n(t_{i+1}) = x^n(t_i) + h_n f(t_i, x^n(t_i)) + h_n\alpha_i.$$

Remark 1. The Euler quasibroken lines can also be represented in the form

$$x^n(t) = x^n(t_i) + \int_{t_i}^{t} f(s, x^n(t_i))ds + \alpha_i(t - t_i).$$

Since α_i are chosen arbitrarily, we obtain an infinite set of Euler quasibroken lines even for fixed h_n.

We prove that family (7.41) is uniformly bounded and equicontinuous. For $t_k \le t \le t_{k+1}$ and any n, we get

$$\begin{aligned} \|x^n(t) - x_0\| &= \left\| h_n \sum_{i=0}^{k-1} (f(t_i, x^n(t_i)) + \alpha_i) + (t - t_k)(f(t_k, x^n(t_k)) + \alpha_k) \right\| \\ &\le h_n \sum_{i=0}^{k} \| f(t_i, x^n(t_i)) + \alpha_i \| \le h_n(M + \alpha^m)(k+1) \\ &\le h_n n(M + \alpha^m) \le a(M+1) \le b. \end{aligned}$$

This family is equicontinuous because

$$\| f(t_i, x^n(t_i)) + \alpha_i \| \le M + 1, \quad i = \overline{0, n-1}.$$

Hence, by the Arzelà theorem, for any α^m, one can select a convergent subsequence from the family of quasibroken lines,.

We show that, for any solution $x(t)$ of Eq. (7.40), there exists a sequence of quasibroken lines convergent to this solution. Let $x(t)$ be an arbitrary solution of (7.40). We construct a sequence of inscribed broken lines

$$\tilde{x}^n(t) = x(t_i) + \frac{t - t_i}{h_n}(x(t_{i+1}) - x(t_i)), \quad t_i \le t \le t_{i+1}, \ i = \overline{0, n-1}.$$

Let us show that, by the choice of α_i, one can find a quasibroken line that coincides with $\tilde{x}^n(t)$ at the points of the partition:

$$\begin{aligned} x(t_1) - \tilde{x}^n(t_1) &= x(t_1) - x_0 - h_n(f(t_0, x_0) + \alpha_0) \\ &= \int_{t_0}^{t_1} [f(s, x(s)) - f(t_0, x_0)]\, ds - \alpha_0 h_n. \end{aligned}$$

Since

$$\| f(s, x(s)) - f(t_i, x(t_i)) \| \le \alpha^m$$

for sufficiently small h_n (this follows from the uniform continuity of the function $f(t, x(t))$, there exists α_0 such that $\|\alpha_0\| \le \alpha^m$ and $\tilde{x}^n(t_1) = x^n(t_1)$.

Similarly, there exist α_i, $\|\alpha_i\| \le \alpha^m$, such that

$$\tilde{x}^n(t_{i+1}) = x^n(t_{i+1}), \quad i = \overline{0, n-1}.$$

It is clear that the sequence of inscribed broken lines converges to $x(t)$, i.e., there exists a subsequence of quasibroken lines that converges to an arbitrary solution of Eq. (7.40). However, one can also find subsequences of quasibroken lines that converge not to a solution of Eq. (7.40) but to a solution of the equation

$$\dot{x} = f(t, x) + \beta(t),$$

where $\beta(t)$ is an arbitrary continuous function such that $\|\beta(t)\| \leq \alpha^m$.

Passing to the limit as $\alpha^m \to 0$, we exclude the superfluous solutions.

Remark 2. Consider a differential inclusion

$$\dot{x} \in F(t, x), \quad x(t_0) \in X_0.$$

We now construct the Euler quasibroken lines for the R-solutions and $t \in (t_i, t_{i+1}]$:

$$X^m(t, \alpha) = \bigcup_{x \in X^m_{\alpha_i}(t_i, \alpha)} \left\{ x + \int_{t_i}^{t} F(s, x) ds \right\}, \tag{7.42}$$

$$h(X^m_{\alpha_i}(t_i, \alpha), X^m(t_i, \alpha)) \leq \alpha_i.$$

If $F(t, x)$ is a Lipschitz function, then, for $\alpha_i = 0$, we get ordinary Euler broken lines convergent to a unique R-solution. For $\alpha_i \neq 0$, the quasibroken lines also converge to the unique R-solution as $\alpha \to 0$.

If an R-solution is not unique, then, by analogy with the previous case, we can show that, for any R-solution, there exists a sequence of quasibroken lines (7.42) that converges to this solution.

Appendix A
Some Elements of Set-Valued Analysis

Let $\mathbb{R}^n$ be an n-dimensional real Euclidean space with norm

$$\|x\| = \sqrt{\sum_{i=1}^{n} x_i^2}.$$

By $\mathrm{comp}(\mathbb{R}^n)(\mathrm{conv}(\mathbb{R}^n))$ we denote the space of all nonempty compact (and convex) subsets of $\mathbb{R}^n$ with the Hausdorff metric

$$h(F, G) = \min\{r \geq 0 : F \subset G + B_r(0), G \subset F + B_r(0)\},$$

where $B_r(a)$ is a ball in $\mathbb{R}^n$ of radius $r \geq 0$ centered at a point $a \in \mathbb{R}^n$.

Let $\rho(x, F) = \min_{f \in F} \|x - f\|$ be the distance from a point $x \in \mathbb{R}^n$ to a set $F \in \mathrm{comp}(\mathbb{R}^n)$. The mapping $\rho(\cdot, F)$ satisfies the Lipschitz condition

$$|\rho(x, F) - \rho(y, F)| \leq \|x - y\| \quad \text{for all } x, y \in \mathbb{R}^n.$$

The Hausdorff distance can also be introduced as follows: Assume that a set $D_{F,G}$ consists of numbers of the form $\rho(f, G)$, $\rho(g, F)$, where $f \in F$ and $g \in G$. Then $h(F, G) = \sup D_{F,G}$.

In addition to the ordinary set-theoretic operations, we consider two more operations in the space $\mathrm{comp}(\mathbb{R}^n)$, namely, the operations of summation and multiplication by a scalar.

Definition 1 ([22, 23]). The sum $F + G$ of two sets $F, G \in \mathrm{comp}(\mathbb{R}^n)$ is defined as the set

$$H = F + G = \{h = f + g : f \in F, g \in G\}.$$

Example 1. $B_{r_1}(a_1) + B_{r_2}(a_2) = B_{r_1+r_2}(a_1 + a_2)$.

Definition 2 ([22, 23]). The product λF of a scalar $\lambda \in \mathbb{R}$ by a set $F \in \mathrm{comp}(\mathbb{R}^n)$ is defined as the set

$$G = \lambda F = \{g = \lambda f : f \in F\}.$$

Example 2. $\lambda B_r(0) = B_{|\lambda| r}(0)$.

Definition 3 ([22, 23]). The image of the set $F \in \text{comp}(\mathbb{R}^n)$ under a linear transformation specified by a matrix $A \in \mathbb{R}^{n \times n}$ is defined as the set

$$G = AF = \{g = Af : f \in F\}.$$

Lemma 1. *If sets $F, G \in \text{comp}(\mathbb{R}^n)(\text{conv}(\mathbb{R}^n))$, a scalar $\lambda \in \mathbb{R}$, and a matrix $A \in \mathbb{R}^{n \times n}$, then the sets $F + G, \lambda F, AF \in \text{comp}(\mathbb{R}^n)(\text{conv}(\mathbb{R}^n))$.*

Lemma 2. *If $F, G, H \in \text{comp}(\mathbb{R}^n), \alpha, \beta \in \mathbb{R}$, and $A, B \in \mathbb{R}^{n \times n}$, then the following equalities are true:*

1. $(F + G) + H = F + (G + H)$;
2. $\alpha(\beta F) = (\alpha\beta)F$;
3. $F + G = G + F$;
4. $1 \cdot F = F$;
5. $F + \{0\} = F$;
6. $A(F + G) = AF + AG$;
7. $\alpha(F + G) = \alpha F + \alpha G$;
8. $A(BF) = (AB)F$;
9. $EF = F$.

However, the space $\text{comp}(\mathbb{R}^n)(\text{conv}(\mathbb{R}^n))$ is not a linear space with respect to the operations presented above because, in the general case, it is impossible to introduce the notion of element opposite to $F \in \text{comp}(\mathbb{R}^n)(\text{conv}(\mathbb{R}^n))$ and the following property of linear spaces is not true:

$$(\alpha + \beta)F = \alpha F + \beta F. \tag{A.1}$$

In the general case, instead of this property, we have the following inclusion:

$$(\alpha + \beta)F \subset \alpha F + \beta F.$$

Thus, for $\alpha = 1,\ \beta = -1$, and $F = B_r(0)$, we find

$$(1 + (-1))\, B_r(0) = \{0\},$$
$$1 \cdot B_r(0) + (-1) \cdot B_r(0) = B_r(0) + B_r(0) = B_{2r}(0).$$

However, equality (A.1) is true in the case where $\alpha > 0, \beta > 0$, and the set $A \in \text{conv}(\mathbb{R}^n)$.

Let $F \in \text{comp}(\mathbb{R}^n)$ and let $M_m(F) = \frac{1}{m}\sum_{i=1}^{m} F$. Then the following inequality holds:

$$h(\text{co}\, F, M_m(F)) \le \frac{|F|(n+1)}{m},$$

where $|F| = \max_{f \in F} \|f\|$ is the modulus of the set F.

Definition 4 ([127]). A sequence of sets $\{A_n\}_{n=1}^{\infty}$, $A_n \in \text{comp}(\mathbb{R}^n)$, $n = \overline{1, \infty}$, is called convergent to $A \in \text{comp}(\mathbb{R}^n)$ if the sequence $\{h(A_n, A)\}_{n=1}^{\infty}$ converges to zero.

Theorem 1 ([127]). *The metric space* $\text{comp}(\mathbb{R}^n)$ *is a complete space.*

Definition 5 ([23]). A support function of the set $A \in \operatorname{comp}(\mathbb{R}^n)$ is defined as a scalar function $c(A, \psi)$ specified by the condition

$$c(A, \psi) = \max_{a \in A}(a, \psi),$$

where (a, ψ) is the scalar product of the vectors $a, \psi \in \mathbb{R}^n$.

The support set of the set $F \in \operatorname{comp}(\mathbb{R}^n)$ in the direction of the vector $\psi_0 \in \mathbb{R}^n$ is defined as the set of all vectors $f_0 \in F$ on which the maximum is attained in the definition of the support function

$$U(F, \psi_0) = \{f_0 \in F : (f_0, \psi_0) = c(F, \psi_0)\}.$$

The hyperplane Γ_{ψ_0} in the space $\mathbb{R}^n$ specified by the relation $\Gamma_{\psi_0} = \{x \in \mathbb{R}^n : (x, \psi_0) = c(F, \psi_0)\}$ is called the support hyperplane for the set f in the direction of the support vector ψ_0.

The following representation is true for the support set $U(F, \psi_0)$:

$$U(F, \psi_0) = F \cap \Gamma_{\psi_0}.$$

The hyperplane Γ_{ψ_0} splits the entire space $\mathbb{R}^n$ into two half spaces $\mathbb{R}^+$ and $\mathbb{R}^-$. The set F lies in the negative half space $\mathbb{R}^-$ relative to the vector ψ_0, i.e., the inequality

$$(f, \psi_0) \le c(F, \psi_0)$$

holds for all points $f \in F$.

Definition 6. The set $F \in \operatorname{comp}(\mathbb{R}^n)$ is called strictly convex in the direction of the vector $\psi_0 \in \mathbb{R}^n$ if its support set $U(F, \psi_0)$ is formed by a single point. The set F is called strictly convex if it is strictly convex in any direction.

We now present the main properties of the support function. Let $F, G \in \operatorname{comp}(\mathbb{R}^n)$ and let $\psi, \psi_1, \psi_2 \in \mathbb{R}^n$. Then

(1) $c(F, \lambda\psi) = \lambda c(F, \psi)$ for $\lambda \ge 0$;

(2) $c(F, \psi_1 + \psi_2) \le c(F, \psi_1) + c(F, \psi_2)$;

(3) $c(F + G, \psi) = c(F, \psi) + c(G, \psi)$;

(4) $c(\lambda F, \psi) = c(F, \lambda\psi)$;

(5) $c(AF, \psi) = c(F, A^T\psi)$;

(6) $c(\operatorname{co} F, \psi) = c(F, \psi)$;

(7) $\operatorname{co} F = \bigcap_{\psi \in S_1(0)}\{x \in \mathbb{R}^n : (x, \psi) \le c(F, \psi)\}$;

(8) if $F = G$, then $c(F, \psi) = c(G, \psi)$ for all $\psi \in \mathbb{R}^n$; if $c(F, \psi) = c(G, \psi)$ for all $\psi \in S_1(0)$, then $\operatorname{co} F = \operatorname{co} G$;

(9) if $F \subset G$, then $c(F, \psi) \le c(G, \psi)$ for all $\psi \in \mathbb{R}^n$; if $c(F, \psi) \le c(G, \psi)$ for all $\psi \in S_1(0)$, then $\operatorname{co} F \subset \operatorname{co} G$;

(10) if $F \bigcap G \neq \emptyset$, then $c(F, \psi) + c(G, -\psi) \ge 0$ for all $\psi \in \mathbb{R}^n$; if $c(F, \psi) + c(G, -\psi) \ge 0$ for all $\psi \in S_1(0)$, then $\operatorname{co} F \cap \operatorname{co} G \neq \emptyset$;

(11) $|c(F, \psi_1) - c(G, \psi_2)| \le |F| \cdot \|\psi_1 - \psi_2\| + \|\psi_1\| h(F, G) + 2\|\psi_1 - \psi_2\| h(F, G)$;

(12) $h(\operatorname{co} F, \operatorname{co} G) = \max_{\psi \in S_1(0)} |c(F, \psi) - c(G, \psi)| \le h(F, G)$;

(13) the set F is strictly convex in the direction of the vector $\psi_0 \in \mathbb{R}^n$ if and only if its support function $c(F, \psi)$ is differentiable at the point ψ_0.

Definition 7. A set-valued mapping is defined as an arbitrary function $F : \mathbb{R}^m \to \operatorname{comp}(\mathbb{R}^n)$, i.e., a function whose argument is a vector $x \in \mathbb{R}^m$ and values are elements of the space $\operatorname{comp}(\mathbb{R}^n)$, i.e., nonempty compact sets from the space $\mathbb{R}^n$.

Definition 8 ([24]). A set-valued mapping $F : \mathbb{R}^m \to \operatorname{comp}(\mathbb{R}^n)$ is called measurable if, for any nonempty compact set K, the set $\{x \in \mathbb{R}^m : h(F(x), K) \le \varepsilon\}$ is Lebesgue measurable.

Definition 9. A set-valued mapping $f : \mathbb{R}^m \to \mathbb{R}^n$ is called a measurable section (a single-valued measurable branch or a measurable selector) of the set-valued mapping $F : \mathbb{R}^m \to \operatorname{comp}(\mathbb{R}^n)$ if $f(x)$ is measurable and $f(x) \in F(x)$ for almost all $x \in \mathbb{R}^m$.

Theorem 2 ([49]). *If a set-valued mapping $F : \mathbb{R}^m \to \operatorname{comp}(\mathbb{R}^n)$ is measurable, then it has a measurable single-valued branch.*

Theorem 3 ([48]). *If $F : \mathbb{R}^m \to \operatorname{comp}(\mathbb{R}^n)$ is a measurable set-valued mapping, $\psi_0 \in \mathbb{R}^n$, then there exists a measurable single-valued branch $f(x)$ of the mapping $F(x)$ that belongs to the set $U(F(x), \psi_0)$.*

Theorem 4 ([49]). *Let a set-valued mapping $F : \mathbb{R}^m \to \operatorname{comp}(\mathbb{R}^n)$ and a function $v : \mathbb{R}^m \to \mathbb{R}^n$ be measurable. Then there exists a measurable branch $f(x)$ of the mapping $F(x)$ such that the condition*

$$\rho(v(x), F(x)) = \|v(x) - f(x)\|$$

is satisfied for almost all $x \in \mathbb{R}^m$.

Theorem 5 ([48]). *Assume that the function $f : \mathbb{R}^m \times \mathbb{R}^p \to \mathbb{R}^n$ is measurable with respect to $x \in \mathbb{R}^m$ and continuous in $u \in \mathbb{R}^p$. Moreover, suppose that a mapping $U : \mathbb{R}^m \to \operatorname{comp}(\mathbb{R}^p)$ and a function $v : \mathbb{R}^m \to \mathbb{R}^n$ are measurable and, in addition, $v(x) \in f(x, U(x))$. Then there exists a measurable branch $u(x)$ of the mapping $U(x)$ such that $v(x) = f(x, u(x))$.*

We now fix a segment $I = [t_0, t_1]$ and a set-valued mapping $F : I \to \operatorname{comp}(\mathbb{R}^n)$.

Definition 10 ([10]). The Aumann integral of the set-valued mapping $F(t)$ on the segment I is defined as the set

$$G = \int_{t_0}^{t_1} F(t)dt = \left\{ \int_{t_0}^{t_1} f(t)dt : f(t) \in F(t) \right\}.$$

Here, the Lebesgue integral on the right-hand side is taken over all single-valued branches of the mapping $F(t)$, where it exists.

Theorem 6 (Lyapunov [78]). *Assume that the set-valued mapping $F(t)$ is measurable and satisfies the estimate $|F(t)| \le k(t)$, where $k(t)$ is summable on I. Then $G = \int_{t_0}^{t_1} F(t)dt$ is a nonempty convex compact set in the space $\mathbb{R}^n$.*

Note that this integral may exist even in the case where a set-valued mapping is not measurable on I because the condition of its existence is the presence of a single-valued Lebesgue integrable branch of the set-valued mapping.

Thus, the set-valued mapping

$$F(t) = \begin{cases} S_1(0), & t \in J, \\ \{0\}, & t \in I \setminus J, \end{cases}$$

where J is a nonmeasurable subset of I, is not measurable on I. However, this set-valued mapping contains a Lebesgue integrable single-valued branch $f(t) \equiv 0, t \in I$, and therefore,

$$0 \in \int_{t_0}^{t_1} F(t)\,dt.$$

Theorem 7 ([10, 78]). *Assume that a set-valued mapping $F(t)$ is measurable and satisfies the estimate $|F(t)| \le k(t)$, where $k(t)$ is summable on I. Then*

$$c\left(\int_{t_0}^{t_1} F(t)dt, \psi \right) = \int_{t_0}^{t_1} c(F(t), \psi)dt.$$

Definition 11 ([23]). A set-valued mapping $F : \mathbb{R}^m \to \mathrm{comp}(\mathbb{R}^n)$ is called upper semicontinuous at a point $x_0 \in \mathbb{R}^m$ if, for any number $\varepsilon > 0$, one can find a number $\delta > 0$ such that

$$F(x) \subset F(x_0) + S_\varepsilon(0)$$

for $\|x - x_0\| < \delta$.

Definition 12 ([23]). A set-valued mapping $F : \mathbb{R}^m \to \mathrm{comp}(\mathbb{R}^n)$ is called lower semicontinuous at a point $x_0 \in \mathbb{R}^m$ if, for any number $\varepsilon > 0$, there exists a number $\delta > 0$ such that the inclusion

$$F(x_0) \subset F(x) + S_\varepsilon(0)$$

holds for $\|x - x_0\| < \delta$.

Definition 13 ([23]). A set-valued mapping $F : \mathbb{R}^m \to \mathrm{comp}(\mathbb{R}^n)$ is called continuous at a point $x_0 \in \mathbb{R}^m$ if it is both upper and lower semicontinuous at this point.

As one of the most important properties of set-valued mappings extensively used in applications, we can mention the Michael theorem [85]. We now present one of its numerous interpretations.

Theorem 8 ([8]). *Let X be a metric space and let Y be a Banach space. Assume that a set-valued mapping $F(\cdot)$ from X into a closed convex subspace of Y is lower semicontinuous. Then there exists a continuous selector $f : X \to Y$ from $F(\cdot)$.*

Remark. The condition of convexity of $F(x)$ for all $x \in X$ is essential because if $F(x)$ is not convex, then even its continuity does not guarantee the existence a continuous selector for $F(\cdot)$.

Theorem 9 ([52]). *Assume that a set-valued mapping $F : \mathbb{R} \times \mathbb{R}^m \to \mathrm{conv}(\mathbb{R}^n)$ is measurable with respect to t and continuous in x. Then there exists a single-valued branch $f(t,x) \in F(t,x)$ measurable with respect to t and continuous in x.*

Definition 14 ([51]). A set-valued mapping $F : I \times \mathbb{R}^m \to \mathrm{comp}(\mathbb{R}^n)$ satisfies the Carathéodory conditions if

(a) for any fixed $x \in \mathbb{R}^m$, the set-valued mapping $F(\cdot, x)$ is measurable;

(b) for almost all fixed $t \in I$, the set-valued mapping $F(t, \cdot)$ is upper semicontinuous.

Definition 15 ([51]). A set-valued mapping $F(t,x)$ satisfies the improved Carathéodory conditions if it satisfies condition (a) and the condition

(b′) for almost all fixed $t \in I$, the set-valued mapping $F(t, \cdot)$ is continuous.

Theorem 10 ([28]). *If a set-valued mapping $F(\cdot,\cdot)$ satisfies the improved Carathéodory conditions, then, for any measurable set-valued mapping $Q : I \to \mathrm{comp}(\mathbb{R}^n)$, the set-valued mapping $F(\cdot, Q(\cdot)) : I \to \mathrm{comp}(\mathbb{R}^n)$ is measurable.*

Note that this property is not true if the improved Carathéodory conditions are replaced by the ordinary conditions.

Definition 16 ([3]). A set-valued mapping $F : \mathbb{R} \to \mathrm{comp}(\mathbb{R}^n)$ is called absolutely continuous if, for any $\varepsilon > 0$, there exists a number $\delta > 0$ such that, for any natural N, the following inequality is true:

$$\sum_{i=1}^{N} h(F(b_i), F(a_i)) < \varepsilon$$

for $a_1 < b_1, \ldots, a_N < b_N$ and $\sum_{i=1}^{N}(b_i - a_i) < \delta$.

Definition 17 ([4]). A set-valued mapping $F : \mathbb{R}^m \to \text{comp}(\mathbb{R}^n)$ is called locally Lipschitz if, for any $x_0 \in \mathbb{R}^m$, there exist a neighborhood $U(x_0) \subset \mathbb{R}^m$ and a constant $L \geq 0$ such that

$$h(F(x''), F(x')) \leq L\|x'' - x'\|$$

for any $x', x'' \in U(x_0)$. Moreover, this mapping is called Lipschitz if there exists $L \geq 0$ such that

$$h(F(x''), F(x')) \leq L\|x'' - x'\|$$

for any $x', x'' \in \mathbb{R}^m$.

Appendix B
Differential Inclusions

Consider a differential inclusion

$$\dot{x} \in F(t, x), \tag{B.1}$$

where $t \in I \subset \mathbb{R}$ is time, $x \in \mathbb{R}^n$ is the phase vector, and $F : I \times \mathbb{R}^n \to \mathrm{comp}(\mathbb{R}^n)$.

Definition 1. An absolutely continuous function $x(t)$ defined on a segment (in an interval) $J \subset I$ is called an ordinary solution of the differential inclusion (B.1) on J if

$$\dot{x}(t) \in F(t, x(t))$$

almost everywhere on J.

It is known that, in the theory of differential equations, the transitions from differential equations to integral equations, and vice versa, are equivalent. For differential inclusions, this is not true, i.e., a solution of inclusion (B.1) is a solution of the integral inclusion

$$x(t) \in x(t_0) + \int_{t_0}^{t} F(s, x(s))ds \tag{B.2}$$

but not all solutions of the integral inclusion (B.2) are solutions of the differential inclusion (B.1).

Example 1 ([34]). Let $F(t, x) \equiv [0, 1]$, $x_0 = 0$, $I = [0, 2]$ and let

$$x(t) = \begin{cases} 0, & t \in [0, 1), \\ 2t - 2, & t \in [1, 2]. \end{cases}$$

It is clear that $x(t)$ is not a solution of the differential inclusion

$$\dot{x} \in [0, 1], \quad x(0) = 0$$

on $[0, 2]$ because $\dot{x}(t) = 2 \notin [0, 1]$ for $t \in (1, 2)$ but is a solution of the corresponding integral inclusion

$$x(t) \in 0 + \int_0^t [0, 1]ds$$

because $\int_0^t [0, 1]dt = [0, t]$ and $0 \le x(t) \le t$ for $t \in [0, 2]$.

For this reason, the theory of differential equations deals with a different form of integral inclusions for which it is possible to obtain a result similar to the corresponding result in the theory of differential equations.

Definition 2 ([34]). A continuous function $x(t)$ is called a generalized solution of inclusion (B.1) on J if the integral inclusion

$$x(t'') - x(t') \in \int_{t'}^{t''} F(t, x(t))dt \tag{B.3}$$

is valid for all $t' < t'' : t', t'' \in J$.

Theorem 1 ([34]). *Assume that* $F : I \times \mathbb{R}^n \to \text{conv}(\mathbb{R}^n)$ *satisfies the following conditions:*

(1) $F(\cdot, x)$ *is measurable for all* $x \in \mathbb{R}^n$*;*

(2) $F(t, \cdot)$ *is continuous for almost all* $t \in I$*;*

(3) $|F(t,x)| \le m(t)$, $(t,x) \in I \times \mathbb{R}^n$, $m(t)$ *is summable on* I.

Then the set of ordinary solutions of inclusion (B.1) *coincides with the set of generalized solutions.*

Corollary 1. *Let* $F : I \times \mathbb{R}^n \to \text{comp}(\mathbb{R}^n)$ *be a set-valued mapping satisfying conditions* (1)–(3) *of Theorem* 1.

Then the set of ordinary solutions of inclusion (B.1) *is contained in the set of generalized solutions.*

We now present some other definitions of solutions of the differential inclusion (B.1).

Definition 3 ([65]). A function $x(t)$ is called a quasisolution of the differential inclusion (B.1) if there exists a sequence of functions $\{x_k(t)\}_{k=1}^{\infty}$ such that

(1) $x_k(t)$ is absolutely continuous on J;

(2) $|\dot{x}_k(t)| \le m(t)$, $t \in J$, $m(t)$ is summable on J, $k = 1, 2, \ldots$;

(3) $\lim_{k\to\infty} x_k(t) = x(t)$, $t \in J$;

(4) $\lim_{k\to\infty} \rho(\dot{x}_k(t), F(t, x_k(t))) = 0$ almost everywhere on J.

Definition 4 ([65]). A function $x(t)$ is a called a Riemannian solution of the differential inclusion (B.1) if $\dot{x}(t)$ is Riemann integrable and $\dot{x}(t) \in F(t, x(t))$ for all $t \in J$.

Definition 5 ([65]). A function $x(t)$ is called a classic solution of the differential inclusion (B.1) if $x(t)$ is continuously differentiable on J and $\dot{x}(t) \in F(t, x(t))$ for all $t \in J$.

Since differential inclusions are obtained as a result of generalization of differential equations, all problems typical of the theory of ordinary differential equations appear in the theory of differential inclusions, namely, the problems of existence of solutions, their extendability, boundedness, continuous dependence on the initial conditions and parameters, etc. At the same, for differential inclusions, a family of trajectories originates from every initial point. This set-valuedness leads to the appearance of various specific problems, including the closedness and convexity of the family of solutions, existence of boundary solutions, selection of solutions with given properties, and many others.

First, we present some results concerning the conditions of existence of ordinary solutions to the differential inclusion (B.1) with initial condition $x(t_0) = x_0$.

Theorem 2 ([51]). *Assume that, at every point (t, x) of the domain $D = \{t_0 \le t \le t_0+a,\ \|x-x_0\| \le b\}$, a set-valued mapping $F(t, x)$ satisfies the following conditions:*

(1) *the set $F(t, x)$ is nonempty and closed;*

(2) *$F(\cdot, x)$ is measurable for all x;*

(3) *$F(t, \cdot)$ is continuous for all t;*

(4) *$|F(t, x)| \le m(t)$, where $m(t)$ is summable on $[t_0, t_0 + a]$.*

Then, for $t_0 \le t \le t_0 + d$, there exists a solution of problem (B.1)*, where*

$$d \le a, \quad \varphi(t_0 + d) \le b, \quad \varphi(t) = \int_{t_0}^{t} m(s)ds.$$

Definition 6 ([23]). A function $\omega(t, r) \ge 0$ $(t \ge t_0,\ 0 \le r \le b)$ is called a Kamke function if it is continuous in r, measurable with respect to t, $\omega(t, r) \le m_0(t)$, where $m_0(t)$ is summable on the segment $[0, c]$ for any c, and the function $r(t) \equiv 0$ is a unique solution of the problem

$$\dot{r}(t) = \omega(t, r(t)), \quad r(t_0) = 0,$$

for $t \ge t_0$.

Thus, if the function $k(t)$ is summable, then $k(t)r$ is a Kamke function $(0 \le r \le b)$.

Theorem 3 (Filippov [23, 105]). *Assume that, at every point (t, x) of the domain $D = \{t \in [t_0, T],\ \|x-x_0\| \le b\}$, a set-valued mapping $F(t, x)$ satisfies the following conditions:*

(1) *the set $F(t, x)$ is nonempty and closed;*

(2) *$F(\cdot, x)$ is measurable for all x;*

(3) *the set $F(t, x)$ is convex;*

(4) *for any $r > 0$, $\|x - y\| \le r$, and almost all t,*

$$h(F(t, x), F(t, y)) \le w(t, r), \tag{B.4}$$

where $w(t, r)$ is a Kamke function.

In addition, let the function $y(t)$ be absolutely continuous for $t \in [t_0, T]$, let its graph be contained in D, $y(t_0) = y_0$, and let, for almost all $t \in [t_0, T]$,

$$\rho(\dot{y}(t), F(t, y(t))) \leq \eta(t),$$

where $\eta(t)$ is summable on $[t_0, T]$.

Then, for $(t_0, x_0) \in D$, one can find a solution $x(t)$ of the problem

$$\dot{x} \in F(t, x), \quad x(t_0) = x_0,$$

such that

$$\|x(t) - y(t)\| \leq r(t), \quad \|\dot{x}(t) - \dot{y}(t)\| \leq w(t, r(t)) + \eta(t) \tag{B.5}$$

for almost all $t \in [t_0, t^*]$, where $r(t)$ is the upper solution of the problem

$$\dot{r} = w(t, r) + \eta(t), \quad r(t_0) = \|x_0 - y_0\|,$$

and t^* is an arbitrary number such that $(t, x(t)) \in D$ for $t_0 \leq t \leq t^*$.

Remark 1. If, in Theorem 3, condition (4) is replaced by the Lipschitz condition, i.e.,

$$h(F(t, x), F(t, y)) \leq k(t)\|x - y\|, \quad k(t) \text{ is summable on } [t_0, T],$$

then condition (3) can be removed and, in inequalities (B.5), we have

$$r(t) = \|x_0 - y_0\| e^{\int_{t_0}^{t} k(s)ds} + \int_{t_0}^{t} \eta(s) e^{\int_s^t k(\tau)d\tau} ds.$$

Definition 7. An integral funnel of the point (t_0, x_0) (of the set K) is defined as a set of points lying on the graphs of all solutions passing through this point (resp., through the points of the set K).

Definition 8. A section $t = t'$ of the funnel of the point (t_0, x_0) is defined as a set of attainability at time t', i.e., as the set of points $\{x(t')\}$ that can be attained at time t' by moving along all possible solutions originating at time t_0 from the point x_0.

A section of the funnel of the set K is defined similarly.

Theorem 4 ([23,33]). *Assume that the following conditions are satisfied in a bounded domain D:*

(1) *the set $F(t, x)$ is nonempty and closed;*

(2) *$|F(t, x)| \leq m(t)$, where $m(t)$ is a function summable on $[t_0, t_1]$;*

(3) *$F(t, \cdot)$ is upper semicontinuous on D;*

(4) *$F(\cdot, x)$ is measurable on D;*

(5) *the set $F(t, x)$ is convex.*

If all solutions of (B.1) *on the segment $[t_0, t_1]$ exist and are contained in D, then the set $H_F(t_0, x_0)$ of these solutions is a compact set in the space $C[t_0, t_1]$. The same is true for the set $H_F(K)$ of all solutions with all possible initial conditions $(t_0, x_0) \in K$, where K is a compact set and $K \subset D$. If K is a connected compact set (and, in particular, if K is a point), then the set $H_F(K)$ is connected.*

If the set $F(t, x)$ is not convex, then both the funnel and the set of attainability can be nonclosed.

Example 2 ([48]). Consider a system

$$\dot{x} = -y^2 + u^2, \quad \dot{y} = u, \quad -1 \le u(t) \le 1. \tag{B.6}$$

Here, the set F is an arc of the parabola

$$v_1 = v_2^2 - y^2, \quad -1 \le v_2 \le 1,$$

i.e., is not convex (v_1 and v_2 are the projections of points of the set F onto the coordinate axes).

For $0 \le t \le 1$, we consider the set of solutions with initial conditions $x(0) = y(0) = 0$. If $y(t) \equiv 0$, then $u(t) = 0$ everywhere, $\dot{x} = -y^2 + u^2 = 0$, and $x(t) \equiv 0$. If $y(t)$ is not identically equal to zero ($0 \le t \le 1$), then $\dot{x} = -y^2 + u^2 \le 1$. Moreover, $\dot{x} < 1$ in the intervals where $y(t) \ne 0$. Hence, $x(1) < 1$ for all solutions and the point $t = 1,\ x = 1,\ y = 0$ belongs neither to the graphs of solutions, nor to the segment $0 \le t \le 1$ of the integral funnel.

We now consider a solution $x_k(t),\ y_k(t)$ for which $x_k(0) = 0$, $y_k(0) = 0$, and

$$u = \begin{cases} 1, & \frac{2i}{k} \le t < \frac{2i+1}{k}, \\ -1, & \frac{2i+1}{k} \le t < \frac{2i+2}{k}, \end{cases} \qquad i = 0, 1, 2, \ldots.$$

In this case,

$$0 \le y_k(t) \le \frac{1}{k}, \quad \dot{x}_k(t) \ge 1 - \frac{1}{k^2}, \quad x_k(1) \ge 1 - \frac{1}{k^2}.$$

Thus, points of the graphs of solutions with trivial initial conditions lie arbitrarily close to the point $t = 1,\ x = 1, y = 0$, whereas the point itself does not lie on the graph of this solution. Hence, the set of these points and the segment $0 \le t \le 1$ of the funnel are not closed.

Equations (B.6) can be regarded as equations of a controlled system, i.e., a system whose motion can be controlled by an arbitrary choice of the function $u(t)$ within the prescribed limits. This means that, for a unit period of time, this system cannot be transferred from the state $x = y = 0$ into the state $x = 1,\ y = 0$ but can be transferred into a state arbitrarily close to $x = 1,\ y = 0$ by changing the function $u(t)$ sufficiently rapidly from 1 to -1 and back (sliding mode).

In the absence of the condition of convexity of the set $F(t, x)$, the relations between the sets of solutions of the inclusion $\dot{x} \in F(t, x)$ and the inclusion

$$\dot{x} \in \operatorname{co} F(t, x) \tag{B.7}$$

were studied, e.g., in [31, 157].

Theorem 5 ([31, 104]). *Assume that a set-valued mapping $F(t, x)$ satisfies the conditions:*

(1) *the set $F(t, x)$ is nonempty and closed;*

(2) *$|F(t, x)| \le m(t)$, where $m(t)$ is a summable function;*

(3) *$F(t, \cdot)$ is upper semicontinuous in x;*

(4) *for any $r > 0$, $\|x - y\| \le r$, and almost all t,*

$$h(F(t, x), F(t, y)) \le w(t, r),$$

where $w(t, r)$ is a Kamke function.

Then each solution of inclusion (B.7) *with the initial condition $x(t_0) = x_0$ is the limit of a uniformly convergent sequence of solutions of the inclusion $\dot{x} \in F(t, x)$ with the same initial condition.*

In this case, the indicated limit may be not a solution of the inclusion $\dot{x} \in F(t, x)$ if the set $F(t, x)$ is not convex.

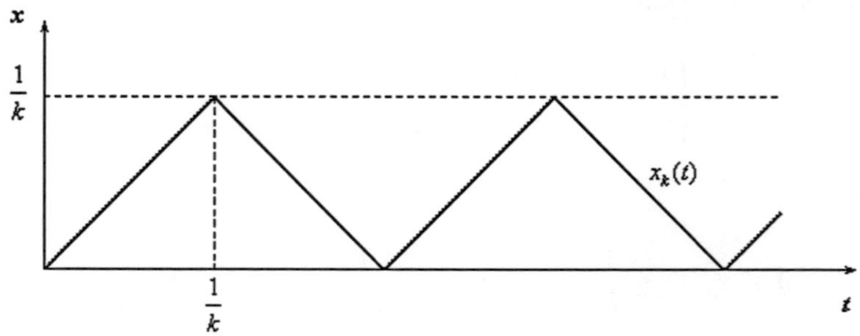

Figure 1.

Thus, if $x \in \mathbb{R}$ and the set $F(t, x)$ consists of two points 1 and -1, then the sequence of solutions $\{x_k(t)\}$ uniformly converges to the function $x(t) \equiv 0$, which is not a solution of inclusion (B.1) (Figure 1).

Condition (4) cannot be removed [104] and replaced by the Hölder condition.

Example 3 ([104]). Assume that the set $F(t,x)$, $t \in \mathbb{R}$, $x \in \mathbb{R}^2$, does not depend on t and consists of two points $(-1, x_1^2 + \sqrt{|x_2|})$ and $(1, x_1^2 + \sqrt{|x_2|})$. Then $\operatorname{co} F(t,x)$ is the segment connecting these points.

The vector function $x(t) \equiv 0$ satisfies the inclusion $\dot{x} \in \operatorname{co} F(t,x)$ but does not satisfy the inclusion $\dot{x} \in F(t,x)$. In [104], it is shown that none of the sequences of solutions of the inclusion $x \in F(t,x)$ has the limit $x(t) \equiv 0$, i.e., the solution of the inclusion $\dot{x} \in \operatorname{co} F(t,x)$.

We now consider inclusion (B.1), where $F : D \to \operatorname{conv}(\mathbb{R}^n)$ is a set-valued mapping continuous in D. The section of the integral funnel by the plane $t = \text{const}$ is a closed set $R(t)$ depending on t. Thus, the funnel is the graph of the set-valued function $R(t)$. The following approach to the determination of this function is proposed in [99–102]:

Definition 9 ([23, 101, 102, 144]). A set-valued function $R(t)$ is called an R-solution generated by the differential inclusion (B.1) if, for any t, the set $R(t)$ is closed, the function $R(t)$ is continuous, and, for all t,

$$\frac{1}{\Delta} h\Big(R(t+\Delta), \bigcup_{x \in R(t)} \{x + \Delta F(t,x)\} \Big) \to 0 \quad \text{as } \Delta \downarrow 0. \tag{B.8}$$

Theorem 6 ([101]). *Assume that, for any t, x, $F(t,x)$ is a convex compact set continuous in the collection of its variables as a set-valued mapping. Then there exists $\eta > 0$ such that the R-solution generated by the set-valued function $F(t,x)$ exists in the half interval $[t_0, t_0 + \eta)$.*

Theorem 7 ([101]). *Assume that $F(t,x)$ satisfies the Lipschitz condition in a certain neighborhood $S(R_0)$ of the set $R_0 \in \operatorname{comp}(\mathbb{R}^n)$. Then the indicated solution is unique for all $t \ge t_0$ for which the R-solution $R(t)$ $(R(t_0) = R_0)$ is defined and $R(t) \subset S(R_0)$. Moreover, this solution continuously depends on the initial set R_0.*

Theorem 8 ([101]). *Let $R(t) \subset W$ for $t \in [t_0, T]$, where the set W is open and bounded, and let $F(t,x)$ be a Lipschitz function in W. Then, for $t \in [t_0, T]$, the set $R(t)$ is the set of attainability from $R(t_0) = R_0$ at time t.*

Theorem 9 ([23, 144]). *For any compact set $K \subset \mathbb{R}^n$, there exists an R-solution with the initial condition $R(t_0) = K$. The integral funnel is the graph of the R-solution $R(t)$. If condition (B.4) is satisfied, then the R-solution with the initial condition $R(t_0) = K$ is unique and continuously depends on K, and the graph of $R(t)$ is an integral funnel.*

Example 4. Let $F(t,x) = 2[\alpha, \beta]\sqrt{x}$, $x(0) = x_0$, $0 \le \alpha \le \beta$.

For $x = 0$, the mapping $F(t,x)$ does not satisfy the Lipschitz condition.

Let $x_0 = 0$. We now show that a set-valued mapping

$$R(t) = \begin{cases} 0 & \text{for } 0 \le t \le t_1, \\ [0, \beta^2(t - t_1)^2] & \text{for } t_1 \le t \le t_2, \\ [\alpha^2(t - t_2)^2, \beta^2(t - t_1)^2] & \text{for } t_2 \le t \end{cases}$$

is an R-solution of the differential inclusion

$$\dot{x} \in 2[\alpha, \beta]\sqrt{x}, \quad x(0) = 0,$$

i.e., satisfies the equation

$$\lim_{\Delta \downarrow 0} \frac{1}{\Delta} h\Big(R(t + \Delta), \bigcup_{x \in R(t)} \{x + \Delta \cdot 2[\alpha, \beta]\sqrt{x}\}\Big) = 0. \tag{B.9}$$

For $0 \le t < t_1$, it is clear that $R(t) \equiv 0$ satisfies (B.9).
Let $t_1 \le t < t_2$. Then

$$\begin{aligned} &\lim_{\Delta \downarrow 0} \frac{1}{\Delta} h\Big([0, \beta^2(t + \Delta - t_1)^2], \bigcup_{x \in [0, \beta^2(t - t_1)^2]} \{x + \Delta \cdot 2[\alpha, \beta]\sqrt{x}\}\Big) \\ &= \lim_{\Delta \downarrow 0} \frac{1}{\Delta} h([0, \beta^2(t - t_1)^2 + 2\Delta\beta^2(t - t_1) + \beta^2\Delta^2], \\ &\qquad [0, \beta^2(t - t_1)^2 + 2\Delta\beta^2(t - t_1)]) = 0. \end{aligned}$$

Assume that $t_2 \le t$. Then

$$\begin{aligned} &\lim_{\Delta \downarrow 0} \frac{1}{\Delta} h\Big([\alpha^2(t + \Delta - t_2)^2, \beta^2(t + \Delta - t_1)^2], \\ &\qquad \bigcup_{x \in [\alpha^2(t - t_2)^2, \beta^2(t - t_1)^2]} \{x + \Delta \cdot 2[\alpha, \beta]\sqrt{x}\}\Big) \\ &= \lim_{\Delta \downarrow 0} \frac{1}{\Delta} h([\alpha^2(t - t_2)^2 + 2\Delta\alpha^2(t - t_2) + \alpha^2\Delta^2, \\ &\qquad \beta^2(t - t_1)^2 + 2\Delta\beta^2(t - t_1) + \beta^2\Delta^2], \\ &\qquad [\alpha^2(t - t_2)^2 + 2\Delta\alpha^2(t - t_2), \beta^2(t - t_1)^2 + 2\Delta\beta^2(t - t_1)]) = 0. \end{aligned}$$

For $t_1 = 0$ and $t_2 = \infty$, the R-solution $R(t)$ coincides with the integral funnel $X(t)$. All other R-solutions corresponding to arbitrary values $t_2 \ge t_1 \ge 0$ are such that $X(t) \subset R(t)$.

If $x_0 > 0$, then the R-solution

$$R(t) = \left[\alpha^2\Big(t + \frac{\sqrt{x_0}}{\alpha}\Big), \beta^2\Big(t + \frac{\sqrt{x_0}}{\beta}\Big)\right]$$

is unique and coincides with the integral funnel.

A similar example is constructed in [9].

Later, Panasyuk generalized the notion of R-solutions to the case of right-hand sides $F(t,x)$ measurable with respect to t and continuous in x.

Definition 10 ([23, 101, 102, 144]). A set-valued function $R(t)$ is called an R-solution generated by the differential inclusion (B.1) if, for any t, the set $R(t)$ is closed, the function $R(t)$ is absolutely continuous, and

$$\frac{1}{\Delta}h\Big(R(t+\Delta), \bigcup_{x\in R(t)}\{x+\int_t^{t+\Delta}F(s,x)ds\}\Big)\to 0 \quad (\Delta\downarrow 0) \tag{B.10}$$

for almost all t.

There are several approaches used for the investigation of stability of differential inclusions. These approaches differ by the objects of investigation. Thus, by analogy with the theory of ordinary differential equations, the first approach is based on the analysis of stability of separate trajectories [49, 51]. At present, there exists another approach aimed at the description of dynamics of the sets specified by differential inclusions. Within the framework of this approach, the R-solutions are used for the investigation of stability.

Definition 11 ([107]). An R-solution $R(t)$ $(t_0 \le t < +\infty)$ of the differential inclusion

$$\dot{x}\in \operatorname{co}F(t,x) \tag{B.11}$$

is called Lyapunov stable if, for any $\varepsilon > 0$, there exists $\delta(\varepsilon) > 0$ such that

(1) all R-solutions $X(t)$ of inclusion (B.11) satisfying the condition

$$h(Y(t_0), F(t_0)) < \delta \tag{B.12}$$

are defined for all $t > t_0$;

(2) the following inequality holds for all solutions satisfying inequality (B.12):

$$h(Y(t), F(t)) < \varepsilon.$$

Definition 12 ([107]). An R-solution $R(t)$ $(t_0 \le t < +\infty)$ of the differential inclusion (B.11) is called asymptotically stable if:

(1) it is Lyapunov stable;

(2) for any R-solution $X(t)$ satisfying the inequality

$$h(Y(t_0), F(t_0)) < \delta,$$

the following relation is true:

$$\lim_{t\to\infty} h(Y(t), F(t)) = 0.$$

Definition 13 ([23]). A solution $\psi(t)$ ($t_0 \le t < +\infty$) of the differential inclusion (B.1) is called stable if, for any $\varepsilon > 0$, one can find $\delta > 0$ such that, for any $\tilde{x}_0$ satisfying the inequality $\|\tilde{x}_0 - \psi(t_0)\| < \delta$, every solution $\tilde{x}(t)$ with the initial condition $\tilde{x}(t_0) = \tilde{x}_0$ exists for $t_0 \le t < +\infty$ and satisfies the inequality

$$\|\tilde{x}(t) - \psi(t)\| < \varepsilon.$$

Definition 14 ([23]). A solution $\psi(t)$ ($t_0 \le t < +\infty$) of the differential inclusion (B.1) is called weakly stable if, for any $\varepsilon > 0$, there exists $\delta > 0$ such that, for any $\tilde{x}_0$ satisfying the inequality $\|\tilde{x}_0 - \psi(t_0)\| < \delta$, a solution $\tilde{x}(t)$ with the initial condition $\tilde{x}(t_0) = \tilde{x}_0$ exists for $t_0 \le t < +\infty$ and satisfies the inequality

$$\|\tilde{x}(t) - \psi(t)\| < \varepsilon.$$

Definition 15 ([23]). A solution $\psi(t)$ ($t_0 \le t < +\infty$) of the differential inclusion (B.1) is called asymptotically stable if

(1) it is stable and

(2) satisfies the condition

$$\|\tilde{x}(t) - \psi(t)\| \to 0 \quad \text{as } t \to \infty.$$

Definition 16 ([23]). A solution $\psi(t)$ ($t_0 \le t < +\infty$) of the differential inclusion (B.1) is called weakly asymptotically stable if

(1) it is weakly stable and

(2) satisfies the condition

$$\|\tilde{x}(t) - \psi(t)\| \to 0 \quad \text{as } t \to \infty.$$

Example 5. Consider a differential inclusion

$$\dot{x} \in \alpha x + [-1, 1], \quad x(0) = x_0,$$

where α is an arbitrary parameter. In this case, the R-solution can be represented in the form

$$R(t) = [x_1(t), x_2(t)],$$
$$x_1(t) = \left(x_0 + \frac{1}{\alpha}\right) e^{\alpha t} - \frac{1}{\alpha}, \quad x_2(t) = \left(x_0 - \frac{1}{\alpha}\right) e^{\alpha t} + \frac{1}{\alpha}.$$

Any solution $x(t)$ of this differential inclusion satisfies the relation $x_1(t) \le x(t) \le x_2(t)$.

For $\alpha < 0$, the R-solution is asymptotically stable. For any value of α, the stable ordinary solution $x(t)$ does not exist. At the same time, for any value $\alpha < 0$, every solution $x(t)$ is weakly asymptotically stable.

Example 6 ([23]). Consider a differential inclusion

$$\dot{x} \in [\alpha, \beta]x, \quad x(0) = x_0,$$

where α and β are arbitrary constants. For the solution of the differential inclusion $x(t)$, we can write the following inequality:

$$x_0 e^{\alpha t} \le x(t) \le x_0 e^{\beta t}, \quad x_0 \ge 0.$$

If $x_0 = 0$, then the solution $x(t)$ is asymptotically stable for $\alpha \le \beta < 0$, stable for $\alpha \le \beta = 0$, weakly asymptotically stable for $\alpha < 0 < \beta$, weakly stable for $\alpha = 0 < \beta$, and unstable for $0 < \alpha \le \beta$. The R-solution is asymptotically stable for $\alpha \le \beta < 0$ and stable for $\alpha \le \beta = 0$ for any x_0.

References

[1] V. N. Afanas'ev, V. B. Kolmanovskii, and V. R. Nosov, *Mathematical Theory of Construction of Control Systems* [in Russian], Vysshaya Shkola, Moscow (1999).

[2] M. A. Aizerman and E. G. Pyatnitskii, "Foundations of the theory of discontinuous systems," *Avtomat. Telemekh.*, No. 7, 33–47 (1974); No. 8, 33–61 (1974).

[3] Z. Arstein, "On the calculus of closed set-valued functions," *Indiana Univ. Math. J.*, **24**, No. 5, 433–441 (1974).

[4] Z. Arstein and J. A. Burne, "Integration of compact set-valued functions," *Pacific J. Math.*, **58**, No. 2, 297–307 (1975).

[5] L. T. Ashchepkov, *Optimal Control over Discontinuous Systems* [in Russian], Nauka, Novosibirsk (1987).

[6] J.-P. Aubin, "Mutational equations in metric spaces," *Set-Valued Anal.*, **1**, No. 1, 3–46 (1993)

[7] J.-P. Aubin, *Impulse Differential Inclusions and Hybrid Systems: A Viability Approach*, Lecture Notes, Université Paris–Dauphine (2002).

[8] J.-P. Aubin and A. Cellina, *Differential Inclusions. Set-Valued Maps and Viability Theory*, Springer, Berlin (1984).

[9] J.-P. Aubin and H. Frankovska, *Set-Valued Analysis*, Birkhäuser, Boston (1990).

[10] R. J. Aumann, "Integrals of set-valued functions," *J. Math. Anal. Appl.*, **12**, No. 1, 1–12 (1965).

[11] D. D. Bainov and A. B. Dishliev, "Population dynamic control in regard to minimizing the time necessary for the regeneration of a biomass taken away from the population," *Math. Model. Numer. Anal.*, **24**, No. 6, 681–692 (1990).

[12] D. D. Bainov and S. D. Milusheva, "Justification of the averaging method for a system of differential equations with fast and slow variables with impulses," *J. Appl. Math. Phys.*, **32**, 237–254 (1981).

[13] D. D. Bainov and S. D. Milusheva, "Application of the partially-multiplicative averaging for a class of functional-differential equations with impulses," *Rend. Sem. Mat. Univ. Politech*, Torino, **40**, No. 1. 139–161 (1982).

[14] C. Baiocchi and A. Capelo, *Variational and Quasivariational Inequalities. Applications to Free Boundary Problems*, Wiley, New York (1984).

[15] H. T. Banks and M. Q. Jacobs, "A differential calculus of multifunctions," *J. Math. Anal. Appl.*, No. 29, 246–272 (1970).

[16] N. N. Bautin, "Theory of pointwise transformations and the dynamical theory of clocks," *Tr. ICNO–5, Akad. Nauk Ukr. SSR,* Vol. 2, Kiev (1963), pp. 29–54.

[17] M. Benchohra, "Initial-value problems for first-order impulsive differential inclusions in Banach spaces," *Nonlin. Oscil.*, **4**, No. 2, 146–154 (2001).

[18] M. Benchohra and A. Boucherif, "On the first-order initial value problems for impulsive differential inclusions in Banach spaces," *Syst. Appl.,* **8**, No. 1, 119–126 (1999).

[19] M. Benchohra and A. Boucherif, "An existence result for first-order initial-value problems for impulsive differential inclusions in Banach spaces," *Arch. Math. (Brno),* **36**, No. 3, 159–169 (2000).

[20] M. Benchohra and A. Boucherif, "Initial-value problems for impulsive differential inclusions of first order," *Different. Equat. Dynam. Syst.,* **8**, No. 1, 51–66 (2000).

[21] V. I. Blagodatskikh, "Some results in the theory of differential inclusions (a survey)," in: *Summer School on Ordinary Differential Equations*, Brno (1975), pp. 29–67.

[22] V. I. Blagodatskikh, *Introduction to Optimal Control. Linear Theory* [in Russian], Vysshaya Shkola, Moscow (2001).

[23] V. I. Blagodatskikh and A. F. Filippov, "Differential inclusions and optimal control," in: *Collection of Surveys "Topology, Ordinary Differential Equations, and Dynamical Systems," Vol. 2, On the Fiftieth Anniversary of the Institute (Proc. of MIAN SSSR, **169**)* [in Russian], Nauka, Moscow (1985), pp. 194–252.

[24] F. S. de Blasi, "On the differentiability of multifunctions," *Pacific J. Math.,* **66**, No. 1, 67–81 (1976).

[25] F. S. de Blasi and F. Iervolino, "Equazioni differentiali con soluzioni a valore compatto convesso," *Boll. Unione Mat.Ital.,* **2**, No. 4–5, 491–501 (1969).

[26] F. S. de Blasi and F. Iervolino, "Euler method for differential equations with set-valued solutions," *Boll. Unione Mat.Ital.,* **4**, No. 4, 941–949 (1971).

[27] N. N. Bogolyubov and Yu. A. Mitropol'skii, *Asymptotic Methods in the Theory of Nonlinear Oscillations* [in Russian], Nauka, Moscow (1974).

[28] Yu. G. Borisovich, V. D. Gel'man, A. D. Myshkis, and V. V. Obukhovskii, "Set-valued mappings," in: *Itogi VINITI, Ser. Mathematical Analysis* [in Russian], Vol. 19, VINITI, Moscow (1987), pp. 127–130.

[29] L. P. A. J. Brandao, F. S. de Blasi, and F. Iervolino, "Uniqueness and existence theorems for differential equations with compact convex-valued solutions," *Boll. Unione Mat. Ital.,* **4**, 534–538 (1970).

[30] F. L. Chernous'ko, *Estimation of the Phase State of Dynamical Systems. Method of Ellipsoids* [in Russian], Nauka, Moscow (1988).

[31] P. I. Chugunov, "On the proper solutions of differential inclusions," *Differents. Uravn.,* **17**, No. 4, 660–668 (1981).

[32] E. A. Coddington and N. Levinson, *Theory of Ordinary Differential Equations*, McGraw-Hill, New York (1955).

[33] J. L. Davy, "Properties of the solution set for a generalized differential equation," *Bull. Austral. Math. Soc.,* **6**, No. 3, 379–398 (1972).

[34] M. Dawidowski, “On some generalization of Bogolubov averaging theorem,” *Funct. Approx. (PRL),* No. 7, 55–70 (1979).

[35] G. Debreu, “Integration of correspondences,” in: *Proc. of the Fifth Berkeley Symposium on Mathematical Statistics and Probability,* University of California (1966), pp. 351–372.

[36] K. Deimling, *Multivalued Differential Equations,* de Gruyter, Berlin (1992).

[37] V. P. Demidovich, *Lectures on the Mathematical Theory of Stability* [in Russian], Nauka, Moscow (1967).

[38] G. Diniz, J. F. R. Fernandes, J. F. C. A. Meyer, and L. C. Barros, “A fuzzy Cauchy problem modelling the decay of the biochemical oxygen demand in water,” in: *Proc. of the Joint 9th IFSA World Congress and 20th NAFIPS Internat. Conf.* (2001), pp. 512–516.

[39] A. B. Dishliev and D. D. Bainov, “Dependence upon initial conditions and parameters of solutions of impulsive differential equations with variable structure,” *Int. J. Theor. Phys.,* **26**, No. 6, 655–675 (1990).

[40] T. Donchev, “Functional differential inclusions involving dissipative and compact multifunctions,” *Glas. Mat.,* No. 33 (53), 51–60 (1998).

[41] A. L. Dontchev, *Perturbations, Approximations and Sensitivity Analysis of Optimal Control Systems*, Springer, Berlin (1983).

[42] B. I. Dzyra, “Application of the method of averaging to the investigation of one-frequency oscillations excited by instantaneous forces,” in: *Analytic and Qualitative Methods in the Theory of Differential Equations* [in Russian], Institute of Mathematics, Ukrainian Academy of Sciences, Kiev (1972), pp. 43–50.

[43] L. Erbe and W. Krawcewicz, “Existence of solutions to boundary-value problems for impulsive second-order differential inclusions,” *Rocky Mountain J. Math.,* **22**, No. 2, 519–539 (1992).

[44] A. N. Filatov, *Averaging in Systems of Differential and Integrodifferential Equations* [in Russian], FAN, Tashkent (1971).

[45] A. N. Filatov, *Asymptotic Methods in the Theory of Differential and Integrodifferential Equations* [in Russian], FAN, Tashkent (1974).

[46] A. N. Filatov and L. V. Sharova, *Integral Inequalities and the Theory of Nonlinear Oscillations* [in Russian], Nauka, Moscow (1976).

[47] O. P. Filatov and M. M. Khapaev, *Averaging of the Systems of Differential Inclusions* [in Russian], Moscow University, Moscow (1998).

[48] A. F. Filippov, “Some problems in the theory of optimal regulation,” *Vestn. Mosk. Univ., Ser. Mat., Mekh.,* No. 2, 25–32 (1959).

[49] A. F. Filippov, “Classic solutions of differential equations with set-valued right-hand sides,” *Vestn. Mosk. Univ., Ser. Mat., Mekh.,* No. 3, 16–26 (1967).

[50] A. F. Filippov, “On the existence of solutions of set-valued differential equations,” *Mat. Zametki,* **10**, No. 3, 307–313 (1971).

[51] A. F. Filippov, *Differential Equations with Discontinuous Right-Hand Sides* [in Russian], Nauka, Moscow (1985).

[52] A. Fryszkowski, "Carathéodory-type selectors of set-valued maps of two variables," *Bull. Acad. Pol. Sci., Ser. Sci. Math., Astron., Phys.,* **25**, No. 1, 41–46 (1977).

[53] A. T. Fuller, "Study of an optimum nonlinear control system," *J. Electron. Contr.,* **15**, No. 1 (1963).

[54] A. T. Fuller, "Absolute optimality of nonlinear control system with integral-square error criterion," *J. Electron. Contr.,* **17**, No. 3, 301–317 (1964).

[55] I. I. Gikhman and A. V. Skorokhod, *Introduction to the Theory of Random Processes* [in Russian], Nauka, Moscow (1965).

[56] A. Halanay and D. Wexler, *Teoria Calitativă a Sistemelor cu Impulsuri*, Editura Academiei Republicii Socialiste România, Bucureşti (1968).

[57] P. Hartman, *Ordinary Differential Equations*, Wiley, New York (1964).

[58] M. Hukuhara, "Integration des applications mesurables dont la valeur est un compact convexe," *Funkcial. Ekvac.,* No. 10, 205–223 (1967).

[59] T. Janiak and E. Luczak-Kumorek, "The theorem of middling for functional-differential equations of neutral type," *Discuss. Math.,* No. 11, 63–73 (1991).

[60] B. S. Kalitin, "On the oscillations of a pendulum with shock pulses," *Differents. Uravn.,* **5**, No. 7, 1267–1274 (1969).

[61] B. S. Kalitin, "On the oscillations of a pendulum with shock pulses. Part 2," *Differents. Uravn.,* **6**, No. 12, 2174–2181 (1970).

[62] B. S. Kalitin, "On the limiting cycles of pendulum systems with pulsed perturbations," *Differents. Uravn.,* **7**, No. 3, 540–542 (1971).

[63] I. Karkinbaev and N. A. Perestyuk, "On the substantiation of application of asymptotic methods to the investigation of impulsive systems," in: *Asymptotic Methods in the Theory of Nonlinear Oscillations* [in Russian], Naukova Dumka, Kiev (1979), pp. 43–50.

[64] R. Z. Khas'minskii, *Stability of Systems of Differential Equations Under Random Perturbations of Their Parameters* [in Russian], Nauka, Moscow (1969).

[65] N. Kikuchi, "On contingent equations," in: *Lecture Notes Math.,* No. 243 (1971), pp. 169–181.

[66] M. Kisielewicz, "Method of averaging for differential equations with compact convex-valued solutions," *Rend. Math.,* **9**, No. 3, 397–408 (1976).

[67] M. Kisielewicz, *Differential Inclusion and Optimal Control*, PWN, Warszawa (1991).

[68] A. E. Kobrinskii and A. A. Kobrinskii, *Vibroimpact Systems* [in Russian], Nauka, Moscow (1973).

[69] A. F. Kononenko, "Structure of the optimal strategy in controlled dynamical systems," *Zh. Vychisl. Mat. Mat. Fiz.,* **20**, No. 5, 1105–1116 (1980).

[70] M. A. Krasnosel'skii and S. G. Krein, "On the principle of averaging in nonlinear mechanics," *Usp. Mat. Nauk*, **10**, No. 3(65), 147–152 (1955).

[71] N. N. Krasovskii and A. I. Subbotin, *Differential Games* [in Russian], Mir, Moscow (1974).

[72] N. M. Krylov and N. N. Bogolyubov, *Introduction to Nonlinear Mechanics* [in Russian], Ukrainian Academy of Sciences, Kiev (1937).

[73] A. B. Kurzhanski and I. Valyi, "Ellipsoidal techniques for dynamic systems: Control synthesis for uncertain systems," *Dynam. Control,* No. 2, 87–111 (1992).

[74] V. Lakshmikantham, D. D. Bainov, and P. S. Simeonov, *Theory of Impulsive Differential Equations*, World Scientific, Singapore (1989).

[75] V. Lakshmikantham, S. Leela, and A. S. Vatsala, "Interconnection between set and fuzzy differential equations," *Nonlin. Anal.,* **54**, 351–360 (2003).

[76] V. Lakshmikantham and A. A. Tolstonogov, "Existence and interrelation between set and fuzzy differential equations," *Nonlin. Anal.,* **55**, 255–268 (2003).

[77] P. Lancaster, *Theory of Matrices*, Academic Press, New York (1969).

[78] A. A. Lyapunov, "On completely additive vector functions," *Izv. Akad. Nauk SSSR, Ser. Mat.,* No. 6, 465–478 (1940).

[79] C. Marchal, "Chattering arcs and chattering controls," *J. Optim. Theor. Appl.,* **11**, No. 5, 441–468 (1973).

[80] A. Marchaud, "Sur les champs continus de demi-cones convexes et leurs integrales," *C.R. Acad. Sci. Paris,* No. 199, 1278–1280 (1934).

[81] A. Marchaud, "Sur les champs de demi-cones et equations differentielles du premier order," *Bull. Soc. Math. France,* No 62, 1–38 (1934).

[82] A. Marchaud, "Sur les champs continus de demi-cones convexes et leurs integrales," *Comput. Math. Ser.,* No 8, 89–127 (1936).

[83] A. Marchaud, "Sur les champs de demi-cones convexes," *Bull. Sci. Math.,* **62**, No. 2, 229–240 (1938).

[84] S. M. Markov, "Existence and uniqueness of solutions of the interval differential equation $X' = F(t, X)$," *C. R. Acad. Bulgare Sci.,* **31**, No. 12, 1519–1522 (1978).

[85] E. A. Michael, "Continuous selections, I," *Ann. Math.,* **63**, No. 2, 361–381 (1956).

[86] V. D. Mil'man and A. D. Myshkis, "On the stability of motion in the presence of pulses," *Sib. Mat. Zh.,* **1**, No 2, 233–237 (1960).

[87] V. D. Mil'man and A. D. Myshkis, "Random pulses in linear dynamical systems," in: *Approximate Methods for the Solution of Differential Equations* [in Russian], Ukrainian Academy of Sciences, Kiev (1963), pp 64–81.

[88] Yu. A. Mitropol'skii, *Method of Averaging in Nonlinear Mechanics* [in Russian], Naukova Dumka, Kiev (1971).

[89] Yu. A. Mitropol'skii and G. N. Khoma, *Mathematical Substantiation of the Asymptotic Methods of Nonlinear Mechanics* [in Russian], Naukova Dumka, Kiev (1983).

[90] Yu. A. Mitropol'skii, A. M. Samoilenko, and N. A. Perestyuk, "On the substantiation of the method of averaging for the second-order equation with pulse action," *Ukr. Mat. Zh.,* **29**, No. 6, 750–762 (1977).

[91] B. S. Morduhovich, "Discrete approximations and refined Euler–Lagrange conditions for nonconvex differential inclusions," *SIAM J. Control Optim.,* **33**, 882–915 (1995).

[92] A. D. Myshkis, "On the domains of uniqueness of the solution of a system of linear partial differential equations," *Mat. Sb.,* **19(61)**, No. 3, 489–520 (1946).

[93] A. D. Myshkis, "General theory of differential equations with delayed argument," *Usp. Mat. Nauk,* **4**, No. 5, 99–141 (1949).

[94] A. D. Myshkis and A. M. Samoilenko, "Systems with pushes at given times," *Mat. Sb.,* **74**, No. 2, 202–208 (1967).

[95] S. P. Negoda, "Linear impulsive systems with 'pushes' of general location," in: *Analytic Methods for the Investigation of Nonlinear Differential Systems* [in Russian], Kiev (1992), pp. 60–62.

[96] M. S. Nikol'skii, "On one method of approximation of the attainability set for a differential inclusion," *Zh. Vychisl. Mat. Mat. Fiz.,* **28**, No. 8, 1252–1254 (1988).

[97] O. D. Nurzhanov, "On periodic solutions of nonlinear integrodifferential equations with pulse action," in: *Analytic Methods in the Theory of Differential Equations* [in Russian], Institute of Mathematics, Ukrainian Academy of Sciences, Kiev (1977), pp. 88–103.

[98] A. I. Ovseevich, "Extremal properties of ellipsoids approximating the domains of attainability," *Probl. Upravl. Teor. Inform.,* **12**, No. 2, 1237–1240 (1983).

[99] A. I. Panasyuk, "On the dynamics of sets specified by differential inclusions," *Sib. Mat. Zh.,* **27**, No. 5, 155–166 (1986).

[100] A. I. Panasyuk, "Qualitative dynamics of sets specified by differential inclusions," *Mat. Zametki,* **45**, No. 1, 80–88 (1989).

[101] A. I. Panasyuk and V. I. Panasyuk, *Asymptotic Optimization of Nonlinear Control Systems* [in Russian], Belor. Univ., Minsk (1977).

[102] A. I. Panasyuk and V. I. Panasyuk, "On one equation generated by a differential inclusion," *Mat. Zametki,* **27**, No. 3, 429–437 (1980).

[103] A. I. Panasyuk and V. I. Panasyuk, *Basic Asymptotic Optimization of Controlled Systems* [in Russian], Nauka i Tekhnika, Minsk (1986).

[104] G. Piangiani, "On the fundamental theory of multivalued differential equations," *J. Different. Equat.,* **25**, No. 1, 30–38 (1977).

[105] A. Plis, "Trajectories and quasitrajectories of an orientor field," *Bull. Pol. Acad. Sci., Ser. Sci. Math., Astron. Phys.,* **11**, No. 6, 369–370 (1963).

[106] A. Plis, "On trajectories of orientor fields," *Bull. Pol. Acad. Sci., Ser. Sci. Math., Astron. Phys.,* **13**, No 8, 571–573 (1965).

[107] V. A. Plotnikov, "Averaging of differential inclusions," *Ukr. Mat. Zh.,* **31**, No. 5, 573–576 (1979).

[108] V. A. Plotnikov, *Method of Averaging in Control Problems* [in Russian], Lybid', Kiev–Odessa (1992).

[109] V. A. Plotnikov, *Investigation of Some Differential Equations with Set-Valued Right-Hand Sides* [in Russian], Doctoral-Degree Thesis (Physics and Mathematics), Odessa (1994).

[110] V. A. Plotnikov and P. M. Kitanov, "Bogolyubov's theorem for quasidifferential equations with pulses," *Ukr. Mat. Zh.,* **49**, No. 11, 1504–1511 (1997).

[111] V. A. Plotnikov and N. M. Kitanov, "Continuous dependence of the solutions of impulsive differential inclusions and impulsive control problems," *Kibernet. Sist. Anal.,* No 5, 71–85 (2002).

[112] V. A. Plotnikov and L. I. Plotnikova, "Averaging of differential inclusions with multivalued pulses," *Ukr. Mat. Zh.,* **47**, No. 11, 1526–1532 (1995).

[113] V. A. Plotnikov, R. Ivanov, and N. Kitanov, "Differential inclusions with finite number of impulses in fixed moments," *Discr. Math. Appl., Research Math., Blagoevgrad,* No. 5, 246–254 (1995).

[114] V. A. Plotnikov, R. P. Ivanov, and N. M. Kitanov, "Method of averaging for impulsive differential inclusions," *Pliska Stud. Math. Bulgar.,* No. 12, 43–55 (1998).

[115] V. A. Plotnikov, A. V. Plotnikov, and A. N. Vityuk, *Differential Equations with Set-Valued Right-Hand Side. Asymptotic Methods* [in Russian], AstroPrint, Odessa (1999).

[116] N. V. Plotnikova, "Approximation of a bundle of solutions of linear impulsive differential inclusions," *Visn. Kharkiv. Nats. Univ., No. 645, Ser. Mat., Prikl. Mat., Mekh.,* Issue 54, 67–78 (2004).

[117] N. V. Plotnikova, "Stability of solutions of linear impulsive differential inclusions," *Nelin. Kolyv.,* **7**, No. 1, 121–131 (2004).

[118] N. V. Plotnikova, "Periodic solutions of linear impulsive differential inclusions," *Nelin. Kolyv.,* **7**, No. 4, 495–515 (2004).

[119] N. V. Plotnikova, "Linear differential equations with pulse actions at indefinite times," in: *Abstr. of the Internat. Sci. Conf. on Differential Equations and Their Applications (June 6–9, 2005),* Shevchenko Kyiv National University, Kyiv (2005), p. 88.

[120] N. V. Plotnikova, "Linear differential equations with multivalued trajectories," in: *D. A. Ovsyannikov and L. A. Petrosyan (editors), Proc. of the Internat. Conf. on 'Stability and Control Processes' (June 26–29, 2005)* [in Russian], Vol. 2, St. Peterburg University, St. Peterburg (2005), pp. 1137–1147.

[121] N. V. Plotnikova, "Systems of linear differential equations with π-derivative and linear differential inclusions," *Mat. Sb.,* **196**, No. 11, 127–140 (2005).

[122] N. V. Plotnikova, "Krasnosel'skii–Krein theorem for differential inclusions," *Differents. Uravn.,* **41**, No. 7, 997–1000 (2005).

[123] N. V. Plotnikova, "Averaging of impulsive differential inclusions," *Mat. Stud.,* **23**, No. 1, 52–56 (2005).

[124] N. V. Plotnikova, "Approximation of a bundle of solutions of linear differential inclusions," *Nelin. Kolyv.,* **9**, No. 3, 386–400 (2006).

[125] N. V. Plotnikova, "Linear differential equations with multivalued trajectories," *Vestn. St. Petersburg Univ., Ser. 10, Prikl. Mat., Inform., Prots. Upravlen.,* Issue 1, 57–63 (2006).

[126] N. V. Plotnikova, "Differential equations with discontinuous right-hand sides," *Tr. Odessa Politekh. Univ.,* Issue 1 (25), 172–178 (2006).

[127] E. S. Polovinkin, *Theory of Set-Valued Mappings* [in Russian], MFTI, Moscow (1983).

[128] B. N. Pshenichnyi, *Convex Analysis and Extremal Problems* [in Russian], Nauka, Moscow (1980).

[129] H. Radström, "An embedding theorem for spaces of convex sets," *Proc. Amer. Math. Soc.,* No. 3, 165–169 (1952).

[130] L. I. Rozenoér, "Pontryagin's maximum principle in the theory of optimal processes," *Avtomat. Telemekh.,* No. 9 (1441) (1959).

[131] A. A. Samarskii and A. V. Gulin, *Numerical Methods* [in Russian], Nauka, Moscow (1989).

[132] A. M. Samoilenko, "Application of the method of averaging to the investigation of oscillations excited by instantaneous pulses in self-oscillating second-order systems with small parameter," *Ukr. Mat. Zh.,* **13**, No. 3 (1961).

[133] A. M. Samoilenko, "Some problems of investigation of the oscillations of differential equations with irregular right-hand sides," *Bul. Instill. Politeh.,* **11 (15)**, No. 3–4, 85–92 (1965).

[134] A. M. Samoilenko, "Numerical-analytic method for the investigation of periodic systems of ordinary differential equations. I," *Ukr. Mat. Zh.,* **17**, No. 4, 82–93 (1965).

[135] A. M. Samoilenko, "Numerical-analytic method for the investigation of periodic systems of ordinary differential equations. II," *Ukr. Mat. Zh.,* **18**, No. 2, 50–59 (1966).

[136] A. M. Samoilenko, "On the substantiation of the method of averaging for the investigation of oscillations in systems subjected to the pulse actions," *Ukr. Mat. Zh.,* **19**, No. 5, 96–104 (1967).

[137] A. M. Samoilenko, "Method of averaging in systems with pulses," *Mat. Fiz.,* Issue 9, 101–117 (1971).

[138] A. M. Samoilenko and N. A. Perestyuk, "Invariant sets of systems with instantaneous changes in the standard form," *Ukr. Mat. Zh.,* **25**, No. 1, 129–134 (1973).

[139] A. M. Samoilenko and N. A. Perestyuk, "On the substantiation of the method of averaging for second-order differential equations with pulses," in: *Nonlinear Oscillations and Stability of Motion* [in Russian], Institute of Mathematics, Ukrainian Academy of Sciences, Kiev (1973), pp. 273–289.

[140] A. M. Samoilenko and N. A. Perestyuk, "Second Bogolyubov's theorem for systems of differential equations with pulse actions," *Differents. Uravn.*, **10**, No. 11, 2001–2010 (1974).

[141] A. M. Samoilenko and N. A. Perestyuk, "On the method of averaging in systems with pulse actions," *Ukr. Mat. Zh.*, **24**, No. 3, 411–418 (1974).

[142] A. M. Samoilenko and N. A. Perestyuk, *Impulsive Differential Equations* [in Russian], Vyshcha Shkola, Kiev (1987).

[143] A. M. Samoilenko and N. A. Perestyuk, *Impulsive Differential Equations*, World Scientific, Singapore (1995).

[144] A. A. Tolstonogov, "On the equation of integral funnel for a differential inclusion," *Mat. Zametki,* **32**, No. 6, 841–852 (1982).

[145] A. A. Tolstonogov, *Differential Inclusions in Banach Spaces* [in Russian], Nauka, Novosibirsk (1986).

[146] Yu. N. Tyurin, "Mathematical statement of a simplified model of production planning," *Ékonom. Mat. Met.,* **1**, No. 3, 391–409 (1965).

[147] V. I. Utkin, *Sliding Modes in the Problems of Optimization and Control* [in Russian], Nauka, Moscow (1981).

[148] A. B. Vasil'ev, "On the continuous dependence of solutions of differential inclusions on the parameter," *Ukr. Mat. Zh.,* **35**, No 5, 607–611 (1983).

[149] V. Veliov, "Second-order discrete approximations to linear differential inclusions," *SIAM J. Numer. Anal.,* **29**, 439–451 (1992).

[150] T. Vogel, *Theorie des Systèmes Evolutifs,* Villons, Goutnier (1965).

[151] S. T. Zavalishchin and A. N. Sesekin, *Impulsive Processes. Models and Applications* [in Russian], Nauka, Moscow (1991).

[152] S. T. Zavalishchin, A. N. Sesekin, and S. E. Drozdenko, *Dynamical Systems with Impulsive Structure* [in Russian], Sredne-Ural'skoe Knizh. Izd., Sverdlovsk (1983).

[153] M. Z. Zgurovskii and V. S. Mel'nik, *Nonlinear Analysis and Control over Infinite-Dimensional Systems* [in Russian], Naukova Dumka, Kiev (1999).

[154] V. I. Zhukovskii and M. E. Salukvadze, *Optimization of Guarantees in Many-Criteria Control Problems* [in Russian], Metsnieryba, Tbilisi (1996).

[155] J. Warga, *Optimal Control of Differential and Functional Equations,* Academic Press, New York (1972).

[156] P. J. Watson, "Impulsive differential inclusions," *Nonlin. World,* **4**, No. 4, 395–402 (1997).

[157] T. Wazewski, *Selected Papers*, PWN, Warsawa (1990).

[158] S. K. Zaremba, "Sur une extension de la notion d'equation differentielle," *C. R. Acad. Sci. Paris,* No. 199, 545–548 (1934).

[159] S. K. Zaremba, "Sur les equations au paratingent," *Bull. Sci. Math.,* **60**, No. 2, 139–160 (1936).

Index